CIVIC EMPOWERMENT IN AN AGE OF CORPORATE GREED

CIVIC EMPOWERMENT IN AN AGE OF CORPORATE GREED

Edward C. Lorenz

Michigan State University Press

East Lansing

♾ The paper used in this publication meets the minimum requirements of ANSI/NISO Z39.48-1992 (R 1997) (Permanence of Paper).

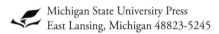 Michigan State University Press
East Lansing, Michigan 48823-5245

Printed and bound in the United States of America.

18 17 16 15 14 13 12   1 2 3 4 5 6 7 8 9 10

LIBRARY OF CONGRESS CATALOGING-IN-PUBLICATION DATA

Lorenz, Edward C.
Civic empowerment in an age of corporate greed / Edward C. Lorenz.
   p. cm.
   Includes bibliographical references and index.
   ISBN 978-1-61186-025-2 (hardback) 1. Fruit of the Loom (Firm) 2. Hazardous waste sites—United States. 3. Wealth—Moral and ethical aspects—United States. 4. Social responsibility of business—United States. I. Title.

   HD9940.U6F785 2012
   338.7'68720973—dc23   2011040291

Cover design by Erin Kirk New
Cover photo is of the Velsicol Chemical plant, ca. 1980, along the Pine River in St. Louis, Michigan taken from Penny Park. Photo is used courtesy of the St. Louis Historical Society.
Book design by Scribe Inc. (www.scribenet.com)

green press INITIATIVE Michigan State University Press is a member of the Green Press Initiative and is committed to developing and encouraging ecologically responsible publishing practices. For more information about the Green Press Initiative and the use of recycled paper in book publishing, please visit www.greenpressinitiative.org.

Visit Michigan State University Press at www.msupress.org

Contents

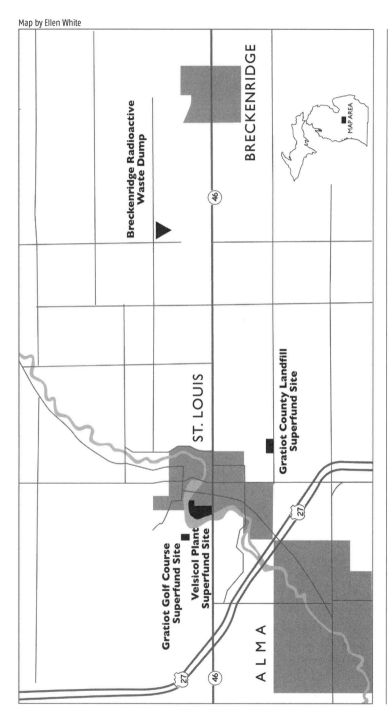

Central Michigan Velsicol Superfund and Nuclear Waste Sites

Preface

IN THE SUMMER OF 1990, WHEN I HAD LIVED FOR ONLY A YEAR IN Alma, Michigan, I attended a seminar at MIT on the "Myth and Reality of American Decline." On the first morning, the leader, Richard Vallely, asked each of us to introduce ourselves. Since I had sat in the front right of the room, he asked me to go first. After I said my name and residence, another member, Haul Riddick, an agricultural chemist from North Carolina, interrupted to say, "If we want to study what is wrong with the American economy, we should go to this guy's hometown." I thought, "Did he misunderstand me? I'm not from Flint or Detroit." But then he explained the terrible mismanagement at Velsicol Chemical, formerly known as Michigan Chemical, in St. Louis, Michigan, a sister or twin city of Alma.

For the remainder of the seminar, I spent most of my time learning from him about Michigan Chemical. Quickly, I learned why colleagues in Alma had discouraged my family from looking for a house in St. Louis when we had moved to the region in 1989. Upon returning home, I went to the college library and asked Larry Hall, the college archivist, about the company. Soon Larry pulled from the archives a collection of material documenting a tragic food-contamination incident in the 1970s that led to thousands of dairy cattle being shot and buried in hazardous-waste landfills. He had stories about the long-term member of Congress from the district, who neglected the crisis while attending to important national affairs, only to be ousted by an enraged dairy farmer. He even helped me see the links of what happened at Velsicol in St. Louis to my longtime interest in global labor policy.

In 1991, Alma College was launching a new "service learning" program, aiming to get students engaged in communities in need of their skills. With support from the Kellogg Foundation, we were to develop an annual focus for the program. Driven by my new knowledge of the local environmental problems, I suggested we make our first case study "The Environment." Old hands on the faculty who remembered the events in the '70s, such as Tracy Luke and Burnie Davis, were intrigued and agreed. Soon a committee of students, working with a chemistry professor, Melissa Strait, and myself, took advantage of the twentieth anniversary of the cattle-feed contamination to review the policy lessons that had been learned, or, as seemed more to be the case, forgotten. What the students found is that little learning had taken place, but that local residents felt abandoned by a process that had allowed the company and public officials implicated in the mistake to escape the consequences.[1] As the late state's attorney Stu Freeman observed when meeting the students, "We wanted a corporate execution. The problem is we shot at Velsicol and missed. We killed the community." The symbol of this failure rested in the heart of St. Louis. There, on 54 acres, the former plant site was demolished and buried next to the heavily contaminated Pine River. Adding insult to injury, the final legal settlement with the company, negotiated by far-off regulators from the U.S. Environmental Protection Agency (EPA), required that Velsicol mark the site forever with a granite tombstone warning of the serious contaminants buried beneath.

The interviews conducted in 1992 in St. Louis had several beneficial results. They seemed to serve as a cathartic for some people in town, restoring hope that something might be done to correct past mistakes. Second, they linked Alma College and me with the community in a way that changed our academic focus. While still interested in global labor standards, I began a crash course in environmental policymaking, with a subfield in public health. When the US EPA was compelled to return to town in 1997 to inform citizens of record levels of DDT found in river fish, the impact of the interview process became clear. On an October evening in 1997, almost two hundred angry citizens packed the senior-citizen center in town to confront EPA staff reporting on their findings. A few weeks later, residents took advantage of new EPA regulations and formed a community advisory group (CAG) for the site, electing me as chair.

Beginning in early 1998, the CAG took advantage of the special town-gown relationship to find the expertise to force officials to launch a massive remediation of the river. Residents relied on college faculty and students as their public research team. By 1999, they had found that Velsicol had indirect links to Fruit of the Loom, a firm that for some reason had set aside $100 million for environmental liabilities. At the end of the year, after learning that Fruit of the Loom had filed for bankruptcy, the CAG filed a $100 million claim in bankruptcy court. Banned from using a small technical assistant grant from EPA for any legal work, the CAG turned to the Alma faculty, who found help from the Law School at the University of Michigan for pro bono assistance. Following up on the CAG's claim, the U.S. Department of Justice entered the Fruit suit, winning in May 2002 access to the $100 million insurance and even the shares in Velsicol.

As these matters unfolded, we became aware of the larger story of Fruit of the Loom. The settlement with Fruit in 2002 linked St. Louis with five other communities with the worst Velsicol Superfund sites. At a Department of Justice hearing held in St. Louis to solicit public comments on the settlement, the most dramatic event came when a delegation from Memphis, Tennessee, suddenly arrived, thanking the CAG for informing them of their rights and castigating federal officials for not inviting them.

As a result of the pioneering work of the CAG, Christy Todd Whitman appointed me to the Superfund subcommittee of the National Advisory Council for Environmental Policy and Technology (NACEPT). There, again, we were exposed to links beyond St. Louis with Fruit of the Loom. As I traveled the country on committee work, I learned both of other communities with some of the 166 Velsicol contaminated sites, and more generally of the similarities between the Fruit of the Loom story and that of other firms in similar communities. On sabbatical in 2003, I also had an opportunity to visit a number of communities where Fruit of the Loom had left negative legacies not related to the environment. In town after town, there were stories of lost pensions, of world-class companies milked of assets, and, in Central America, of grossly exploited workers.

Generally, these experiences not only provided massive amounts of information about repeated egregious corporate exploitation of places and people, but also created an obligation to tell the story. From CAG members in St. Louis to former workers in Georgia to waitresses in Tennessee to fellow members of NACEPT, and most importantly to my wife and kids, person after person said, "You must write the book." They did not say that because I am the best writer or the wisest, they said it simply because I ended up in the middle of the story—meeting people in West Point, Georgia; Campbellsville, Kentucky; and Clarksville, Tennessee; as well as Chicago, mid-Michigan, and even human-rights advocates from Honduras. NACEPT members emphasized that this case study was not so much unique as typical, important both because of the seriousness of the pollution and the extent of job loss.

Oddly, there also were personal links to the leaders of Fruit of the Loom and its predecessors. Long before I moved to Michigan, I had attended the University of Chicago and worked on the staff of the marvelous Joseph Regenstein Library. One of the trustees of our great university had been

Ben Heineman. Upon investigating the background of the pollution in St. Louis, I was shocked to learn that Joseph Regenstein's fortune, which he had graciously shared with my alma mater and innumerable other institutions—from hospitals to museums to zoos in Chicago—came from a firm that left a toxic legacy around the nation. Ben Heineman, whom I recalled as a model business and civic leader, lending his time and talent to all sorts of good causes, especially the civil rights movement, presided over Velsicol during its major food-contamination accident.

While this story had to be written, I could only do it with great help. First, of course, is my wife, Marilyn, who has lived this story from going to Cambridge in 1990 through critiquing the various drafts. All of my kids and now eleven grandkids have been impacted, perhaps negatively, by living too close to the legacy of Velsicol and to my obsession with it: Pete, Karen, Steve, Erin, Dave, Mary, Beth, and Jose. Two of the youngest have been especially involved. Karen became the conduit to the University of Michigan Law School. Teresa was young enough to be a CAG member while she still lived at home. On the CAG as well have been a host of committed citizens, not the least of whom are the various officers: Jane Keon, my successor as chair; Gary Smith; and Carol Layman. Likewise are the many selfless members of the Superfund subcommittee of NACEPT and supporting EPA staff, not to mention the long-serving mayor of St. Louis George Kubin, and his successor Joe Scholtz, and the city manager Bob McConkie, and councilman Bill Shrum. There were two remarkable retirees from Dow Chemical, Drs. Fred Brown and Gene Kenaga, both of whom died shortly before publication. There also are the many attorneys and law students who have helped us, especially Nina Mendelson. Government officials helped out, including Tom Alcano from the US EPA and Scott Cornelius from the Michigan Department of Environmental Quality. While often seemingly criticized in the manuscript, they are greatly to be respected for their openness to advice from volunteers. Dave Dempsey read the manuscript at an early stage. Then there have been a group of dedicated student assistants, especially Shannon Finnegan (who also served as CAG secretary in 2001?2002), Cardell Johnson, Salina Maxwell, Margaret Hewitt, Ben Roberts, and Alyssa Walters. Last but not least are my faculty colleagues who also have served as key CAG officials: Dr. Melissa Strait, the CAG secretary after 2002, and especially Murray Borrello, Technical Committee chair.

Introduction

They were careless people . . . they smashed up things and creatures and then retreated back into their money or their vast carelessness, or whatever it was that kept them together and let other people clean up the mess they had made.

— *THE GREAT GATSBY*

WHILE INNUMERABLE HISTORIC EXAMPLES EXIST OF ABUSE OF INDI-vidual power and excessive self-interest, early twenty-first century global financial crises illustrate that such abuses can impact large numbers of people and communities. Individual excess transitioned into a fundamental societal problem when its pathologies were magnified in corporations no longer properly controlled by either civic processes or cultural norms. As the new century's financial crises unfolded, a common error of analysis focused on relatively recent changes in law or policy that encouraged imprudent behavior, such as the repeal of bank regulations in the previous decade.[1] The problems that became evident in the fall of 2008 had been brewing in American business and civil society for decades and were grounded not merely in contemporary leadership mistakes, however grievous, but in much longer-term civic, economic, and environmental ideologies and practices.

There is no better way to see this than through a case study of a group of firms that at one time or another were linked to Fruit of the Loom. Politically and economically, they often modeled exploitation if not contempt for local communities, their workers, and their investors. Environmentally, the former chemical subsidiary Velsicol left a multistate legacy of carelessly dumped chemical and radioactive wastes. Rather than be distracted by flagrant recent financial misjudgments, this case study reveals the need for widespread and systematic reforms ranging from U.S. institutional leadership practices to renewed citizen empowerment if the sources of the problems are to be addressed. The case of Fruit of the Loom exposes two primary sources of the economic and cultural maladies of the new millennium, as well as effective ways of curing them.

First, the firm's history illustrates how poor corporate and civic leadership can negatively affect workers and residents of host communities, the natural world, and ultimately the economic viability of companies themselves. Advocates of sustainable capitalism have called these three dimensions of corporate impact the "triple bottom line."[2] Velsicol's behaviors, which first appeared confined to egregious environmental practices, demonstrated that disregard for any of the three dimensions likely will undermine continued success in the other two. The business media, as well as the scientific and technical professionals who worked for and with the firm, blindly dismissed early concerns with the company's environmental impact, or later the movement of Fruit's jobs out of historic host communities as inevitable adjustments to allow the company to remain competitive. The firm's bankruptcy in late 1999 proved the error of such compartmentalized economics.

Reviewing the history of this company is especially helpful because of its links to some of the most flawed leaders in finance and government. As the national economy unraveled in 2008 and 2009, even casual observers of U.S. leadership recognized Fruit's collaborators as failures. Whether junk-bond promoters such as Michael Milken, insurance schemers at AIG, or officials of the Federal Reserve, all had helped Fruit flounder, yet had survived its collapse for another decade. The core danger of the 2008–09 collapse, as with the earlier problems in savings and loans, was not the loss of jobs and wealth, but that the response would be the use of short-term public subsidies to allow failed leaders and practices to continue. The public therefore urgently needs to review the longer-term history of the behaviors that underlay the crises, rather than focus only on mistakes tied to mortgages or auto technology. The study of Fruit of the Loom and the companies that became part of its complex structure shows, for example, that the insurance giant AIG did not merely speculate carelessly in the "subprime" mortgage market, but speculated more generally in other high-risk insurance schemes.[3] Examining the history of Fruit of the Loom also shows the long-term consequences of deregulation of the financial sector and promotion of excessive debt financing.[4] While detrimental to the firm's workers, host communities, natural environment, and outside investors, these schemes transferred massive amounts of wealth, especially savings accumulated over many years of productive business activity, to the private accounts of speculators, much as the 2008–09 bank bailout transferred public subsidies to the private accounts of bank owners.[5]

Second, beyond the details of corporate skullduggery and administrative incompetence revealed by the Fruit of the Loom case, the firm's history, especially that of the Velsicol subsidiary, chronicles the evolution of an ideology and methodology for exploiting the environment and people. Firms like Velsicol relied heavily on the power arising from professional expertise, and effectively used it to control citizen objections to their egregious behavior. As with so many mid-twentieth-century U.S. manufacturers, respect for the leaders of Fruit of the Loom arose from a long history of innovation and entrepreneurship within components of the Fruit "empire." The firm's experts pioneered the production of ubiquitous products such as underwear and the window envelope, and made technological advances in aluminum casting and fluorescent lighting. A Nobel laureate praised one of its chief scientists. Community leaders celebrated the philanthropy of its owners. Yet, several of the firm's great philanthropists equally symbolized the pursuit of naked self-interest.

Environmentally, the history of the firm exemplifies what Thomas Berry called "the central human issue and the central Earth issue of the twenty-first century. . . . that after . . . centuries of industrial efforts to create a wonderworld we are in fact creating wasteworld, a nonviable situation."[6] Socially, the history of Fruit demonstrates how separation of corporate ownership and leadership from the communities in which the firm operated encouraged the transfer of wealth from the producers. It also imposed excess risk on the workers and their workplace, exorbitantly enriching managers and their institutional collaborators.[7] Finally, the firm's behavior illustrates the timeless need for civic institutions to adopt and enforce regulations to protect communities and the natural world from pillage and abuse. With the rise of what political scientists labeled "interest group liberalism," in the middle of the century, Fruit exemplified how firms undermined, even corrupted, civic capacity. They retrained citizens to pursue individual interests in jobs and cheap consumer goods rather than seek the more amorphous but much more important long-term public interest.[8]

The leadership at Fruit of the Loom and its affiliates created numerous problems that parallel fundamental troubles in the U.S. policymaking process. While some link these difficulties to the Reagan administration's advocacy of "deregulation" and neoliberalism, this study shows that those problems had earlier manifestations, more deeply rooted in the dominant American theory of "political economy."[9] The company's controversial abuses of the environment and workers began in periods often marked by progressive regulatory regimes. This investigation suggests that the widely shared free-market ideology that dominated both political parties in the second half of the twentieth century facilitated these abuses. The transition to an era of "deregulation" did not correspond to intensification of these wrongs; in fact, some progress in addressing the firm's problems occurred after the appearance of "Reaganomics." The "Fruit" case is important because it clarifies that neither the liberal nor neoliberal ideologies behind the dominant U.S. policy process effectively protects the public interest, and it also suggests a procedural approach and supporting ideology that do identify and defend that common good.[10]

The story of the firm's contamination and later financial collapse did not end with a fully negative result. In St. Louis, Michigan, the only community where the firm left multiple highly contaminated sites, and one where the local plant was shuttered and demolished, citizens modeled a response that might well be copied by those facing similar fates. Not only did St. Louis citizens join together to aggressively confront the firm and docile regulators, but they displayed concern and support for the other communities and global civic culture threatened by the firm and its allies elsewhere. They concluded that if a humane policy process, marked by restraint and integrity, is to emerge in the world, communities familiar with the failure of current corporate and civic leadership must learn from their local experience but act globally, confronting related abuses of the natural world, workers, communities, and investors.

TWO EXAMPLES

Two cases may serve to illustrate the chronology of the problems on which this book will focus. They substantiate that corporate irresponsibility and the ensuing damage originated years before late twentieth-century deregulation. Both cases show the variety of destructive impacts that resulted from the absence of civic responsibility among corporate leadership—one primarily costing jobs, the other contaminating the natural world. The first example occurred at a factory of West Point-Stevens, the large apparel firm acquired by Fruit of the Loom in 1988. This company, like the subsequent example, demonstrates how firms that passed in and out of the Fruit of the Loom empire were adversely affected by that experience.

West Point-Stevens, especially the old J.P. Stevens part of the firm, had been infamous for hostility to organized labor. In the 1960s and 1970s, Stevens fought a protracted battle with the Amalgamated Clothing and Textile Workers Union (ACTWU). During what some would characterize as the height of progressive U.S. liberalism, Stevens held off the ACTWU organizing campaigns for two decades, a confrontation immortalized in the Oscar-winning movie *Norma Rae*.[11] The movie, which appeared in 1979, described the successful labor-organizing activity of Crystal Lee Jordan at the O.J. Henley plant in Roanoke Rapids, North Carolina. In fact, it took nineteen months after the release of *Norma Rae* before the company signed a comprehensive union agreement, a few weeks before Ronald Reagan won the White House.[12] For the next two decades, during the peak of what many consider anti-progressive neoliberalism, the union endured. What doomed the ACTWU was not hostility from the Reagan administration, but the excesses of corporate raiders, who perfected a scheme that permitted speculators with modest means, such as Fruit of the Loom's Bill Farley, to acquire companies. He procured Fruit in 1984 by borrowing billions, and then attempted to pay down his debt using assets of its various affiliates, such as West Point. While Farley's efforts to acquire West Point-Stevens transpired during the era of deregulation, he, as so many of his junk-bond compatriots, refined the techniques long before 1980.

While 1980s deregulation of banking and the stock market facilitated such transactions, the rise of institutional investors provided the capital needed by the deregulated market. Roanoke Rapids suffered because these investors focused only on short-term growth in stock value, resulting from debt-financed takeovers of cautious old firms. Investors like Bill Farley understood that companies with a large market share and a reputation for innovation could be milked for short-term profits if new investment were halted and production moved to low-wage countries. A firm raided for its cash and trademark could be sold shortly after purchase, with much of the acquisition debt following the firm like an albatross. For plants such as the Roanoke Rapids facilities of West Point-Stevens, multiple debt-financed leveraged buyouts burdened them with pressure for reduced investment in technology and wages.

In the mid-1980s, West Point-Pepperell had borrowed to buy out Stevens, and then in 1988 Farley borrowed $1.6 billion to buy West Point. The massive indebtedness that grew from these consolidations cursed the surviving firms with unsustainable interest payments. The subsequent reduction in technical investment initiated a continuous decline in global competitiveness. At one point during his takeover of West Point, Farley's cash flow covered only 70 percent of his interest payments.[13] What had been West Point's comfortable productivity advantage over foreign competitors disappeared, and the firm could only compete by moving production to places where reduced labor costs compensated for low productivity. To pay the soaring debt, Farley squandered assets not only from West Point but from highly profitable subsidiaries, such as his aluminum-casting operation, Doehler-Jarvis (see chapter 5). As the empire collapsed around him, Farley lost control of Fruit of the Loom a few months before it entered bankruptcy. However, the indisputable tragedy of such mismanagement was the closure in spring 2003 of facilities such as the "Norma Rae" factory in North Carolina.

The purpose of investigating these cases is to validate the long-term nature of the problems at Fruit of the Loom. It was not decimated by 1980s-era deregulation or by other short-term policy changes, such as new trade laws. These were not causes but effects of a long-term rise of self-interested irresponsibility among business leaders, institutional investors, policy experts, and the media, which celebrated rather than critiqued the behaviors. The "Norma Rae" factory did not close in 2003 because of a labor battle; it folded after two decades of labor peace because of the incompetence of corporate leaders and the disinterest of the company's investors in actual production. This situation did not originate in the 1980s; it emerged earlier in what this book contends is a version of U.S. liberalism that welcomed excessive focus on self-interest.[14]

The second case, which receives greater attention later, is that of Velsicol Chemical, a company whose owners egregiously impacted the environment in their pursuit of self-interest. Velsicol Chemical and its predecessors had a long history of environmental degradation and explicit contempt for critics of their irresponsibility. Contrary to those who concentrate only upon the deleterious effect of deregulation in the 1980s, Velsicol faced its first effective regulation after 1980 and made its worst environmental impacts in the early 1970s, the era often identified as the pinnacle of modern environmental policy. The history of Velsicol is important because it shows that the problems inherent in the political economy of the last century were not rectified by new government environmental regulations. Some corrective action only occurred when citizens were willing to identify the public interest and to demand accountability from civic institutions and civil society.

With and without federal environmental laws, Velsicol defied public law and regulators for decades. Velsicol had been linked to massive fish kills in the Mississippi River, yet escaped responsibility because of its effective lobbying (see chapter 1). Company leaders convinced local community leaders, in an era of prosperity, that jobs were more critical than environmental stewardship. After the appearance of Rachel Carson's *Silent Spring*, Velsicol, joined by petrochemical leaders and their collaborators, initiated an attack vilifying Carson and her allies. While one could argue that the attacks on *Silent Spring* came before modern environmental regulation, the firm's most infamous environmental catastrophe occurred in the spring of 1973 at the height of federal environmental regulation and shortly after creation of the U.S. Environmental Protection Agency (EPA). Velsicol's St. Louis, Michigan, factory mistakenly shipped tons of a highly toxic fire retardant made of polybrominated biphenyls (PBBs) in place of an animal feed supplement, resulting in PBB's entry into the food chain. While the details of this incident are explained in chapters 2 and 3, it reveals the failure of the regulatory state to prevent accidents.

Remediation of the consequences of the PBB accident took place under the Superfund law. Officially the Comprehensive Environmental Response, Compensation, and Liability Act (CERCLA), passed by the lame-duck Congress in December 1980 after the election of Ronald Reagan, Superfund authorized the EPA and state partners to identify the worst hazardous waste sites (the National Priorities List) and establish a mechanism to assure funding for their cleanup.[15] When the company settled claims after the PBB blunder, agreeing to close and contain contaminants at the St. Louis facility, the small town had three Superfund sites: the former plant site, a dumpsite in the middle of a golf course, and the former county landfill. In the 1980s and 1990s, decades supposedly characterized by deregulation, Superfund evolved, with strengthening amendments in 1986 and later reform efforts empowering communities to resist corporate opposition to vigorous enforcement.

While some communities took advantage of support for informed engagement in the regulatory process, the media increasingly became an accomplice of corporate interests. For example, in 2003, after the loss of 50,000 U.S. jobs and bankruptcy, CBS-RADIO provided Fruit's lobbyist, John Albertine, with a forum to defend deregulation.[16] Similarly, the *Detroit News* published a series on the twentieth anniversary of Velsicol's contamination of the food chain, ignoring the long-term peer-reviewed epidemiological studies documenting a number of serious health problems. Despite this evidence, the *News* editorialized, "The PBB case is especially interesting because it was one of the first of a long string of chemical scares caused by rising environmentalist concerns about industrial pollution. As Mr. Tobin [the *News*'s reporter] noted, nearly all of those scares . . . have turned out to be medical busts."[17]

The study of Fruit of the Loom documents how ideology tended to blind journalists and many applied scientists. On the one hand, they tended to approach all technical issues with a general faith in scientific progress, seeing use of pesticides, for example, as inherently better than past "primitive" procedures. Additionally, the persistent faith in individualism and the free market implanted a bias in favor of unregulated "progress."[18] The *Detroit*

News account of the PBB legacy, and Dr. Albertine's comments on deregulation both arise from trust in the ability of market forces not only to allocate resources but also to determine the "public" interest. This faith has origins in the early history of the United States and, in its excess form, is distinctly "American."[19]

Confidence in the quest of self-interest burdened the United States with special challenges in protecting human and natural resources. The exploitation of people or the environment had to become extreme before civil society accepted restrictions on market freedom. The disasters at places like St. Louis or Roanoke Rapids inescapably resulted from what the great Austrian economist Joseph Schumpeter called the creative destruction that inevitably arose from the success of creative entrepreneurs.[20] As scholars of the abolition movement have recorded, the United States long had difficulty determining the balance between the self-interest of the slave owner and the slave.[21] Similarly, although excessive dumping of poisons into the environment at Donora, Pennsylvania, in 1948 were condemned, the economic benefits of emitting tiny amounts of endocrine disruptors were not censured, despite the costs of altered reproductive behavior.[22] The policy process habitually rejected development restraint, overturning local regulations as "takings" of individual rights.[23] Albert Hirschman explained how the American experience—moving to a continent vacant of all but Indians, who were easily pushed aside and forgotten—instilled in the civic culture a sense that the challenges arising from excessive individualism were forgotten by "going West," or otherwise abandoning devastated communities.[24] Even when Americans claimed to be guided by higher principles, these values were blatantly phrased in individualistic or libertarian terms.[25]

As the new millennium approached, with the frontier vanished, the nation nostalgically renewed the traditional liberal faith in individualism. Contemporary neoliberals claimed that innovation required the rejection of efforts to force protection of the amorphous community interests. They merged fatalism with fear to condemn any deviation from the principles of laissez-faire political economy.[26] Nobel Prize–winning economists, such as Gary Becker, rationalized growing income inequality, not as an affliction or predicament, but as proof of the value of education. Under the concept of "human-capital formation," they converted the mission of schooling, especially higher education, from imparting wisdom into a hedonistic quest for wealth in the brutal global economy. Becker and his associates avoided examining the dubious contributions of leaders at firms like Fruit of the Loom, who procured excessive compensation while failing to manufacture products efficiently in nations with strong civil societies.[27]

The unprecedented productivity of the U.S. free market provided powerful reinforcement for an economic ideology justifying minimal regulation of the environment and employment. Leaders emerged in politics and the professions who no longer regarded the economy as a subsystem of the larger civil society, but as the total system. This trend was especially true in agriculture and its growing use of the latest science, including Velsicol's pesticides, to produce abundant and nearly identical foods. Citizens became consumers, directed not to worry about the wider or longer-term consequences of production methods, but to enjoy cheap abundance. As a leading agricultural economist, John Ikerd, observed:

> To the extent that neoclassical economics includes any remaining element of happiness, it most clearly is *hedonistic* in nature rather than *eudaimonic.* Economists commonly refer to overall well-being as the ultimate objective of economic activity, and in contemporary psychology, "the terms *well-being* and *hedonism* are essentially equivalent." The current pursuit of economic wealth is a pursuit of individual, hedonistic, sensory pleasure. And the pursuit of individual wealth, within this context, inevitably encourages the exploitation of others for individual gain and thus degrades the integrity of personal relationships.[28]

In the last decades of the twentieth century, the reappearance of staggering income differences, the decline in political participation, and the weakening of independent institutions, such as churches that previously sanctified the "habits of the heart," compelled many to worry about the American prospect.[29] Individualism spread not only from boardrooms to colleges and universities, it also thrived in the churches, undermining their leavening role that Tocqueville saw as nurturing those civic "habits of the heart." Many churches shifted their message from the transcendent and the sacred to individual adjustment, personal therapy, and the blessing of "success."[30] The media popularized other movements toward hedonism, as did modern public-relations practices that promoted unregulated individualism and rationalized its egregious conduct.[31]

ALTERNATIVES

Since the early republic, some of the great works of social analysis and literature have noted the tension between the excesses of individual freedom and the need to define and protect community. For example, after touring the United States during the 1830s, Alexis de Tocqueville produced a classic assessment of these tensions in a large federal democracy. He understood that all societies walked a fine line between the twin scourges of anarchy and oligarchy. On both sides, a few powerful or violent leaders emerged, bringing oppression, exploitation, and at its worst, pillage and rape to the many. After observing the young U.S. democracy, Tocqueville concluded his visit hopeful that the U.S. political culture controlled the worst tendencies inherent in individual freedom through its "habits of the heart."[32] These habits included a distinctive mix of commitment to republican principles and personal self-restraint, reinforced especially by the country's vigorous and competing churches. Although people in the United States sought individual success without hesitation or shame, they qualified that pursuit by "explaining almost all the actions of their lives by the principle of self-interest rightly understood." Tocqueville added:

> The principle of self-interest rightly understood produces no great acts of self-sacrifice, but it suggests daily small acts of self-denial. By itself it cannot suffice to make a man virtuous; but it disciplines a number of persons in habits of regularity, temperance, moderation, foresight, self- command; and if it does not lead men straight to virtue by the will, it gradually draws them in that direction by their habits.[33]

In 1985, *Habits of the Heart* shifted from Tocqueville's concept to become the title of a best-selling study of the decline of community and the rise of individualism in America.[34] *Habits of the Heart* was neither the first nor the only investigation of growing concern about personal interest. What came to be called "communitarianism" produced numerous studies during the last third of the twentieth century that critiqued the abandonment of "rightly understood" as a modifier for self-interest, and its replacement with "naked" or "radical."[35] Of course, even before the communitarians, the general culture had described the extreme pursuit of wealth at all costs as "greed," and listed it as a deadly sin. A number of leading business scholars had expressed concern with the negative consequences of entrepreneurship as much as they celebrated the innovation. William Kapp once summarized his research as presenting "a detailed study of the manner in which private enterprise under conditions of unregulated competition tends to give rise to social costs which are not accounted for in entrepreneurial outlays but instead are shifted to and borne by third parties and the community as a whole."[36] When the management guru Peter Drucker examined the topic in the 1980s, he likewise emphasized the need for observing warning signs of entrepreneurship gone wrong, rather than focusing on "success stories."[37] The prevailing justification for entrepreneurial creative destruction was progress. Communities like St. Louis in the late twentieth century experienced a contrary fate. They witnessed job loss and environmental pollution arising from earlier enterprise with no replacement innovation, only abandonment.

REACTION

The Fruit of the Loom case is worthy of study because in one community, St. Louis, the only town with three of the firm's highly contaminated sites, citizens awakened and found their "habits of the heart." At the turn of the new millennium, they organized to challenge irresponsible corporate leadership, and the media and experts who defended such behavior. The uniqueness of their action is visible through a review of the problems, both environmental and economic, in many of the other communities that hosted company facilities. A number of these communities also tried to confront the closures and contamination left by the company, only to fail. St. Louis took advantage of all possible structures and processes to sustain corrective efforts. The community's story can be emulated, and that makes the story worth telling.

Reforms of Superfund administered under the Reagan administration and the first Bush administration provided an initial structure and funding, which empowered St. Louis citizens. Once empowered, however, they continued to insist on a wider role for citizens in policymaking than those reforms envisioned. The community's contemporary response began in 1997 after EPA and state officials returned to town in 1997 to report that earlier Superfund remediation had failed. Within weeks, outraged residents took advantage of the new reforms of Superfund to create a community advisory group (CAG) to advise the EPA—the Pine River Superfund Citizen Task Force—and to secure a modest amount of funding: a technical assistance grant (TAG). The TAG allowed the citizens to hire their own experts to secure second opinions on technical and scientific issues.[38]

When the task force started in 1998, it only raised some technical challenges to the government and companies (see chapter 7). By the end of the next year, it sought large funding for a sophisticated health study, and in 2000 filed a $100 million claim against Fruit of the Loom in bankruptcy court. The details of that history are at the core of this book, providing a model to address the problems from modern corporate irresponsibility and individual excess. Partnering with a local college for free technical expertise, the task force did not stop with aggressive legal tactics.[39] Slowly, it pioneered ways to directly challenge elite expertise in and out of government and in the press. Early in the millennium, CAG leaders forced the Department of Justice to hold three hearings in the community on environmental settlements with the local polluters.[40] At the second of these, the local community expanded its focus beyond its region, inviting people from Memphis, Tennessee, who faced Velsicol contamination there, to come to the St. Louis hearing. This step inaugurated the community's awareness that local groups needed to join together to confront national problems created by irresponsible global leadership.[41] The challenge that remained was to link with communities outside the United States where firms such as Fruit of the Loom have found few requirements to report their impacts on what John Elkington has called the triple bottom line: people, plant, and profit.[42] Relying on weak international regulatory mechanisms, the leadership of firms can exploit workers, abuse the environment, and ignore investor protections. Consequently, in 2007 and 2008, the task force pioneered an approach to global corporate irresponsibility.

Long advised by policy experts to "think globally and act locally," local environmental activists usually focus on reviewing general environmental issues and then apply those lessons to their specific circumstances. By implication, the local groups lack the expertise or resources to shape the national or global policy process. In 2006–07, the Pine River Task Force learned it was being outmaneuvered in complex policy negotiations by the perceived wisdom developed by global experts and the mechanisms they controlled for justifying policy—conferences, peer-reviewed technical publications, and global forums. The catalyst for the task force's new strategy materialized when global petrochemical lobbyists began to undermine the global consensus on the restricted use of DDT. After the *New York Times* published a story in its science section that argued for use of DDT against malaria in Africa and critiqued the waste of funds on Superfund cleanups of DDT in the United States, the CAG determined that it needed a global response.[43] The Pine River group teamed with Alma College to sponsor the Eugene Kenaga International DDT Conference in March 2008. They captured the endorsement of the Society for Environmental Toxicology and Chemistry, and the International Society of Environmental Toxicologists, bringing together African and First World health experts to directly challenge the World Health Organization and urge support for the Stockholm Convention on Persistent Organic Pollutants.[44]

CASE STUDIES AND SOLUTIONS

This study is guided by the assumption that the case method is an appropriate way to understand the functioning of the national policy process and the vibrancy of an ideology, such as neoliberalism. In the social sciences, there is debate about the value of case studies. As one political scientist observed, "Most researchers who work with quantitative methods are averse to doing case studies. They presume that case studies cannot test or inspire theory because they suffer from selection bias."[45] Only work built on large data sets and analyzed dispassionately is acceptable. The goal is to create unifying theories. Based on such theorizing, neoliberal political economists have long proposed that regulation inhibits innovation. Since the late 1970s, they have had an opportunity to test their theories, beginning with the transportation deregulation of the Carter administration and intensifying under President Reagan and his successors.[46] The country has experienced exceptional levels of economic growth since deregulation began. A neoclassical economist can say, based on data for the total economy, that rational analysis proves that much of the old regulation was bad. They boast of the precision of the laws of market economics, proven valid by mathematical models.

A few argue against this methodology. Nancy Cartwright of the London School of Economics has warned, "Economists simply do not know enough to fill in their law claims sufficiently."[47] The gaps in mathematical models reveal the need for alternative approaches, such as the case method used in this study. Almost a century ago, Pierre Duhem, the French mathematician, writing about seemingly more precise physical theory, extolled the value of studying specific cases: "Between an abstract symbol and a concrete fact there may be a correspondence, but there cannot be complete parity; the abstract symbol cannot be the adequate representation of the concrete fact."[48] While the limits of our theories may be minimized in much of the natural and social sciences, our legal system is quite conscious of them. Kim Scheppele, professor of comparative law at the University of Pennsylvania, observed:

> Social scientists of a statistical bent often dismiss as "anecdota" the sort of experiences that [the victim of an accident] had, while lawyers and judges often distrust the statistics that social scientists provide because the statistics can't ever be said to be "true" in a particular case. . . . Nonetheless, the statistical backdrop makes it harder to argue that each accident taken alone is really as bizarre as it sounds. To the judge or the lawyer, accidents are possible because each case is unique. But to the social scientist, one could have "law without accidents"

because from the bird's-eye view of the patterns, very little is truly unpredictable when taken in the aggregate.[49]

Although on the national level, theories about deregulation seem correct, the case for individuals can be quite different. As the country has decreased regulation in the last quarter century, the nation has experienced fresh economic activity. Yet, most studies of income distribution show growing inequality during the era.[50] Without careful analysis of the impact upon specific subgroups in the population, macroeconomic data crunching can mask negative impacts upon the majority of the population behind massive benefits arising for a few. Additionally, short-term economic gains for a few may fail to measure the long-term legacy of unsustainable resource use, abandoned communities, egregious social dysfunction, and individual injustice. Such conditions may be rampant in a place with immense aggregate economic growth, such as the U.S.-Mexico border region.[51] The fundamental error arises from forgetting that people function not with what Scheppele called a bird's-eye view, but with a very human one. They do not only scan the world from the sky to get the big picture, they land and feel the complexity on the ground.

One of the challenges in assessing ideologically driven movements, such as neoliberalism, is the difficulty of understanding actual, as opposed to theoretical, impacts. The epistemological difficulty with analyzing neoliberalism is that it subsumes the political under the economic. Work, culture, even family relations can be reduced to market issues, as when reference is made to the "labor market." Yet, many people consider work as a source of meaning for life—a vocation. Similarly, many citizens assume policymaking is a process of providing equity or justice, not merely profits. If Jefferson was right that one of the three core purposes of government is to promote happiness, neoliberalism is woefully inadequate, centered on life devoted to economic maximization. In real communities, numerous citizens assess civic efficacy with standards beyond economic efficiency, including equal enjoyment of the environment.

David Rosenbloom captured this complexity of good public policy by saying it must balance three nearly contradictory concerns: what he called the managerial, the political, and the legal.[52] The neoliberals fixate on the managerial emphasis upon economic efficiency. This approach relies on experts who view individuals impersonally and emphasize the scientific method and rationalism to achieve knowledge. By contrast, the political approach depends on elected representatives and their agents to address the concerns of society, seeing individuals primarily as parts of groups and communities. Civic leaders ascertain citizen needs and expectations through elections, lobbying, and other forms of interaction with constituents. This is a messy process, for on most issues only an intense minority—on both sides of a debate—has an opinion. The third position is explained by Scheppele. As a civil society based on law, citizens expect case-by-case adjudication, reacting to people as discrete individuals with rights, who demand procedural due process no matter how inefficient.[53]

The PBB accident can be approached in each of these three ways. From the managerial perspective, studies of accidents reveal how often, in comparable situations, companies and their employees make mistakes. The actuarial tables of insurance companies grow from this process. However, the politician cannot respond to angry constituents by saying, "It was bound to happen to someone, just accept it." No, the residents of the contaminated watershed, or the workers, or the farmers, expect their political representatives to respond to their problems regardless of the odds. Finally, the judge must decide if the injury from the accident to a specific individual warrants application of a legal sanction. Neoliberals envision citizens taking the first approach, trusting the occasional accident to be essential for economic growth. People living in real places, not ideologically isolated neoliberal "think tanks," want the three approaches balanced and applied at the same time. Unlike a Washington lobbyist, the citizen grasps that the regulation of self-interest is superior to the deregulated economic growth promised by neoliberalism.

The PBB accident introduced a known carcinogen into the bodies of approximately eight million people. Worse, PBB is not the contaminant that makes the Velsicol site in Michigan one of the most expensive Superfund cleanups. Massive DDT dumping caused the current remediation, but that is not the end of the firm's environmental legacy. In filings with the U.S. Justice Department in 2002, Velsicol admitted responsibility for 165 other sites, at least two of which rival the cleanup costs of the one in Michigan.[54] However, they are not the sole disasters that unfolded at the Fruit of the Loom empire. At one point in the late twentieth century, Fruit's leadership controlled U.S. businesses employing over 75,000 people.[55] By 2004, that employment had fallen to 12,800. While a modest number of those jobs survived abroad, human-rights groups challenged their value.[56] Furthermore, the leaders of Fruit of the Loom and its predecessors and descendants repeatedly manipulated corporate finances, costing thousands of investors and pensioners hundreds of millions of dollars. That these problems happened repeatedly in many places—Roanoke Rapids; Clarksville, Memphis, and Toone, Tennessee; Toledo, Ohio; and elsewhere—not just St. Louis, proves that the PBB mistake was not simply an accident but a warning of what should have been prevented.

A CIVIC GOAL

This study begins and ends with advocacy of building sustainable communities by promoting alternative forms of policymaking and responsible leadership. The evidence from this case study attests to the urgency of the task. That John Albertine could be a leader at Fruit of the Loom and still be solicited by the media to comment on regulation demonstrates the need for a renewed civic consciousness and commitment to integrity and sustainability. While the word is used too frequently, including in the context of this book, sustainability means, at a minimum, that "the present generation has the obligation to pass on to future generations an average capital stock,—of goods, services, knowledge, raw materials—that is equivalent to today's." A stronger version assumes that "some natural resources and ecological processes are critical; they cannot be depleted below a certain level without dramatic ramifications."[57] While people, economic institutions, and technological tools and practices change, building a sustainable community embraces permanence and not short-term exploitation. Good stewards of such a community struggle to reject the short-term and destructive, while laboring to promote the general welfare by preserving resources, maintaining public law, or encouraging vigorous civic life. The hope in producing this book is that it will motivate readers to organize their communities and inspire them to promote sustainable public policies.

One step toward sustainability is to keep Rosenbloom's three approaches in balance. At its core, this study documents the consequences of poor leadership at Velsicol; its last parent, Fruit of the Loom; and other firms linked to them. Their negative impact upon management practices, finance, employment, and the environment was so damaging to humans as well as to the natural world that this story cries out for a conscious response. It proves the need for renewal of civic life as a defense against those who were formerly called robber barons. Despite the complacent neoliberal demands for less regulation, this case makes it clear that that ideology is fatally flawed.

The pattern for this new civic vitality complies with Tocqueville's "habits of the heart." The community advisory group process of the US EPA modeled this approach. Without glorifying a rather minor governmental procedural change, the CAG was one example of empowered citizens, a rare species at the start of the new millennium in America. The CAG at the Velsicol site in Michigan regularly objected to legal interpretations and decisions that failed to acknowledge the equity implicit in protecting the environment for future generations. In the mid-twentieth century, political scientists Grant McConnell and Theodore Lowi warned that U.S. civil society was in peril because "justice" had come to be considered as the product of a deal between powerful interests.[58] This study concludes with an appeal for emphasis upon Rosenbloom's third approach: restoring public law to its prominence. A core ingredient of a sustainable society is widely respected and fairly enforced public law.

As a corrective to the emotions and fears they believe are emboldened by modern environmental regulation, one school of modern legal scholars calls for injecting natural and social-science expertise to the law. Chicago's Cass Sunstein has modeled this solution, advocating precise, scientific calculations of risks and benefits.[59] He assumes that with enough data, the experts can determine accurately the perils of alternative government decisions.[60] While tempting to an academic, no amount of research can detect precisely all risks, since the long-term ones may be assessed incorrectly or remain obscure to contemporary scientists.[61] There are innumerable precedents of more general failures of what James Scott of Yale University terms "high-modernist" planning.[62] Thus, this study advocates that the experts serve as consultants to an active citizenry.

In 1978, as the Velsicol problems escalated in Michigan, an environmental aide to Governor William Milliken, an expert employed by a thoughtful public figure of the era, voiced concern about local citizens' input into the environmental policy process.[63] In a letter to a county commissioner, he said, "I have reservations about the kind of citizen committee that [the state senator] was proposing. We have learned from past experience that it's awfully easy to frighten the public to a degree which may be unwarranted about toxic chemical problems."[64] Considering that the expert-dominated process yielded a $38 million solution that not only failed to contain the pollution but exposed inhabitants to additional toxins for at least another quarter century, the citizens could not have done worse.

This study teaches that the most basic principle of democracy is right: the collective citizenry must control the leaders, even their highly trained experts. The average citizen is not always correct, and there clearly is the need for professionals employed in the civil service to monitor policy implementation; however, experts and self-interested individualists also can be wrong and can be more dangerous if mistaken than a single ordinary citizen. Whatever the merits for specific environmental policy, reinvigorated public engagement is essential for the good of civil society. In order to escape Hirschman's pathological excuse of "going West," the United States needs a corrective to excessive individualism. The St. Louis case, especially when contrasted to the other communities harmed by Fruit of the Loom, demonstrates processes and structures to control neoliberal excess and its professional allies in the media and the natural and social sciences. This case therefore highlights the endless need to control "robber barons," and also to regain control of the policy process and mark it with restraint and integrity.

This study does not end as a pessimistic account of the triumph of neoliberal cynicism, but with descriptions of a community-empowerment process that has had at least modest success. Such empowerment resembles other historical examples. As this investigation concludes, the St. Louis, Michigan, example is linked, hopefully not presumptuously, to the battle for humanistic civic life throughout the modern world, beginning with the Renaissance. That linkage illustrates the value of pondering the struggles of the past that pitted special interests against the public good, and more importantly, restores confidence that communities can achieve a good society.[65] As citizens seek to restore expectations and methodologies that support their vision of the long-term public interest, they can perceive that this is an old fight that must be renewed periodically to restore control over elite economic leaders and their accomplices. It is about civic empowerment to direct innovation and enterprise toward the creation of a good society.

This study celebrates the historic strengths of the American system, such as the multilayered federal policy structure with a preference for public decisions, which most impact daily lives, legislated at the local level. Another attribute has been the distrust of the corruption and hunger for power of far-away experts, beginning with the minions of George III. This examination does not reject the civic tradition; rather, it challenges the twentieth-century policy innovation that empowered professionals, linked to special interests, at the expense of those with a vision of the long-term public interest. As Frank Fischer phrased the current debate: "Already talk of democracy all too often serves as little more than a thinly veiled guise for elite government. The question is: Can the democratic process be rescued from the increasingly technocratic, elitist policy-making processes that more and more define our present age?"[66] Max Weber—one of the great defenders of the virtues of the modern bureaucratic state, but also a perceptive critic of its vices—warned citizens to find mechanisms to assert their values.[67]

While celebrating vigorous local participation in technocratic policymaking, this study is not guided either by simplistic adulation of local wisdom or a bias against either business or national political leadership. Local people often can be foolish, primitive, superstitious, and bigoted. They seldom possess the expert knowledge of modern professionals, whether physicians, accountants, or economists. Nor do they understand global and national political imperatives, as do better-trained leaders. Yet, this study maintains that local weaknesses must be balanced by concern with the potential immorality of the specialist or the corruption of the Washington official. The only checks on experts and insiders is either a benevolent dictator, or a vigorous public-policy process that allows outspoken or outraged citizens to say, "No!" The locals can do a lot of irrational things, but those may be not nearly as evil in intent or consequence as the health problems spawned by the irresponsible chemists at Velsicol, or the theft of worker and investor wealth by the devious accountants in the pocket of Fruit of the Loom's officials. An unqualified local manager may cost a handful of his neighbors work when his incompetence causes his business to fail. Such behavior pales by comparison to the impact of the leadership at Fruit of the Loom, whose actions caused 65,000 workers to lose their jobs, and more to surrender their pensions, health insurance, and investments.

Welcoming business creativity has given the United States its great wealth, permitting most residents to live more affluent lives than almost anyone else in the world. This study is not oblivious to that achievement. In fact, many communities adversely affected by the behavior of Fruit of the Loom wish they had an innovative entrepreneur to bring good jobs to their hometowns. Nothing written in this study seeks to inhibit innovation, some of which inevitably results in failures and errors. The cases examined here are not examples of random business failures, inevitable in a free economy. They are not random accidents, such as the unavoidable odds that for every mile driven by millions of drivers, some misjudgments will lead to injury and death. These cases are more analogous to the accidents of the chronically drunk. The public response is no-fault insurance for the random driver errors, and criminal prosecution for the severely impaired. However, there have been no criminal prosecutions of leaders involved in the events chronicled here, with the exception of one set of charges against Velsicol's officials, punished by token fines after the firm hired one of the best-connected law firms in the country to guide it to victory. Their success in enriching themselves at the expense of their workers is reason to cease protecting the recklessness of the civically and socially impaired. That our media consults them as economic experts rather than exposes their disingenuousness suggests the extent of collusion in their actions and ideology. This story intends to encourage corrective actions and embrace of an ideology that envisages reform of our failing civic processes.

CHAPTER 1

Corporate Leadership Problems

On December 29, 1999, Fruit of the Loom, the underwear giant, filed for bankruptcy protection in Wilmington, Delaware. What had gone wrong to turn the owner of one of the world's best-known icons, the Fruit of the Loom trademark, into a financial failure? Its CEO until early in 1999, William Farley, had once boasted that he would reverse the decline in U.S. manufacturing jobs; yet, in the mid-1990s he had begun moving thousands of jobs out of the United States to Central America. He even shifted the corporation's legal headquarters from the Sears Tower to the Cayman Islands. One business critic nicknamed him "Bill (Off Shore) Farley."[1] One claim filed against the company in bankruptcy court dramatized the cancer that long had eaten at the heart of the firm. The Pine River Superfund Task Force filed the largest claim, requesting $100 million to pay for cleanup of Superfund sites linked to the former subsidiary Velsicol Chemical.[2] For the same reason, the U.S. Department of Justice, U.S. Fish and Wildlife Service, and National Oceanic and Atmospheric Administration intervened, seeking assets for seven major contamination sites around United States, including two in the Pine River watershed.

Deciphering the links of Fruit and Velsicol not only clarifies why such a large claim would be filed from a community that never had an underwear factory, but also reveals much about the pathologies within modern corporations and civic life that allowed what one former worker called "community rape." Fruit of the Loom and its sister firm, Union Underwear, were among the pioneers in divorcing production from specific factories, workers, and communities. The chemical affiliates Michigan Chemical in St. Louis and its parent Velsicol pioneered a different set of corporate behaviors. First, they ignored repeated complaints about emissions from local government, neighbors, and the rare critical expert. Additionally, they relied on allies in the chemical industry, industrial agriculture, land-grant universities, and federal and state agricultural agencies to show contempt for their critics. When necessary, they even threatened retaliation from plant closings to libel lawsuits to intimidate anyone who persisted in challenges to their behavior and the faith of "sound science." To be fair, many of their leaders also could be leading philanthropists and otherwise models of good citizenship, yet there always was a detachment from the specific places and people who produced their wealth.

The histories of the companies began when each were local firms manufacturing underwear and chemicals in communities in which the business owner lived. Long before the end of the century, they had become global conglomerates, hated by many in the former home community. Before the start of the new millennium, Fruit of the Loom and its affiliates were to produce not just underwear and pesticides, but also various textiles, fluorescent light fixtures, auto parts, and boots. Focusing initially on the underwear and chemical origins of the firm helps define the behavior patterns that would bring each part of the conglomerate into disrepute in its original hometown. All the cases raised the most basic questions about leadership failure. What had prevented Fruit and the subsidiaries, despite disproportionate market share in their industries, from being able to pay their bills? How could Fruit of the Loom have responsibility for several dozen highly polluted sites? How had these sites developed in the most scientifically advanced nation in the world?

The failure of individual communities to understand the full structure and strategies of global corporations with facilities in their midst was a core reason that these problems became unmanageable. Much of this study will be devoted to St. Louis, Michigan, because the experiences that led to the filing of the large bankruptcy claim in 2000 grew from the decision in St. Louis to survive, and to break out of the web that had been woven around the town. Many of the communities impacted by the entrepreneurship modeled at Fruit of the Loom did not survive as well. Important lessons arise from the conflict between creative destruction, survival, and renewal. Successfully confronting the web created by Fruit of the Loom depended upon understanding its origins and evolution. Equally, it required understanding of the shared assumptions of liberal and neoliberal ideologies that justified Fruit's behavior because it maximized private wealth accumulation regardless of the social costs.[3]

The citizens in St. Louis began to see the flaws in both the liberal and neoliberal ideologies because they could ignore only with difficulty the real consequences of implementing these ideologies. Meanwhile, the $100 million citizen claim served as a warning that excessive pursuit of corporate self-interest could result in the public questioning all the unconscious assumptions underlying the modern economic system. Unfortunately, few of the other communities in which the firm operated reacted as St. Louis did. In places as varied as West Point, Georgia; Clarksville, Tennessee; and Toledo, Ohio, the general rule was to accept the legacies of the new business practices: massive worker dislocation, investor losses, toxic contamination, and weakened civic capacity. A key reason for the failure to respond was that the new corporate practices and structure were seen in isolation. Most communities had no sense of the web of which they had become a victim.

On June 28, 2005, a jury failed to convict an entrepreneur who allegedly had led HealthSouth into one of the numerous corporate frauds of the era. Whatever the merits of that decision, the post-trial analysis explained that it was difficult to achieve conviction because the structure of the firm, like that of so many modern corporations, was exceptionally complex. The jury, bogged down in trying to understand the structure, eventually threw up its collective hands in frustration. The analysts speculated that some of the structural complexity of the modern firm could even be intentional, shielding leaders from the negative consequences of their actions.[4] Fruit of the Loom and its related companies certainly fit this mold. The corporation that emerged in the 1980s as modern Fruit of the Loom included not merely a host of textile firms and the complex chemical subsidiaries responsible for pollution in Michigan, but a host of other manufacturers as well. The best way to understand this structure is to begin with the model of manufacturing pioneered in Fruit's underwear divisions. Then, reviewing the early criticisms of the chemical subsidiaries, and the responses of their leadership, is the best approach to understanding the origins of modern failures to have economic actors practice responsible institutional leadership that is guided by sustainable pursuit of the long-term public interest.

THE FRUIT UNDERWEAR MODEL

Fruit of the Loom traced its origins to B. B. and R. Knight in Providence, Rhode Island. Founded in 1848, Knight was for nearly a century one example of the first U.S. industrial revolution. Industrialization had come to America in 1791 when Samuel Slater built the first textile mill along the Blackstone River in nearby Pawtucket. Not surprisingly, Bill Farley, who would come to control the modern Fruit of the Loom, was born in Pawtucket. From his youth in the decaying factory town, Farley knew that prospects had changed radically for his home state's textile industry. Knight had been one of the first casualties, filing for bankruptcy protection in the midst of the "Roaring Twenties." However, Knight had one asset of exceptional value, one of America's oldest trademarks—"Fruit of the Loom."[5] Proper use of that asset, Farley would learn, became the skill that would allow the firm to survive where others failed, but only as a shell of its former self.

Generally, the decline of the New England textile industry in the 1920s presented Americans with several lessons related to business. In a free-market system with modern transportation, opportunities to avoid one region's strict labor standards and the taxation that supported moderate public services would tempt and eventually compel firms to seek out new host communities. In the 1920s, the region offering a low-cost deal to the textile industry was the American South. It took only limited insight to know that after the South, the whole world waited in the wings to lure manufacturers offshore, with labor and environmental exploitation and low taxes.[6]

B. B. and R. Knight learned the lessons of the New England textile decline about as well as could be expected. In 1928 the company still had 2,180 workers in its seven mills in Rhode Island.[7] Once the Great Depression hit and business conditions moved from decline into crisis, Knight remade itself. In 1934 there were catastrophic labor disputes and desperate efforts to stabilize textile employment under the National Recovery Administration.[8] When those efforts failed, the company closed its six textile mills, retaining only its finishing operation, and took its trademark as its name. It would get others to produce its former products and pay for its one distinctive asset, the "Fruit" label.

In many ways, Fruit began to follow the practice first pioneered by Union Underwear (as in "union suit"). Union, with whom Fruit merged in the 1960s, pioneered being a "manufacturer" without a factory. The company's founder, Jacob Goldfarb, purchased cloth from suppliers, turned the cloth over to subcontract manufacturers, and sold union underwear at the end of the process. He demonstrated that a business owner could extract wealth with little or no involvement or responsibility in the communities or with the workers who produced the product.

However, Goldfarb was not as irresponsible as later manufacturers. He moved from licensing into ownership of factories. Perhaps it was his early life as a Polish immigrant trying to make it in America without advanced education that made him willing to commit to communities from where his wealth came. After he succeeded as an underwear tycoon, he gave lavishly to religious charities and to higher education. Yet, later business leaders also gave to universities; what was distinctive about Goldfarb was that he regularly visited the towns where his factories were located. He especially became known in Kentucky, where he helped communities recover from the Depression.

In an effort to counteract the era's unemployment, the city fathers of Frankfort, Kentucky, offered to provide Goldfarb with a factory if he would come there and start producing "union suits." A few years later, in 1938, Goldfarb moved even further from licensee into manufacturer. He negotiated a twenty-five-year license from Fruit of the Loom to put the Fruit label on Union Underwear. As a result of growing demand, he worked with local officials in western Kentucky to open a second factory in Bowling Green. The former licensee had become a major producer, seemingly abandoning its licensing model.[9]

Before Goldfarb died, however, he contributed to the renewal of the manufacturing process he had pioneered to start Union. In 1955 he sold out to Philadelphia and Reading Corporation, a holding company that was transitioning into a new-style conglomerate. Philadelphia and Reading was searching for a place to put the assets that it had accumulated in the previous century from railroading and anthracite mining. The year after acquiring Union Underwear, Philadelphia and Reading added Acme Boot of Clarksville, Tennessee. A few years later, in 1961, it fought a bruising battle with another budding conglomerate to purchase Fruit of the Loom.

By that time, Fruit of the Loom had moved fully away from manufacturing. It had not focused on loyalty to host communities since 1934. Instead, Fruit had developed a mode of operation that fit well into the new world of conglomerates. While places such as Frankfort and Bowling Green, Kentucky, rejoiced at their links to a company, a new generation of business leaders, including those at Fruit of the Loom, was in search of ways to minimize geographic responsibilities. Even before the end of the century, when this behavior pattern became chronic, the growth industries demonstrated that a new era had dawned, marked by contempt for local wishes. That this change had begun should have been clear in Rhode Island in 1934 with the closing of Knight's factories. It became crystal clear in the two chemical companies that ultimately would become part of the Fruit empire.

MICHIGAN CHEMICAL AND ITS COMMUNITY

In the midst of the Great Depression, few might expect to find the creation or growth of very many new factories, such as those of Union Underwear.

But some new resource-exploiting companies, such as the petrochemical industry, hardly noticed the downturn.[10] Michigan Chemical and Velsicol Chemical, two later subsidiaries of what would be Fruit of the Loom, were incorporated during this era. Joe Regenstein founded Velsicol in 1931 to produce petroleum-derived chemicals.[11] Four years later, two brothers from St. Louis, Michigan—Walter and Donald Wilkinson—founded Michigan Chemical to turn brine lying under the middle of the state into salable chemicals. Thirty years later, the two firms would merge as they moved through an infamous decade during which they became linked to some of the most well-known environmental and worker-health crises in modern America.

One of the challenges to responsible business leadership comes whenever financial resources are scarce and managers are tempted to cut corners on standards to reduce costs. According to one apologetic manager, Michigan Chemical regularly faced such temptations.[12] From its founding in 1935, the company did not make a profit until 1937, and then only $7,485. It would not be until 1945, when its DDT production for the military was booming, that it made over $100,000, and in 1947 and 1948 combined profits totaled $882,919. Proof of success came in the next decade when the firm opened another Michigan plant, and one each in Arkansas, Florida, Ohio, and Idaho.

Throughout the company's thirty years of independent existence, sales rose steadily, with periodic readjustments. In 1940 they totaled just over $100,000. By the end of the war they approached a half million. As discoveries from the wartime chemical revolution became available for civilian use, the company boomed. Sales jumped tenfold in the four postwar years. In 1951, sales were nineteen times larger than in 1945. The directors summarized the reason for the company's success on the cover of the 1952 annual report, where they illustrated the process for producing dichlorodiphenyltrichloroethane or DDT. In 1944 the company had begun DDT production to supply the U.S. military and, as the war wound down, the United Nations Relief and Rehabilitation Administration (UNRAA). The United Nations used Michigan Chemical's DDT to kill body lice on refugees.

After the war, the company reformulated the DDT into a host of specialized pesticides—many called Pestmaster, but some with more obvious target victims, such as Weedmaster, Roachmaster, and a generic Bugmaster.[13] Soon the company was riding a postwar fascination with pesticides. Former servicemen were the perfect targets for Michigan Chemical ads. If they had served in the Pacific Theater or Italy, they had seen DDT in action. Now the company reminded them of a "Vacation Necessity—A Can of Pestmaster."

They added, "So, don't spend your first peace-time vacation in five years, shooting flies and slapping mosquitoes. The answer is to take a can of PEST-MASTER 5% Residue Spray with you to the lake, the beach, the mountains or wherever you like to vacation." They also reminded the reader that PEST-MASTER 10% Powder would be good to have along "in case of bedbugs, ants, and roaches." The company provided advertisements directed at "leading drug chains," as they launched their "springtime insecticide campaigns."[14]

Not only did the company promote DDT for Americans on vacation. There was an aggressive effort to meet the needs of international health agencies, from continued use by UNRAA to the ministries of public health in places as varied as Egypt, Holland, Mexico, China, Italy, and Cuba. The company boasted that Pestmaster not only eliminated mosquitoes, but also flies, ticks on sheep, and lice on hogs. To overcome the resistance of traditionalists in the public-health community, the company asserted in their promotional materials that "Up to Date Thinking Prompts Use of Pestmaster in Large Scale Insect Control." Their advertising material showed people appearing to be nurses and physicians using DDT. For Americans with less-pressing survival needs, the company marketed Weedmaster 2, 4-D for lawns. Best of all, Weedmaster was water-soluble for easy spraying. The company pointed out it was harmless to skin and did not discolor clothes.[15]

DDT CONCERNS

There was much early evidence that the positive claims of DDT safety were not true. As early as 1937 there had been studies questioning the safety of some chlorinated hydrocarbons, such as DDT. Workers seemed to develop hepatitis and a rare and fatal liver disease from exposure.[16] After the mass production and use of DDT during the war, critics appeared. Before the war's end, E. B. White, writing in the "Talk of the Town" section of the *New Yorker*, quoted Edwin Teale of the New York Entomology Society warning, "A spray as indiscriminate as DDT can upset the economy of nature as much as a revolution upsets social economy." White also interviewed Richard Pough of the Audubon Society saying, "We are definitely alarmed over the possibilities of DDT. It might conceivably eliminate all insect-eating birds as well as shrews, moles, bats, and skunks. The insect eaters make up one half the base that supports the whole natural economy. . . . Mind you, we don't object to its use to save lives now. What we're afraid of is what might happen when peace comes."[17]

The next year the *New Republic* carried a story on the "Dynamite of DDT." It reported on the death of insects and birds following spraying in Pennsylvania. The magazine reached conclusions that were not only remarkable for summarizing the wisdom of the ages, but also diametrically opposed to the newer arrogance of "science."

> DDT's greatest defect for use out of doors is its nonselective killing power. Even in experienced hands, its use may be likened to firing a broadside at a throng of people in which we have both enemies and friends. Fortunately for us, there are more kinds of insects that are helpful than harmful. The destructive ones make all insects seem baneful, but it is probable that we could not live without the services of beneficial insects any more than without the birds.[18]

A year later, the *New York Times* reported that gardeners faced an increasing problem with red spider mites because DDT killed the mites' predators.[19] In the earlier *New Republic* story, they reported on the same phenomenon and warned as well of several cases of previous pesticides creating a "resistant race" of pests.[20]

The criticisms of indiscriminate DDT use in the popular press and academic forums during and immediately after World War II prove two things. First, later claims by pesticide manufacturers that in the early years no one knew about problems resulting from the chemical were false. Second, scientists had no excuse for lack of caution, for there was a robust literature in peer-reviewed science journals raising fundamental questions about DDT. Garth Fitzhugh and Arthur Nelson of the Food and Drug Administration in 1946 documented the chronic toxicity of DDT for laboratory rats, noting all the problems later cited for the pesticide. They found nervous-system problems, growth of liver tumors, and evidence of what later would be called endocrine system disruption.[21]

Nelson joined another U.S. Food and Drug Administration (FDA) scientist, Geoffrey Woodard, three years later to report on research on dogs, where DDT exposure caused damage to the adrenal glands much like that found in humans with Addison's disease.[22] The same year, four scientists from the University of California's citrus experiment station reported at the American Chemical Society's national meetings that DDT was absorbed by various fruits.[23] In 1951, three other FDA scientists, led by Edwin Laug, reported widespread presence of DDT in humans. Of seventy-five people tested, only fifteen did not have DDT in their fat. Eleven of their samples had over 10 parts per million. Only two women out of thirty-two did not show traces of DDT in breast milk. James Delaney held hearings in Congress on the carcinogenic properties of pesticides in food.[24] In 1955 the Public Health Service produced a large study on the toxicity of DDT and related insecticides.[25] Consequently, in 1958, the so-called Delaney Clause was added to the Federal Food, Drug, and Cosmetics Act, forbidding any additive to food that could cause cancer, potentially also banning certain pesticides if their residue remained on food.[26]

Knowing the apparent risks of DDT as well as the evidence of resistance even among targeted species, the proper response in the late 1940s would seem to have been limited use in public health or agricultural emergencies.[27] Yet, of course, that is not what happened. Chemists for companies such as Michigan Chemical pushed production to the maximum. Why would graduates of the best university science programs not rise up and tell companies, including their own employers, to slow introduction and use of DDT? These studies raised for the first time the core questions at the heart of this book related to private behavior. Why cannot moderation in behavior be required by citizens when, if left unchecked, such behavior can bring harm?

The behavior of technically advanced companies such as Michigan Chemical calls for a new policy model to control the outcomes of the creativity of skilled scientists. Michigan Chemical boasted of a management group in the 1950s that included scientists with terminal degrees in all key positions. Dr. Paul Maney was the scientific director of the pharmaceutical division, Dr. Cleveland Hollabaugh and Dr. Dwight Williams ran the research and patents office, and Dr. Adrian Gammill headed the rare-earths division.[28] The scientists in that division were among the national leaders in rare-earth research, funded by the Battelle Memorial Institute.[29] In the mid-1960s, before the company's worst environmental mistake, the company president, Helmuth Schultze, held a doctorate in chemistry from Rensselaer Polytechnic Institute.[30] Even when not led by an eminent scientist, experienced chemical industry executives, such as Theodore Marvin, headed the company.[31] The company recruited nationally, boasting of its research opportunities.[32] If such people could not control irresponsible production, then the experience with Michigan Chemical suggests that more than a code of professional behavior is needed to protect human health and the environment.[33]

Another source of concern at Michigan Chemical was the role of the board of directors. During most of the DDT era, from 1944 to 1963, the board included a number of nationally recognized business and social leaders. Beginning in 1947 they recruited Charles Stuart Mott, the General Motors leader and philanthropist, to serve on the board. At times, Harding Mott and Gene Tunney, the famed prizefighter, served with him. In 1951 the Dodge family bought into the company, and Frederick Van Lennep, husband of Frances Dodge, served on the board until it gave up control to Velsicol appointees in the early 1960s.[34] Yet, this distinguished board would not oversee the reckless environmental behavior of their company. The company was growing, that is what mattered.[35]

Another puzzle at Michigan Chemical, made all the more dramatic by what was to transpire with Velsicol, was the role of leaders of the firm in philanthropy. The president during the early DDT years was a New York "industrialist and philanthropist," William T. Morris. Morris was especially noted for his concern with the children of workers, cancer research, and helping the communities where his firms were located.[36] In St. Louis his name is remembered for the swimming pool he funded—although, as was his habit, he paid only half the costs and required the community to raise the other half. Likewise, of course, there were the Motts on the board, creators of one of the best endowed and respected foundations in America.

AWARENESS OF PROBLEMS

The reason for concern with the leadership of Michigan Chemical is that in this era, the firm left massive amounts of environmental contaminants in the watershed and community beyond the plant. Even the most casual observer of operations could not help see the noxious emissions into the Pine River and across the community. When finally cleaned up at the end of the century, the Environmental Protection Agency would need to remove more than a half-million tons of contaminants from river sediment. Near the plant there was little aquatic life. As early as 1935, four months before Michigan Chemical began operating its plant in St. Louis, the Saginaw City Council unanimously passed a resolution objecting to the company's failure to "plan for the disposal of its waste brine except to discharge into the Pine river."[37] Since the river was a tributary of the one used as the Saginaw drinking-water supply, the city council unanimously and unsuccessfully petitioned the state Stream Control Commission to prevent the dumping. A few years later (in late 1941), 121 citizens in St. Louis complained about contamination in the river. Public officials made an effort to respond, determining blame but stopping short of confronting the polluters.[38] Repeatedly state environmental officials noted the problems originating at the company's outfalls into the river.[39]

In the 1950s, after ten years of DDT production, the state found large amounts of DDT and other contaminants in the Pine River near the plant.[40] The very fact that the state was measuring these emissions proved the awareness of the problem. At the end of the 1950s, the company began processing rare earths to extract elements used in the new color picture tubes, producing significant quantities of low-level radioactive wastes. Again they followed earlier practices in waste disposal. They simply hauled the fiber drums containing the radioactive waste to a lot they bought by a creek northeast of town. Despite esteemed and well-trained leadership, the company became an accident waiting to happen.

In 1970 the company had one more warning of the potential problems ahead. Citizen complaints, having failed to move the state, now focused on the growing federal interest in environmental regulation that marked the era. Senator Edmund Muskie, the Democratic vice-presidential nominee in 1968, had come to national prominence as an environmentalist.[41] With Muskie the apparent front-runner for the 1972 Democratic presidential nomination, St. Louis citizens approached him for support against what they simply called "the Chemical." Muskie responded by asking the state to investigate. He noted that those who wrote him about the Pine River claimed that the company was "pouring untreated wastes into the river, causing foaming, and rendering the stream poisonous."[42] As in the past, the company's leadership ignored this warning. The failure of the leaders to act demonstrates two core problems in the regulatory process in technological industries: first, the nearly blind confidence of experts in new technologies, and second, the unwillingness to hear the concerns of either average citizens or staff of regulatory agencies. Michigan Chemical's behavior also demonstrated several problems with the behavior of modern corporate leaders. First, they repeatedly showed disregard for their responsibility as stewards of their natural environment. Second, they treated their local community, its civic institutions, and public officials with contempt. The third related to the ties of Michigan Chemical to outside owners, especially Velsicol Chemical. Many residents thought the local Wilkinson brothers owned the company throughout the period up to the mid-1970s. In fact the Wilkinsons had sold their control in the 1940s, and Velsicol Chemical bought a majority of shares by the mid-1960s. This confusion, if not deception, is another problem arising from modern corporate ownership. Clearly there are benefits for a company when neighbors feel an identity of interests with a local firm. While the firm may be treating workers as easily replaceable cogs in the corporate machine, local workers feel the "Wilkinson brothers wouldn't do that."

If the U.S. policy process, with its numerous centers of power on the state and federal level, is to work effectively, every effort must be made to enhance the power of citizens and public servants to second-guess the mistaken decisions of private executives and their tendency to accept and defend the decisions of their managers. Proving that these issues did not solely arise because of the egregious behavior of one rogue company is the reason for the remainder of this study. Whether looking at Michigan Chemical, Velsicol Chemical, or other parts of the Fruit conglomerate, highly educated and respected leaders did harm to their workers, their host communities, and their investors for too long.

VELSICOL'S ORIGINS

Unlike Michigan Chemical, Velsicol was born in controversy, and its development seemed marked by especially bitter quarrels among the founders—Joseph Regenstein and his first cousin, Julius Hyman. Dr. Hyman was the brains behind the company and Regenstein the financier. A chemistry graduate of the University of Chicago, with a Ph.D. from Leipzig, Hyman had returned to the United States to work in the labs of the Pure Oil Company on development of new petroleum-derived products.[43] His work at the company had opened his eyes to the world of hydrocarbon polymers. Pure Oil only saw the polymers as useful in producing oil for the foundry industry. Hyman saw much more.

Within a week of leaving Pure Oil, the thirty-year-old Hyman filed for patents on polymers mixed with paint and varnish that produced light-colored and transparent film. He turned to a cousin, Henry Degginger, to find financing for a company to bring his invention into production. Degginger was a major investor in Arvey Envelope, controlled by another cousin, Joseph Regenstein. Regenstein inherited Arvey from his father, Julius. Sensing the potential of his nephew, Julius had joined with his son Joe and financed Hyman's education. Now Joseph was about to realize the return on the family investment.

In January 1931, Regenstein incorporated Velsicol, called in its first year Varnoil Corporation. Initially he issued 200 shares, at $100 each. Regenstein's firms bought 160 shares; Hyman sold his two polymer patents for $4,000 to the company and reinvested it in 40 shares. The board of the corporation, which included Hyman, employed Hyman as general manager. A quick reading of this history can leave the impression that Velsicol was a happy family affair, with everyone sharing risks and opportunities.[44] However, nothing could have been farther from the truth. Regenstein and

Hyman had a history of legal maneuvering and fundamental legal problems, especially with patents. Transo, while still led by Julius Regenstein, had lost a bitter patent case with another envelope producer. That case showed Julius Regenstein to be a shrewd business leader willing to do anything possible to profit. When served with an injunction in the case, Julius used another of his subsidiaries to store and deliver the products he had been ordered not to produce. The court concluded that "the evidence shows [Regenstein's action] was a plain and deliberate violation of the injunctions."[45]

Not surprisingly, given this patent experience, both Arvey and Velsicol required all employees to sign secrecy and patent agreements, giving all rights to the company. On January 31, 1931, Hyman signed such an agreement.[46] He had signed a similar agreement with Pure Oil and then became embroiled in a lengthy legal dispute over the patents he used to launch Velsicol. It took seven years for Hyman to win control of his patents, with the court finding that Pure Oil had Hyman only work on core oil research, not creation of new polymers.[47] As Hyman later said, "[Velsicol] was organized to do special research work in chemistry and . . . [he] was employed for the specific purpose of making discoveries and developing formulas."[48] In the first fifteen years, he alone assigned fifteen patents to the company. After getting the lab functioning, Regenstein's other firms advanced more than a half-million dollars to build a Velsicol plant in Marshall, in east-central Illinois. Sales increased from $1,021 in 1932 to over $4 million in 1945. The company saw net profits soar in the next decade: from $44,000 in 1940, to $353,347 at the end of the war.[49]

Any computation of Velsicol's profits in the era were complicated by the interrelationship of the firm with Arvey. Not only were corporate finances mixed, and even employees, but also the early inventions at Velsicol greatly helped the envelope business. Regenstein was able to patent the window envelope, with a clear covering of Hyman's patented polymer. However, by the 1940s, Hyman began to work on hydrocarbon-based insecticides not related to envelopes. On March 2, 1945, three of the researchers at Velsicol filed the initial paperwork for a patent on chlordane, the first of a series of exceptionally powerful hydrocarbon-based pesticides. Hyman did not assign the patent to Velsicol. He did the same with a second patent application related to the insecticide. By the following summer, Regenstein confronted Hyman, demanding he turn the patents over to their firm. Instead Hyman resigned and moved to Denver, along with thirty of the fifty Velsicol technical employees. Henry Degginger backed the new Julius Hyman Co., and Hyman hired three of Degginger's sons.

To begin production quickly, Hyman leased facilities formerly used to produce mustard gas at the Rocky Mountain Arsenal. It was appropriate that Hyman moved to a military facility, for war erupted with Regenstein and Velsicol. In the next twelve years, five different cases pitted Hyman against Velsicol or Regenstein. At least twice, these were appealed to the U.S. Supreme Court.[50] While Hyman and his company lost all of these cases, they dragged out sufficiently long enough for Hyman's company to become a major developer of new pesticides. In chemistry, he was one of the world's great innovators, recognized in Nobel laureate Paul Muller's Nobel address for his role in chlordane.[51] One lasting consequence of the production at Rocky Mountain was one of the many Superfund sites linked to the genius of Julius Hyman and Velsicol.[52]

There were at least four common features of Velsicol's early history. Hyman and the firm he helped create were leaders in chemical innovation. Of the twelve persistent organic pesticides (POPs) that would become the subjects of international regulation at the end of the century, one third were discovered by Julius Hyman and patented by Velsicol.[53] As indicated by the number of Velsicol Superfund sites, Hyman and his family left a legacy of environmental contamination near each major production facility. They also began the process followed into the next century of ignoring criticism, and even sponsoring studies showing their operations had minimal impact on the environment and human health.

As early as chlordane production, Velsicol and Julius Hyman funded research published by the American Medical Association (AMA) showing that workers had no ill effects from exposure. As would repeatedly be the case over the next half century, institutions such as the AMA found nothing wrong with the potentially responsible party having intimate involvement in research in which the party had a clear interest. In 1951, Hyman funded the first chlordane research. Two years later, the AMA allowed Dr. Samuel Hyman, who was a Chicago physician, to coauthor research on the impacts of exposure to a chemical made at Julius Hyman and Company.[54] The firm also funded the training of a new generation of scientists, such as with the Carbide and Carbon Chemical Fellowships at Rutgers.[55] Of course, this behavior would become standard in the natural sciences, leading to considerable worry about the integrity of the research process.[56]

Finally, the early history of Velsicol demonstrated the willingness of the firm and its leaders to use the best legal minds of the liberal establishment to assert their rights. For example, Hyman's counsel for his appeals to the Supreme Court included former Florida senator Claude Pepper and Thurmond Arnold, FDR's assistant attorney general.[57] Both were pillars of the post–New Deal liberal establishment.[58] Typically, for what would become known as the "broker state," these progressives had no problem defending the work of a brilliant scientist.[59] After all, none of this behavior, except perhaps the dumping of hazardous wastes, was unethical or illegal. However, the willingness of pillars of the liberal establishment to help Velsicol should have raised questions about the efficacy of the liberal approach to controlling corporate self-interest. The only check upon the worst of the firm's behaviors, such as patent infringement, had been the opposition of other parties suffering direct economic loss. Apparently, there was no similar counterforce to protect human health or the environment from the company's management.

VELSICOL'S GROWTH AND GROWING PROBLEMS

Except for the disputes in the courts, Velsicol for many years operated quietly as a privately held subsidiary of the Regenstein family's Arvey Envelope. After the death of the founder in 1957, his son Joseph Jr. ran the firm. Even more so than his father, Joe Jr. remained away from the public spotlight. His company seemed to be merely a small, specialized firm with only two plants—in Memphis, Tennessee, and Marshall, Illinois—and headquarters and labs in Chicago. However, Velsicol was about to enter an era when its name would loom large as a polluter and an opponent of environmental regulation.

Shortly before the death of Joe Regenstein, the company started an aggressive effort to increase sales. Velsicol's profits soared, with its four potent pesticides marketed for use around the world.[60] Its patents allowed it to monopolize sales of chlordane, heptachlor, endrin, and methyl parathion.[61] If chlordane was potent, heptachlor was a version of chlordane with a special ability to change inside an organism into heptachlor epoxide, four times more fatal than chlordane.[62] Endrin's potency was several times greater than heptachlor.[63]

In the 1950s, Velsicol's ads showing frightening Japanese beetles graced newspapers each spring, boasting that chlordane was "the proven insecticide that safeguards lawn growth!"[64] Furthermore, its pesticides received free coverage in advice sections of garden publications.[65] Most worrisome for the integrity of university science, Velsicol sponsored two films produced by agricultural extension offices promoting its products. Now seen as rather humorous, with ants screaming, "We're hungry! We're hungry!" in one sequence, *Goodbye, Mr. Roach* and *Goodbye, Mrs. Ant* used the credibility of scientists at Clemson and the University of Georgia to promote insecticides.[66]

Probably the pinnacle of success for mass promotion of pesticide use came at the time Joe Jr. took over the firm in the late 1950s. The symbol of this success was the fire-ant eradication program launched by the U. S. Department of Agriculture (USDA) secretary Ezra Taft Benson. The fire ant had come to the United States after World War I, apparently imported through the port of Mobile on ships from South America. For forty years, the fire ant had not attracted much attention, even in Alabama, and there seemed to be few indications that it posed a serious threat to agriculture, wildlife, or people. Yet, with the help of the pesticide industry, the USDA launched a public-information campaign much like the *Goodbye, Mr. Roach* effort. Unlike the roach and ant campaigns, however, instead of creating a groundswell of support for the extermination of insects, the fire-ant campaign energized critics of the industry.[67]

By the fall of 1958, opposition could not be contained. *Chemical and Engineering News* warned the chemical industry generally of the need to build good community relations.[68] Some government experts, such as John George at the Fish and Wildlife Service and others at the Food and Drug Administration, complained and shared their criticisms with independent scientists, including Rachel Carson. When Carson heard that *Reader's Digest* was going to publish an article favorable to pesticides, she wrote the editor, DeWitt Wallace, and shared a report from the National Audubon Society critical of the fire-ant program.[69]

Her letter reversed the *Digest's* approach. When the article appeared in June 1959, it raised fundamental concerns with mass pesticide use. The lead sentence warned, "There is mounting evidence that massive aerial spraying of pesticides may do more harm than good." In six pages, Robert Strother captured the core ecological complaint against the pesticide policies of the era. Reviewing several case studies of wildlife deaths following massive spraying, the digest included a box in the text with the label "What One Bird Can Do." It reported, for example, that "a pair of flickers consider 5,000 ants a mere snack." Focusing on the fire ant, Strother agreed that the ants were pests, but added, "None of the farmers I talked with had suffered any crop damage from fire ants." They could be controlled by direct application to their mounds of small amounts of targeted insecticides.[70]

Just as criticism of the fire-ant program peaked, a bigger pesticide problem, closer to the homes of all Americans, emerged. In the fall of 1959, public health officials announced that cranberries were found to absorb a widely used carcinogenic pesticide, aminotriazole.[71] The danger caused the Food and Drug Administration to seize tainted cranberries. People were shocked that they might have to spend Thanksgiving without a key ingredient of the traditional dinner.

The cranberry crisis had all the components of the later public struggles involving Velsicol and Michigan Chemical. It pitted those concerned with public health and the environment against those from agricultural and industrial interests. The controversy divided members of President Eisenhower's cabinet, with the secretary of Health, Education, and Welfare, Arthur Fleming, on one side, and the secretary of Agriculture, Ezra Taft Benson, firmly backing the pesticide makers.[72] Their arguments directly impacted Velsicol on January 19, 1960, when the Public Health Service banned heptachlor's use where it could be absorbed by foods. Reversing the usual position in which chemical companies found themselves, the FDA spokesman put the burden of proof on Velsicol to prove heptachlor was safe. Benson was incensed.[73] The White House tried to intervene to resolve the dispute between the two secretaries, but to no avail.

Federal and state agricultural officials, symbolized at the time by Ezra Taft Benson, generally minimized the dangers of pesticide use. Before the "cranberry crisis," Benson had resisted judicial interference into USDA DDT programs. In 1957, a group of Long Island residents had brought suit to stop gypsy-moth spraying. While Benson's side won, his rigidity and rudeness catalyzed his critics, especially Rachel Carson.[74]

SILENT SPRING

After the news of the Long Island decision, Carson urged her friend E. B. White to take up his pen in renewed criticism of large-scale spraying. It had been White in 1945 who first authored in the popular media a warning about indiscriminate DDT use. Sensing that Carson was the more prepared and able critic, White turned the tables on her and urged her to undertake the project that would become *Silent Spring*. For several months Carson delayed, but with additional urging from other friends incensed with the deaths of wildlife from uncontrolled spraying, Carson began research and writing. Even before the book appeared, the DDT debate escalated closer to Michigan Chemical's home. That controversy would launch the ruthless defense of DDT that would be conducted for the next half century by petrochemical and agricultural interests and free-market ideologues.

Officials in Michigan had been some of the most aggressive users of DDT. Naturally, there was heavy use on the campus of Michigan State University (MSU), one of the nation's centers of agricultural research. An MSU ornithologist, George J. Wallace, had begun studying robin populations on campus in 1954, the same year the university launched a DDT spraying program to combat Dutch elm disease. By the following year, Wallace and a graduate student, John Mehner, noted dead and dying birds throughout the campus.[75] Over the next few years, Wallace would emerge as one of the early effective critics of indiscriminate DDT use. On May 3, 1960, with national concern about DDT rising, Wallace testified before the U.S. House Subcommittee on Fisheries and Wildlife.

Ever the cautious scientist, Wallace described high levels of DDT in dead robins on campus, but added, "We are frequently asked to prove, by chemical analysis that this high mortality . . . is due to the DDT. . . . This may be asking the impossible."[76]

Wallace's testimony was the second bit of bad news for DDT producers in Michigan and their advocates at MSU in early 1960. A few months before he testified, residents of a rural area in southeast Michigan rejected the advice given at a public meeting by Dr. Raymond James, an MSU entomologist, and C. A. Boyer from the Michigan Department of Agriculture, and voted to halt DDT spraying in their township.[77] Not surprisingly, state and federal agriculture officials sought to purge their critics. They launched a campaign to have MSU fire Dr. Wallace, a tenured professor with a national reputation in ornithology.[78] The chair of the Entomology Department at MSU admitted years later that the only thing that saved Dr. Wallace from being disciplined was the intervention of U.S. Representative John Dingell. Dingell threatened the budgets of the state and federal agriculture departments and the university if they carried out their threats to Wallace.[79]

Velsicol and its allies in university agricultural programs did not learn from this experience. When *Silent Spring* approached publication in 1962, with praise of the research by George Wallace, the extreme threats resumed. The first hint of what was to come followed the serialization of parts of the book in the *New Yorker*.[80] The responses to the publication confirmed the worst fears regarding institutional integrity in the natural sciences, medicine, and higher education. Of course the pesticide manufacturers and their chemical-industry allies responded with hostility—but that would be expected from anyone attacked as they were. What was more disturbing was the support they received from professional and academic scientists and much of the mass media. Worse still, most of this criticism was not motivated by financial concerns, such as the loss of grant support or advertising revenue from pesticide producers. Apparently, the prime motivation was ideological faith in the immunity of professional scientists from lay criticism.

The chemical industry launched the first attacks while the *New Yorker* serialized the book. The National Agricultural Chemicals Association (NACA), the specialized trade group for firms such as Velsicol, and the larger Manufacturing Chemists' Association attacked Carson as one of those people who "lack the information on which to base intelligent judgement."[81] The NACA already had prepared a position paper, "Fact and Fancy," that paired quotations from the *New Yorker* with refutations by the industry. What worried some leaders of the industry was that Carson was the one writer in America with a reputation for making nature a popular subject for books. They pointed out that her 1951 classic *The Sea around Us* "sold 2 million copies in the U.S., stayed on the best seller lists for 86 straight weeks, and sold two million copies in the U.S., and was translated into 30 languages, was issued as a paperback, and was even the title of a movie."[82] Hobart Thomas of Stauffer Chemical said the articles in the *New Yorker* "may mark the beginning of some serious problems for the chemical industry."[83]

Not only did the articles themselves trouble the chemical industry, but also they undermined industry support among traditional allies in government, especially in the USDA.[84] In an editorial, Richard Kenyon of *Chemical and Engineering News* lamented that Carson's work would likely cause the question of pesticide use to get "into the futile circus arena that can evolve in Congressional hearings." Kenyon wanted instead for government to create "an objective panel" to meet in private to study the issue.[85] The editorial also called for risk assessment, which would become the ultimate defensive weapon of industry, and contrasted it with the "sensationalism" of Carson, saying, "In the emotional atmosphere now building, nothing constructive can be expected."[86]

Yet, it was not clear who was most guilty of emotionalism. A week before the above editorial, Kenyon had written the first attack on the book. He began:

> More than 200 years ago, the French writer, Rousseau, pulled the emergency cord on the train of progress and shouted for all to get off and go back. "Return to nature," he urged. "Everything is good as it comes from the hand of the Creator. Everything becomes evil in the hands of man."
>
> Rousseau stirred society, but his ideas did not prevail. Today as never before there is conviction that to gain new knowledge is ennobling. We are dedicated to progress through the use of new knowledge and new ideas. But when the speed of progress carries us into visible danger our emotions naturally urge us to go back.

Rachel Carson, writing about pesticides in the *New Yorker* during June, stirs reactions that suggest retreat in shock.[87]

Representatives of other professions linked to the chemists responded similarly. The American Medical Association ran a mocking review later in 1962, written strategically by a female physician, Therese Southgate. Departing from the measured and respectful tones in accompanying reviews of medical books, Southgate began by saying Carson acted "like a zealous and overanxious mother." Southgate then added:

> That *Silent Spring* has raised a storm, second not even to the cranberry crisis of 3 Thanksgivings ago, is now evident. Part of the irritation that comes in reading the book, outside of the fact that Miss Carson uses emotion-arousing words, is that one feels somehow that she is partly right but cannot know where she is wrong. One is left with the impression that Miss Carson is simply against chemistry and for biology, or against synthetics and for life.[88]

Dr. Southgate even lamented that the predictable hostility in the pesticide industry's responses to the book only had made matters worse, perhaps contributing to its remaining on the best-seller lists for so long. She urged physicians to read it so they could counter its points in talking with patients.

The position of the AMA and the other scientific interests opposed to the book should not be surprising. As Paul Brooks, Carson's editor at Houghton Mifflin, said:

> *Silent Spring* initially offended a relatively small (though very rich) segment of society, the chemical and other related industries (such as food-processing), and—in the federal government—the immensely powerful Department of Agriculture. But the fury with which it was attacked, the attempts to discredit that "hysterical woman" as she was called, have, I believe, deeper roots than a simple concern for profits or power on the part of special interest groups. Her opponents must have realized—as was indeed the case—that she was questioning not only the indiscriminate use of poisons but the basic irresponsibility of an industrialized, technological society toward the natural world.[89]

Much of the popular media joined in this critique. Having also worshipped natural scientists, and especially their chemical concoctions that had changed so much of life, the popular media had difficulty defending the book's shocking attack. Of course, the chemical industry generated significant advertising revenue. The day after the book appeared, *Time* discussed it in its science section. *Time* contrasted the responses of "unwary readers" with "scientists, physicians, and other technically informed people." In a remarkable paragraph, *Time* explained how both types could get a proper interpretation of the book:

> Many of the scary generalizations—and there are lots of them—are patently unsound. "It is not possible," says Miss Carson, "to add pesticides to water anywhere without threatening the purity of water everywhere." It takes only a moment of reflection to show that this is nonsense. Again she says: "Each insecticide is used for the simple reason that it is a deadly poison. It therefore poisons all life with which it comes in contact." Any housewife who has sprayed flies with a bug bomb and managed to survive without poisoning should spot at least part of the error in that statement.[90]

Perhaps most shocking was the formal response within higher education, in institutions one would hope would welcome debate about any topic. In a pattern experienced by George Wallace and later repeated with other problems of Velsicol and the chemical industry, academic institutions were not in the forefront of Carson's supporters. Prominent faculty, such as Frederick Stare at the Harvard School of Public Health and William Darby of Vanderbilt, wrote caustic reviews.[91] However, the strongest institutional effort to oppose Carson came from the University of California at Berkeley. John Martinson, who coordinated the program in "Science and the Citizen" in the extension program at Berkeley, had informed readers of the *Science Guide*, published by his program, that the *New Yorker* was serializing Carson's work. The *Guide* merely referred readers to the series. Martinson did not endorse it. As he reported in a confidential letter to the *New Yorker*:

> The day following the appearance of this issue someone within the University made a confidential call to the University Public Information Officer in charge of science news releases. The substance of the call as relayed to us by P.I.O. (without telling us who the caller was) was to the effect that Rachel Carson is not a chemist, the University has a number of chemists working on insecticides, and this is a controversial question which it would be better not to discuss in a bulletin distributed by the University.[92]

Martinson was not confused about the nature and reasons for the communication from the anonymous caller. It was "censorship." He also guessed that what he faced was "a very minor kind of annoyance compared to the economic pressures that must be exerted by chemical interests when you do a series like 'Silent Spring.'" He concluded, "In a way I'm glad this incident occurred because it provides me with an excellent argument in favor of establishing an independent journal of science criticism."[93]

Less than a month after Martinson's letter informed the *New Yorker* of the censorship, Velsicol launched an unprecedented attack. In a letter dated August 2 from the company's general counsel, Louis McLean, Velsicol asserted that it wished to warn Houghton Mifflin of "legal and ethical" problems with the book. Then it made its legal threat:

> From a legal standpoint, we call to your attention the fact that several of the chemicals named, including, for example, aldrin, chlordane, dieldrin, endrin, and heptachlor, are patented chemicals. Chlordane and heptachlor are manufactured solely by this company. You no doubt are familiar with the fact that disparagement of products manufactured solely by one company creates actionable rights in the sole manufacturer.[94]

The ethical problems with the book, according to McLean, were that the publishers "are willing to publish anything to make a dollar." By contrast, the pesticide industry tests its products to assure they are safe. He added quotations from the *Journal of the American Medical Association*, which stated, "There is no reason to believe that the present use of chemicals in foods is endangering the health of people."[95] Furthermore, he insisted that "pesticides and other agricultural chemicals are essential if we are to continue to enjoy the most abundant and purest foods ever enjoyed by any country of this world."

Of course, these claims were not accepted without question by the critics of careless pesticide use. As early as 1946, the *New Republic* had taken the alternative approach to the impact of pesticides on the food supply. John Terres had asserted: "Birds, along with beneficial insects and weather, are a steady curb on the destructive insects which threaten to consume all of man's green food supplies. If we removed the birds and helpful insects from large areas of the earth, we might soon know a great famine."[96] McLean's perspective on pesticides and food supply was a model statement of the modern confidence in the ability of humans to make the natural world better than it was. Terres's statement reflected the renewal of the more modest assessment of human wisdom that had marked all previous ages and was at the core of the new ecology movement.

McLean's response to this "balance of nature" argument, as he called the ideas of people such as Terres, was that it "overlooks the fact that no benign balance of nature protected the dinosaur, the hairy mammoth, nor does it provide adequate food and protection from disease." He also pointed out that deaths from poisoning, including by chemicals, were at historic lows, according to the Public Health Service. Essentially his conclusion was that there was no valid reason for *Silent Spring*.

McLean then turned to what he saw as the real culprit in this argument. Grouping Carson with the "sincere . . . natural food faddists [and] Audubon groups," he warned that "the chemical industry in this country and in western Europe must deal with sinister influences, whose attacks on the chemical industry have a dual purpose: (1) to create the false impression that all business is grasping and immoral, and (2) to reduce the use of agricultural chemicals in this country and in the countries of western Europe, so that our food will be reduced to east-curtain parity." Writing in a style popularized a decade earlier in the heart of the McCarthy era, he concluded, "Many innocent groups are financed and led into attacks on the chemical industry by these sinister parties."[97]

Given this rhetoric from the attorney for a major firm, the editor at Houghton Mifflin, Paul Brooks, wanted to be on solid legal and scientific ground before proceeding with publication, scheduled for the end of September. He immediately sent copies of Velsicol's letter to Carson's literary agent, Marie Rodell; to Carson's attorney; and to the *New Yorker*. Of course, he called Carson, who was vacationing in Southport, Maine. Orally she provided refutation from memory of most of Velsicol's charges. She also told him for the first time that the *New Yorker* had already received similar threats. She promised to provide him with written support for the points challenged by McLean once she had seen the actual letter, which did not reach her until August 8.[98]

Carson immediately wrote back to Brooks, agreeing with others that Houghton Mifflin should send an acknowledgment of the August 2 letter while awaiting legal opinions from her attorney, Maurice Greenbaum, and the company's attorneys, Choate, Hall and Stewart. Meanwhile Carson referenced several key sources justifying her comments about chlordane and heptachlor. Among the sources was Wolfgang van Oettingen's massive study in 1955 for the Public Health Service; she also referenced a clinical memorandum from the Public Health Service, a study from the Fish and Wildlife Service, and a copy of a letter on contamination of hops that she had received from a sympathetic USDA official.[99]

Brooks had already prepared a firm letter to McLean, asking for specific references to sources proving Carson wrong. Until receipt of such proof, Brooks said they were moving to publication backed by research from the Public Health Service on the health hazards of Velsicol's chlordane. Brooks added that Carson essentially agreed with McLean on heptachlor. Finally, and most importantly, Brooks emphasized Carson's moderation—a point ignored by the chemical industry's ideologues for the next half century: "Miss Carson has recognized in numerous pages throughout the book the value of selective chemical sprays and of the work which has been done in making them and using them effectively."[100]

The first response from Houghton Mifflin to Velsicol was not the letter Brooks drafted, but an innocuous one from the firm's president, William Spaulding. Sent on August 10 to buy time, it requested "a more detailed explanation of the contentions" made in the August 2 letter. Spaulding also asked if McLean's examples from August 2 were the only inaccuracies. If not, Spaulding requested that McLean identify a complete list. Spaulding's only concession to the firmness of Brooks's draft was to conclude with a disclaimer: "We do not, of course, intend by this letter to acknowledge the accuracy of the assertions in your letter."[101]

On August 14, McLean responded to Spaulding with a longer list of alleged inaccuracies. He relied heavily on publications from the American Medical Association, the Food and Drug Administration, the USDA, and state agricultural extension programs.[102] After devoting three pages of the four-and-a-half-page letter to specific sources for his claims of errors, McLean concluded with a general assertion of how well the pesticide industry was regulated. "We have the world's most stringent laws on pesticides and food purity of any country in the world and administrators of those laws take their jobs seriously. It is a mistake to state otherwise. We members of the agricultural chemicals industry do take pride in the contributions pesticides and other agricultural chemicals have made to the increasing abundance and purity of food in this country."[103]

By the time Velsicol's second letter arrived, Houghton Mifflin was well on the way to resolving its discomfort with McLean's threat. They had sent the first Velsicol letter to two pathologists associated with Harvard Medical School, Richard Ford and Arthur McBay, who directed the Massachusetts State Police Chemical Laboratory. The responses confirmed the accuracy of Carson's text, with McBay finding only one of Velsicol's claims meriting a possible qualification in the text. However, he admitted that he had read the section in draft and found nothing wrong with the sentence until McLean attacked the book. With this support, on August 22 Brooks decided to proceed with publication.[104]

Brooks wrote a short letter to McLean on behalf of the company. It acknowledged the letter of August 14 and minimized the sense of concern at the company by saying Brooks was writing in place of the vacationing Spaulding. He then brushed aside Velsicol's criticism and their request for a meeting, concluding, "We have reviewed carefully the sources for the statements in her book, in the light of the points you bring up in your letter. While there may be room for difference of opinion, we still believe, after thorough examination, that Miss Carson's presentation is accurate and fair. Since our concern as well as yours is factual accuracy, we do not believe that a meeting would serve any useful purpose."[105]

Of course, McLean, Velsicol, and the pesticide industry could not tolerate disagreement or debate. In Paul Brooks's unsent first letter to Velsicol, he would have expressed the hope that "this book should act as a useful adjunct to the chemical industry's own warnings that these chemicals must be used carefully."[106] Instead of making the best of the criticism and co-opting it by saying they shared the concerns raised by Carson, McLean and his supporters decided to fight with all means at their disposal, including the threatened lawsuit. They repeatedly exaggerated the extent of Carson's criticisms and sought to turn her into a foolish "food faddist" or other radical. The National Agricultural Chemicals Association (NACA), for example, launched an attack on the book by taking quotations from it and matching them with the pesticide industry's response. By that time, Brooks had had enough and was prepared to counterattack. He discussed the situation with Jeptha Wade and Maurice Greenbaum, suing the NACA for plagiarism. The attorneys advised sending a "stuffy letter and releasing it to the press."

The letter was a tour de force:

> We have been interested in the recent publication of your Association entitled *fact and fancy.* Most of the "allegations" appear to be quotations or paraphrasing of passages in Miss Carson's recent book, *Silent Spring.*
>
> We would be quite willing to authorize such quotations for the purpose of scientific comment, if due reference is made to the source, in order that those who are interested in the subject can check not only the references in your publication referred to as "fact" but also the basis for what you refer to as her "allegations."
>
> It would seem a matter of common courtesy, and certainly a more useful scientific publication, if sources were given on both sides.[107]

The letter apparently delivered the proper message to the NACA. Its president, Parke Brinkley, wrote back to Brooks: "We have no present plans to reprint this publication, but if any future printings are made we will be pleased to specify the title and author of the book for any quotations used."[108] However, while the NACA and McLean backed down over the *Silent Spring* contents, this was not the end of the chlordane and heptachlor debates; they were to continue until the late 1980s, always with Velsicol losing.

Why would the company and its allies fight such a counterproductive battle? Why would McLean and the NACA not see that Carson did not call for abandonment of pesticide use, only restraint? Their behavior clearly contradicts the conventional wisdom about groups seeking political victories. Rational-choice scholars have argued that groups who want a political victory, as well as successful entrepreneurs, know how to build coalitions and creatively adapt in order to win and profit.[109] Carson and all critics of restraint in the use of chemicals in the environment were offering industry an easy way to reach compromise. At the beginning of the book, she placed E. B. White's warning: "I am pessimistic about the human race because it is too ingenious for its own good. Our approach to nature is to beat it into submission. We would stand a better chance of survival if we accommodated ourselves to this planet and viewed it appreciatively instead of skeptically and dictatorially."[110] McLean and Velsicol and the NACA could have said they favored caution. Instead they saw the precautionary approach as one with which they could have no compromise.

After receiving a copy of the book in mid-August, White understood its fundamental challenge to modern corporate ideology and behavior. He wrote to Carson from Maine thanking her for using his quotation and then predicting, "This will be an Uncle Tom's Cabin of a book, I feel—the sort that will help turn the tide." Seeing the coming attacks, to which the *New Yorker* had already been exposed, he added, "In some ways its publication will be a less satisfying literary experience for you than the other books, which were songs of praise, but in the long run I'll bet 'Silent Spring' will be the work you are proudest of." In a remarkable admission of failure for the greatest observer of writing style of the generation, White concluded, "I'm unable adequately to express my gratitude to you for attempting to decontaminate this lovely world." Yet, White did find how to express thanks: "When the thrush sings in my woods again (and I'm sure he will) I will think of you every time, and give thanks."[111]

A few in the chemical profession shared the understanding of White and Brooks. To the credit of Richard Kenyon, editor of *Chemical and Engineering News*, he ran a host of criticisms of both his editorials and his general coverage of pesticides. On July 16 the first two letters to the editor appeared. Paul Kronick of the Franklin Institute, a Yale Ph.D., and another former DuPont employee attacked the text of an article Kenyon had run on July 2 criticizing the "opposition of some life scientists to the indiscriminate use of insect poisons."[112] That article was a report on the increase in pesticide sales in early 1962. Supporters of Carson saw it as a direct challenge, especially an offensive comment from the New Jersey Agriculture Department director F. A. Soraci, saying, "In any large scale pest control program in this area, we are immediately confronted with the objection of a vociferous, misinformed group of nature-balancing, organic gardening, bird-loving, unreasonable citizenry that has not been convinced of the important place of agricultural chemicals in our economy."[113] Kronick complained, "It is unfortunate, though, that you choose to support your position by irrational attacks on members of other scientific professions, whose methods of arriving at conclusions are not so much different from those of chemists."[114]

George V. Caeser joined in Kronick's defense of Carson. A Yale graduate with extensive agricultural-chemistry experience with a milling company in Harbor Beach, Michigan, Caeser, at seventy-one, brought years of experience to his support for Carson.[115] He criticized the American Chemical Society response to Carson, as represented in *Chemical and Engineering News*, and called instead for "respectful recognition by our Society" of Carson's contribution. He warned, "If only a part of this terrible indictment by a distinguished biologist be the truth, the future of mankind is in great danger." He also criticized an earlier story that contained language much like that used by McLean. Caeser found inappropriate that article's "arrogant dismissal of a 'bird-loving unreasonable citizenry' unaware that 'science' is sacrosanct."[116]

After Kenyon ran another editorial on the sensationalism of the pesticide debate, focusing on the scare tactics of the opposition, a French-born organic chemist from Olin Matheson, Peter Muffat, observed, "Your editorial on pesticides (July 30) suffers from the very faults it finds in others: shortage of facts and excess of emotion."[117] Kenyon had focused the editorial on criticisms of the misuse of pesticides, agreeing that "strong measures should be taken against" those who do so. Muffat, using an analogy from the endless wars on illegal drugs, mocked Kenyon, saying, "If the editor knows how to supervise the handling of dangerous poisons by hundreds of thousands of farmers, dairymen, crop dusters, and plain home gardeners, he ought to tell the authorities who are trying to control the use of narcotics." Turning to the argument that the pesticides were needed to avert famine, he reminded Kenyon, "It may cheer you up to think of all the storage bins filled with excess agricultural products." He then asked why so many pesticides were used on ornamental trees, when the concern was famine.[118]

Finally, in early September, Edmund Blau expressed surprise "to find the articles by Rachel Carson in the *New Yorker* have produced the reaction they have in the editorial pages of C&EN." Blau was another exceptional chemist, educated at Cornell, the University of Chicago, and Ohio State. After World War II ordnance work, he had been with the Bureau of Standards for ten years before moving to the Applied Physics Lab at Johns Hopkins University.[119] He noted that after the editorials on July 23 and 30, he reread Carson, "thinking I had missed the 'emotional' quality they are claimed to have . . . but she is neither unreasonable nor unreasoning, and the emotional quality of her articles is no higher than that of the editorials." He concluded by scolding Kenyon: "A scientist of Miss Carson's reputation deserves more careful consideration and discussion than she has received so far in the pages of C&EN."[120]

Unfortunately for the chemical profession, the perspectives of people such as Kronick, Caeser, Muffat, and Blau neither dominated the pesticide debate nor significantly influenced it. By the time many chemists joined forces with pesticide critics, McLean, Velsicol, and NACA had done more harm. For another quarter century they would defend chlordane, heptachlor, and similar compounds and become infamous for a host of other environmental health incidents. The response of the profession seemed inexplicable in a rational or empirical sense. It was not so hard to understand, however, if Caeser's reading of the critics was valid. He saw that they had converted the sciences from fields marked by constant testing of hypotheses, using empirical evidence and welcoming vigorous debate, into a modern theology, so "sacrosanct" it could not be criticized, especially by the laity.

Fortunately, a number of public officials, both appointive and elective, responded more appropriately. Their behavior provided a notable example of why democracy is a better method of resolving issues than the forums controlled by any elite, whether the cloisters of industrial chemistry or those of academic experts. Fortunately for Carson, the country was led at the time by an administration notable for its openness to new ideas, and with a policy process that invited critical input at the highest levels. As soon as the *New Yorker* articles appeared, Jerome B. Wiesner, President Kennedy's science advisor, called a meeting of agency leaders potentially concerned with pesticides. Wiesner then appointed a special panel, chaired by Boisfeuillet Jones. Special assistant to the secretary of the Department of Health, Education, and Welfare, Jones, an attorney, had an exceptional background for this assignment, having served on the advisory panel of the Public Health Service and having been a longtime administrator and faculty member at Emory University. He kept the president so well abreast of the debates about the pending book that at the president's weekly press conference on August 29, Kennedy answered a reporter's question on pesticide policy with reference to "Miss Carson's book."[121]

The president's knowledge of the pesticide issue resulted from an information mechanism developed by the administration, the weekly "Kennedy Seminar." At the seminar, innovative thinkers were invited to meet with the administration's inner circle. Just prior to the Cuban Missile Crisis, Carson appeared before the seminar, without the president in attendance.[122] However, those present briefed the president about her findings. While secretary of Agriculture Orville Freeman remained a critic of Carson, much as Ezra Benson under Eisenhower, this time the support came not merely from Health, Education, and Welfare, but from the secretary of the Interior, Stewart Udall.[123]

OVERREACTION?

Velsicol and McLean followed their initial attacks on Carson with an aggressive defense of pesticides and corporate behavior that seemed out of touch with reality.[124] McLean and Velsicol's lobbyist, Samuel Bledsoe, repeatedly took the lead in attacking critics of pesticides, especially those calling for government regulations. Beginning in 1964, Bledsoe, working with lobbyists from Shell Oil and Geigy Chemical, developed a plan to publish a defense of pesticides. They recruited Rep. Jamie Whitten, the longtime congressman from Mississippi and a staunch defender of big agriculture, as the author. They assembled a text from material submitted to hearings before his committee and added information supplied by the staff at the Library of Congress. Velsicol and its corporate allies first approached M. B. Schnapper of the Public Affairs Press in Washington to publish the book. Schnapper recalled that he found the book "substantively weak and . . . poorly written." However, what worried him more was that the three firms assured him there would be sales of the book, since they planned to buy thousands for distribution. Schnapper said, "I really felt squeamish about what was going on in terms of industrial sponsorship of the book."[125] Unfortunately, few were as cautious as Schnapper.

Eventually, Velsicol got D. Van Nostrand, the scientific publishing firm in Princeton, to produce the book, with the provocative title *That We May Live*.[126] However, their greatest success came with a review in the *Library Journal* written by Harold Bloomquist, a medical librarian from Boston. Instead of seeing through the weak substance, to use Schnapper's phrase, Bloomquist elevated the propaganda piece to official truth, saying, "This is a popularized version of what amounts to the Federal Government's answer to the charges of Rachel Carson. . . . The Government's answer is the scientists' answer." He concluded the review finding Whitten's book to be "a sober antidote to the hysteria surrounding *Silent Spring*."[127] The lobbying and public relations campaign around *That We May Live* fully launched the process that would be used repeatedly over the next half century to discredit Carson as an irresponsible, emotional opponent of sound science.

McLean did not stop with the book. He continued to look for forums to defend DDT and Velsicol's products. In 1967, *Bioscience* published a paper written by Richard Goodwin of Connecticut College calling on biologists and other scientists to become environmental-policy advocates and to take a long-term view of pesticide impacts. Goodwin was an eminent biologist with service on the National Research Council. The core of his argument rejected those who see long-term as twenty-five years, pointing out that biologists know that evolution occurs slowly over millennia, as the gene pool slowly adapts and advances. Goodwin explicitly identified the conflict of interest experienced by scientists, between their concern for the distant future and their need not to offend powerful industrial interests that support public funding for the professions.[128] Focusing on pesticides, Goodwin thanked Rachel Carson "especially for bringing the problem into focus. . . . Previous to [*Silent Spring*] the gravity of the issue was not generally understood either by biologists or by the public."[129]

McLean could not allow Goodwin to go unchallenged. He responded with an article appearing six months later in the journal. Once again, however, McLean fell back upon McCarthyist hyperbole. The foreword to his essay reiterated his characterization of his opponents as extremists:

> If history has taught us but one thing, it is this: no discussion of the environment is meaningful if it is limited to the tangible environment. The extremist Nazi theories of heredity that denied rights to non-Aryans and the extremist views of environment of Marx and Lenin that similarly denied rights to classes of people degraded scientific thought, as typified by the teachings of Lysenko, both were implemented by creating an intangible environment of fear and propaganda. The crescendo of controversy over pesticides a few years ago also polluted the intangible environment with long-range, adverse effects.[130]

In the text, he started with his old argument that pesticides had reduced deaths by starvation. After several pages pointing out that pesticides were insignificant contaminants of air, food, water, and soil, he turned his attention to his critics—first, the ghost of Rachel Carson (she had died in 1964). He criticized her speech before the National Press Club in 1962, saying, "In the colorful language with which she was gifted, she stated that scientific information here was screened to serve the 'gods of profit and production.' She accused the medical professions, control officials, and the scientists of our colleges and universities of having been purchased by industry."[131]

Then he broadened his focus and classified critics of pesticides into two groups. First there were the "purposeful" critics, those who used "the controversy to sell natural foods at unnatural prices . . . or in any way to profit from the controversy." Then there were the compulsive critics, whom he said Freud had described under the heading "neurotics, driven by primitive, subconscious fears to the point that they see more reality in what they imagine than in fact."[132] He spent the remainder of the essay attacking the quackery of his critics:

> If you read medical journals (Stare, 1966; Bernard, 1965; Marmor et al., 1960), you will learn that the same purposeful and compulsive types, the anti-pesticide people, in almost every instance hold numerous beliefs in nutritional quackery and medical quackery and that they oppose public health programs. The compulsive see simplicity as purity, feel rejected by mankind and man-endeavors such as science, medicine, and business. They are not able to adjust to the assaults on ego we all experience: failure to achieve the ultimate socially or in business, and especially the ego-shattering fact that we all grow older. Thus while they seek youth and purity in the simple and primitive, they suffer increasing fear of loss of health and physical powers. While presenting a holier-than-thou attitude, they are actually preoccupied with the subject of sexual potency to such an extent that sex is never a subject of jest.

Finally, in another line seemingly lifted from a McCarthy-era speech, he called upon the chemical industry to develop a list of anti-pesticide leaders with "the number of other variant views they have expressed."[133]

It must be presumed that McLean was speaking for Velsicol, since he made all of these statements with his company affiliation indicated. However, he at times represented not just the firm, but also the entire pesticide industry. A year later he appeared as the designated point man for the Industry Task Force for DDT. Established by the National Agricultural Chemicals Association, which had coordinated the campaign against Carson and her allies since the summer of 1962, the task force faced its first regulatory challenge when Wisconsin citizens forced a state hearing on use of DDT. While he volunteered for the role, McLean came with little notice from Velsicol's Chicago headquarters. He was confronted this time not with hysterical women and food faddists, but with attorneys and scientists from the

Environmental Defense Fund. Almost immediately, EDF attorney Victor Yannacone put McLean on the stand and asked him about his denigration of the critics of DDT. McLean's responses discredited both the DDT case and the pesticide industry generally.[134] But, McLean modeled a standard approach for the industry. Long after the national restrictions on general DDT use, a new generation of free-market critics of environmental policy continued to distort Carson's argument and her ghost.[135]

FISH KILLS

The response of public officials to McLean and Velsicol was not surprising given the negative publicity the company received in the era. In the closing days of the Kennedy administration, Robert LaFleur of the Louisiana Division of Water Pollution Control asked the Public Health Service to help determine the cause of repeated fish kills in the Mississippi. The kills had been occurring for six years, but in 1963 were especially severe. The state had considered disease, changes in water temperature, general alkalinity, all with no evidence. Given the size of the 1963 kills and the lack of evidence of any simple cause, the state had turned to the federal government, believing contamination was coming from upstream. After intense analysis of fish and sediment samples, the officials found the dead fish contained high levels of endrin.[136]

By the spring of 1964, the Public Health Service tracked the likely source of endrin to the Memphis, Tennessee, plant of Velsicol. One report in the *New York Times* said the company refused public-health officials entry.[137] The company denied that. A later account held that the company tightly controlled their sampling on factory grounds.[138] The officials moved off Velsicol property to the sewers leaving the plant area, and to the "Hollywood dump" used by Velsicol. They found byproducts of endrin manufacture in the dump, and 8,000 pounds "caked in deposits up to three-feet thick" in the Memphis sewers.[139]

The fish kill turned into a major national exposé of Velsicol's practices, far greater than the brief attention given to heptachlor in 1959–60 after the cranberry scare. To counteract the attacks, Velsicol used all the political weapons at its control. It turned to the Senate minority leader, Everett Dirksen of Illinois, for criticism of the Public Health Service. Dirksen echoed "a defense offered by the company." In a statement that reversed the roles of company officials and Public Health Service professionals, he said of the

Public Health Service, "Some have become so intoxicated with power that our Federal Government now need not be held accountable for its wild charges."[140] Of course, it was hardly wild that the Health Service had spent six months methodically investigating fish tissue and sediment. What was more wild was that Joseph Regenstein, Velsicol's owner, was one of the leading campaign contributors, to both parties, in 1964.[141] As was often the case with companies caught doing something wrong, the firm used public relations and political maneuvering to extricate themselves from the problems, rather than take the more direct approach of changing behavior.

For example, in the midst of the fish kill, Velsicol ran an ad, "Campaign with Chlordane," showing apparent protestors holding signs that read, "Get Rid of Japanese Beetles" and "Join the Community Campaign."[142] Already, on the local level, Velsicol had laid the groundwork for support. The previous year there had been two related problems at the Memphis plant that threatened to cause trouble for the company. On June 3, 1963, twenty neighbors of Cyprus Creek, which flowed by the Hollywood dump, complained to the city health department about nausea and related symptoms linked to fumes from the stream. Four days later, twenty-six workers in factories near Velsicol's plant had to be rushed to the hospital. When more than forty people filed millions of dollars in lawsuits against Velsicol, the company applied political pressure rather than pollution controls.

John Kirk, second in command to Regenstein, held a dinner for 150 civic leaders and made clear the company's approach to community complaints. Instead of apologizing, he took the offensive, saying, "It came as quite a shock to us to discover that there was some question about whether we were welcome in the City of Memphis."[143] The conservative machine mayor, Henry Loeb, responded as desired, assuring Kirk that "this plant is very much wanted in Memphis."[144] In 1964, Loeb's more populist successor would do little more.[145] The director of the Memphis sanitation bureau lamented the absence of federal regulations that would undermine such corporate bullying of a community.[146] In 1964, it was business as usual in the South. Given that the sewers and dump primarily were in African American neighborhoods, the city did not resist the company's leaders—a textbook case of environmental injustice.

Outside Memphis, Velsicol did not win in 1964. Instead of compromising with overwhelming opposition, incensed by its behavior, it defied the public. The 1963 fish kills had offended scores of people in the lower Mississippi Valley, and Velsicol could not threaten to shut its factory to quiet them. Hunters in Louisiana saw mergansers floating dead in the river from eating endrin-tainted fish. Thousands of dead fish floated by towns, including the fish sought by commercial fishermen. The poisoning especially impacted the catfish that were a prime source of food for poor bayou residents. The fish were not subtly ill, but had visible hemorrhages, and if still living had convulsions before dying. Still, Velsicol resisted the obvious with all its power. Bernard Lorant, Velsicol's vice president for research, contested every criticism. He denied health officials access to the plant and repeatedly insisted that endrin did not kill fish.[147] Meanwhile, McLean insisted it was incomprehensible that a chemical could travel five hundred miles and kill fish.[148]

As was a pattern in other contaminations, the company recruited a variety of allies in the academic, industrial, and agricultural sectors to defend its interpretation of endrin. At USDA hearings in April 1964, the Velsicol supporters were given plenty of opportunity to state their views.[149] William Tompkins, chief aquatic biologist with the Massachusetts Division of Fisheries, claimed that he had tested endrin on fish and had found no problems. Leo D. Newsom of the Louisiana State University Department of Entomology concurred with Tompkins.[150] Velsicol had the scientists challenge the very competence of the Public Health Service that had reported dieldrin as well as endrin in the Memphis sewers. Velsicol asserted that it had not manufactured dieldrin in Memphis and thus the Health Service analysis methods must have been flawed.[151]

While the design of the Agriculture Department hearings assured full defense of Velsicol, things changed in the Senate committee chaired by Abraham Ribicoff. He focused not on Velsicol but on the operations of government agencies in responding. To counteract the Department of Agriculture, Anthony Celebrezze, secretary of Health, Education, and Welfare (HEW), also ordered a conference be held in May in New Orleans on the fish kill. During the conference, HEW concluded that the cause was endrin from Velsicol. Velsicol expressed "amazement that Secretary Celebrezze would accept such a report since the conclusions drawn at the New Orleans conference are not supported by the scientific evidence presented there." Senator Dirksen said the Public Health Service had "unjustly crucified" Velsicol, and demanded more hearings from Senator Ribicoff.[152] As a result of the New Orleans report, Velsicol was forced to stop dumping endrin in Memphis. Later, Murray Stein of the Public Health Service observed, "Until we went after the Velsicol people, the fish kills were persistent. Once the discharges stopped, the big fish kills stopped. Draw your own conclusions."[153]

As in the past, however, Velsicol was not prepared to accept the conclusions. At the renewed hearing in July, Velsicol's attorney, Albert E. Jenner, one of the pillars of the Chicago bar, accused Stein of running a "kangaroo court."[154] The company also renewed the attacks on *Silent Spring.* They got Dirksen to read into the *Congressional Record* statements from the academic critics of the book. Unable to allow Rachel to rest in peace following her death in April 1964, they prepared a new publicity campaign for spring 1965, saying, "Spring busted out all over again in 1965—just as it did in 1964, 1963, and even in 1962. . . . In case you haven't noticed, trees leafed, birds sang, squirrels reconnoitered, fish leaped—1965 was a normal spring, not the 'silent' type of the late Miss Carson's nightmares." However, a *New York Times* writer, Oscar Godbout, pointed out that the advertising campaign ignored the fact that "there is 100 per cent mortality of lake trout fry in Lake George this 'normal' spring from residual DDT action."[155]

THE MERGERS

Seemingly everything changed with Velsicol in 1965. Just as it was completing its new advertising campaign attacking *Silent Spring*, it was being sold by Regenstein along with its subsidiary Michigan Chemical. Regenstein had purchased the Michigan firm in the early 1960s during the controversy over *Silent Spring.* By December 31, 1962, Regenstein owned 295,524 shares of Michigan Chemical's 779,279. In the next six months it added 60,000 more shares to raise its holdings to 46 percent. Longtime owners such as Frederick Van Lennep sold all their shares to Velsicol. Velsicol then asked for a reduction in the number of directors to ten and nominated a slate from Velsicol, including John Kirk and Bernard Lorant, for three of those positions. If the reduction of the board to ten were not approved, Velsicol nominated three other directors, including Louis McLean.[156] On February 13, 1964, the American Stock Exchange announced that Velsicol owned 394,100 shares or 50.5 percent.[157] At the same time that Velsicol completed the purchase of Michigan Chemical, it bought the Chattanooga plant of Tennessee Products and Chemical Corporation, and Wood Ridge Chemical in New Jersey. With the addition of Michigan Chemical, the company had eleven manufacturing facilities in seven states, most with fundamental pollution problems. Meanwhile, at the top of Velsicol was the CEO and president Joseph Regenstein, who, with a massive fortune from the window envelope, showed more interest in philanthropy than chemical battles.

In 1950, his father had created the Joseph and Helen Regenstein Foundation, which by the end of the century had given $100 million to Chicago nonprofits. The recipients were a who's who of Chicago institutions: the

University of Chicago received a massive new library, a wing was added to the Art Institute and then old masters purchased to fill it, and the Field Museum of Natural History and the Museum of Science and Industry received large grants. Many Chicago hospitals benefited from the foundation's largess—Northwestern, Michael Reese, Rush-Presbyterian-St. Luke's—as well as programs for the mentally handicapped. An unassuming man, Regenstein was "an avid sportsman and outdoorsman, . . . enjoyed deep-sea fishing and gardening as well as drawing and reading."[158] Concerned with the learning-disabled and mentally handicapped, he promoted special education, receiving the Lambs Good Shepherd Award in 1986 at a dinner hosted by Betty Ford.[159] At Regenstein's death in 1999, the director of the Lincoln Park Zoo recalled the casual lunches at a diner on the north side of Chicago, where Regenstein acted as just one of the neighbors. "He was so down to earth; there was no airs about him."[160]

What is most ironic, given the legacy of Velsicol in places such as Memphis; St. Louis, Michigan; Marshall, Illinois; and Wood Ridge, New Jersey, is that much of the foundation's concern has been with nature, especially zoos. Regenstein gave the sea-otter exhibit at the Shedd Aquarium, supported the fruit and vegetable garden at the Chicago Botanic Garden, and gave lavishly to the Lincoln Park Zoo, where Joseph served for many years as a board member. At Lincoln Park there are the Regenstein Large Mammal House, the Regenstein Birds of Prey Exhibit, and the Regenstein African Journey. Perhaps the most ironic gift, however, was the Swamp Exhibit at the Brookfield Zoo, where visitors are reminded of the dangers of polluting waterways because of the damage to fish. Again in April 2001, the foundation distributed $50 million to the Lincoln Park Zoo, Art Institute, University of Chicago, and Rush-Presbyterian-St. Luke's Hospital.[161]

The question that is begged by the gifts of the foundation, which led the *Sun-Times* in 1993 to open a story referring to Regenstein as "Homo philanthropus," is how can his concerns be reconciled with the behavior of the corporation that produced the fortune?[162] Of course, this is not a new question; it is just framed so starkly when the company's behavior is contrasted with the environmentally related grants. Perhaps it is a reason that in 1965, having watched Velsicol become infamous for several egregious environmental incidents, Regenstein sold the company.

Writing a few years after the sale about Velsicol's role in opposing *Silent Spring*, Frank Graham believed the changes in corporate leadership were for the better. Apparently not knowing the commitment of the shy Joseph Regenstein Jr. to the natural world, Graham felt the new owner, another Chicago civic leader, would be much better. Graham reported that the

Public Health Service's Murray Stein "noticed an abrupt change in the company's policies."[163] Neither Graham nor Stein could have been blamed for not understanding how difficult it would be to change the management practices and culture that had created the problems ranging from resisting *Silent Spring* to denying responsibility for fish kills. For example, while Velsicol stopped dumping endrin directly into Memphis sewers, it bought land sixty miles east of the city and began dumping untreated toxins there. Graham and Stein did not know about that solution to the fish kills.

Another fundamental problem Graham missed was reflected in the role of Everett Dirksen in defense of Velsicol. Until the mid-1970s, both the Republicans and Democrats claimed a strong commitment to environmental conservation. For example, the leading contenders for the 1972 presidential election, President Richard Nixon and Senator Edmund Muskie, competed for being the most environmentally responsible.[164] Bipartisanship on crucial national concerns sent a message that Americans would find no comfort from their civic leadership if they did not clean up their behavior. The creation of the EPA was a good example of the positive consequences of this bipartisanship.[165] However, the work of Senator Dirksen on behalf of Velsicol signaled a tragic shift in strategy among some leading Republicans. This change would open the door to polluters finding aid and comfort among the free-market ideologues who began to expropriate the Republican label after 1964—a label once owned by environmentalists such as Teddy Roosevelt. As early as the attacks on George Wallace, Rachel Carson, and the critics of Mississippi Valley fish kills, the new ideology defended exploitation of people, communities, and resources as "sound science" and in accord with the laws of economics.

This new ideology would so permeate American business that few industrialists would be able to comprehend at the end of the century when one of their own, Ray Anderson of Interface, would write, "To my mind, and I think many agree, Rachel Carson, with her landmark book, *Silent Spring*, started the *next* industrial revolution in 1962, by beginning the process of revealing that the first industrial revolution was ethically and intellectually heading for bankruptcy."[166]

CHAPTER 2

New Corporations and New Regulations

On June 14, 1965, the Chicago and North Western Railroad (c&nw), headed by Ben Heineman, confirmed rumors of the previous month, announcing that it was buying Velsicol Chemical from the Regenstein family.[1] Velsicol now joined the same type of corporate structure that housed Fruit of the Loom and Union Underwear. Heineman was in the early stages of moving North Western from being a railroad into being the name for a holding company controlling a conglomerate. While there had been holding companies for nearly a century, the new feature of firms such as the North Western was that they controlled subsidiaries in unrelated industries. By contrast, the Regensteins, while they controlled several companies, began with related companies that separated as their products diverged. When Arvey had owned Velsicol in the 1930s, it had produced the windows for Arvey's envelopes. In the case of the Chicago and North Western, Velsicol had no plants on the railroad's lines, and the railroad was unlikely to be a major customer of Velsicol.

Heineman had traditional reasons for moving from railroad owner to head of a conglomerate. He wanted to subsidize the operation of his railroad through the cash from highly profitable subsidiaries. The ideal conglomerate, he saw, would bring together additional resources to maintain weak parts of the firm.[2] By the time Heineman sold control of his conglomerate in the mid-1980s, his purposes had changed, yet neither he nor the business press saw this transition very clearly. In 1965, it appeared that Heineman merely had found a mechanism to fund a transportation revolution at his railroad. The new conglomerate converted a group of small firms in their respective industries into a large corporation, seemingly with the resources to sustain long-term progress. Meanwhile, the conglomerate's CEO, Heineman, emerged as a model of responsible corporate citizenship.

By the 1970s, all this changed. Heineman's conglomerate no longer used extra funds to support the railroad; rather, it milked funds from profitable entities to pay debt from acquisition of new parts of a ballooning conglomerate. Whatever Heineman's goals, the funding of expansions forced subsidiaries to follow management practices that risked accidents and transferred costs to communities that trusted the firms to bring benefits, not impose losses. At Velsicol's facility in St. Louis, Michigan, management practices contributed to an accident that contaminated the food of several million people. With contaminated cows shot in front of TV cameras, the mistake dramatized the limitations of the new liberal environmental regulatory structures that Heineman supported. Simultaneously, the effectiveness of the interest-group system in shielding the firm from responsibility while allowing greater food-system contamination exemplified the reasons for the regulatory failure. The cozy elite deals at the heart of the modern broker state shielded the powerful from responsibility. Yet, the poor policy response to the problems of Velsicol sowed the seeds for a popular critique of the interest-group system.

ELITE INNOVATOR

The recent purchase of Michigan Chemical by Velsicol had required purchase of the shares of a large number of stockholders. In contrast, the Regensteins and their foundation owned 95 percent of Velsicol, and therefore received almost all of the $90 million purchase price.[3] The transaction marked a new stage not only for the railroad but also for the Regensteins. They changed from a private Chicago industrial family to leading regional philanthropists. The major differences between Heineman and Regenstein were in their concerns outside of managing their business. Regenstein was the quiet philanthropist. Ben Heineman was an active participant in public affairs, personifying noblesse oblige. Unlike the shy Regenstein, he served visibly and was glad to be known by the public. Even as corporate managers, the two differed in their early years. Regenstein inherited his position and distanced himself from daily leadership. By contrast, Heineman rose from employee to owner, and spent his first decade as head of the railroad as an active transportation reformer and innovator.

Heineman was a determined man and accepted more than just the challenge of building a conglomerate. He wanted to play a role in public affairs, whether related to transportation or wider issues, especially of urban life and civil rights. As a loyal Democrat with a great interest in social problems, Heineman accepted the responsibility to use his prestige for the public good. However, as with building a successful railroad, making good urban policy became exceptionally difficult, marked by radical new expectations and frequent disappointments. The country moved from consensus on fulfilling the legacy of John Kennedy, as embodied in Lyndon Johnson's Great Society, to racial unrest and fundamental division over the war in Vietnam. More so than Heineman's Democrats, the opposition was reinventing itself.

In Heineman's Illinois, this transformation of the Republicans had caused anguish for the senior senator and Velsicol defender Everett Dirksen. His party of Lincoln now was courting southern segregationists.[4] More importantly for the policies that concerned Heineman, Barry Goldwater's candidacy marked the rise of a new form of conservatism—of neoliberalism. The older ideology of the party had made it quite understandable to Heineman, since it included an expectation that the conservative was to be a person with responsibility for preserving the society and especially the environment. The Republicans, after all, had been the party of Teddy Roosevelt and Gifford Pinchot. Now many Republican leaders questioned this legacy.[5]

This transition was ironic, given that Rachel Carson's work had just brought a new version of conservation to the consciousness of most Americans. The consequences of these changes were hidden for a number of years, since the landslide of 1964 buried the Goldwater wing for the next several elections. This new brand of conservatism would only rise to power when Velsicol needed it most, in the early 1980s. Then, the company, headed by a man who personified the full rejection of laissez faire, would be led by circumstances to advocate being left alone or "deregulated" to avoid the consequences of its mistakes.[6]

POLITICAL REFORMER

Early in his career, Ben Heineman had a fascinating experience in government economic policymaking, serving as an attorney in the World War II Office of Price Administration.[7] Working under Leon Henderson with a team that included future president Richard Nixon and economist John Kenneth Galbraith, Heineman first distinguished himself as a masterful thinker and organizer.[8] Heineman returned to Chicago and established a reputation as a competent liberal attorney. He both lived with and symbolized liberal business integrity, moving to the racially integrated, if upper-middle-class Hyde Park neighborhood on the city's south side.[9]

In the next decade he became known as the most innovative railroad executive in the country. He increased the money plowed back into track and equipment.[10] In 1958 he reached an innovative agreement with the State of Wisconsin to allow abandonment of a number of unprofitable lines in exchange for bringing upgraded double-decker cars to remaining passenger routes.[11] While those lines would eventually fail, he had much better luck with Chicago-area commuter rail service. Bringing similar new equipment and insisting that crews be on time and courteous, he became something of an icon to rail commuters on all other lines. He had done the impossible. Not only were the C&NW commuter trains on time and crews friendly, the company made a profit with them.[12] Heineman again rose to national prominence as a credible spokesperson for a new mass-transit policy.[13] He also played an increasingly visible role on behalf of racial and economic justice. Unfortunately, his efforts brought him into conflict with many white ethnic residents of working-class neighborhoods. He more often had the support of the city's progressive business leaders.

For the next five years, Heineman regularly accepted difficult assignments to help the country move beyond its legacy of racial discrimination. Early in 1966 he chaired Lyndon Johnson's White House Conference on Civil Rights, a conference formed to respond to the conclusions of what was called the Moynihan Report.[14] Here Heineman won the gratitude of Democratic Party regulars in walking a fine line between factions in the civil rights movement. This would not be the only time Heineman worked with a group formed to investigate the ideas of Daniel Patrick Moynihan.[15]

Heineman won praise from leaders of the mainline civil rights organizations and produced the final White House report in collaboration with the old railroad labor leader A. Philip Randolph.[16] In the midst of preparing the report on the White House conference, Heineman had to rush to Chicago to arbitrate a truce between his friend Mayor Daley and Dr. Martin Luther King Jr.[17] The emergency meeting to resolve weeks of low-level conflict between King's peaceful demonstrators and white neighborhood residents called on all of Heineman's negotiating skills. Held at the Episcopal Cathedral and attended by leaders of the major religious denominations in the city, the meeting reached a compromise, with a moratorium on demonstrations matched by a commitment from the Chicago Board of Realtors to equal-opportunity housing.

REFORMING RAILROADS AND CHEMICALS

While winning respect as a liberal political reformer, Heineman's growing business conglomerate was to taint his reform legacy. His railroad reform initiatives necessitated milking cash from the new chemical subsidiaries and promoting production eschewed by more profitable firms. In the mid-1960s Michigan Chemical especially was doing well with its national leadership in "rare earth" processing.[18] In fact, rare earths needed by the color picture tube industry and a variety of new products were so profitable, Michigan Chemical stopped production of some of its old pesticides based on DDT. Its europium was selling for $800 per pound. At one point, yttrium was selling for $1,000 per gram, and even after the company perfected extraction, the price held at $40 per pound, allowing a significant profit.[19] However, the rare-earth extraction process produced tons of low-level radioactive wastes that the company disposed of at the lowest possible cost on farmland northeast of St. Louis. The burial site would become the highest priority remediation site under the Justice Department settlement in 2002. In a classic case of passing on externalities to later generations, it meant the firm recorded record profits in 1965, some 48 percent above 1964.[20]

At Velsicol itself, Heineman replaced Regenstein with the marketing vice president, Norman Hathaway. Hathaway had the promotion and sales qualifications Heineman needed at Velsicol. While he had studied chemistry, as Hathaway later observed, he had "not been the top scholar of my class." Heineman did not need a research scientist; he needed someone who could make sales grow and not worry about the chemistry or its consequences. Hathaway boasted that as soon as he got out of the Navy after World War II, "because he wasn't a 'technical genius,' he decided . . . that his strength would be in marketing." Velsicol was a perfect place for Hathaway in early 1965, when the company needed to recover from the Mississippi fish kills. While some may have hoped they would hire an environmentalist to clean up their production processes, Hathaway provided the alternative solution, to continue to promote the company out of its problems.[21] Hathaway in one interview said, "I'm so pleased that I'm here for life."[22] He was a rare marketer who lived up to his word, dying on the job in 1969.

Next, it was Bob Garrison from Stauffer Chemicals who had to keep profits growing. Ahead of his times as always, Heineman found in Garrison a leader for the early stage of modern globalization. He had once managed Stauffer's Industria Química, and then had risen to head all international operations. His goal in his new position was to have "Velsicol . . . hit international marketing hard." Given the company's environmental track record, he was the ideal choice. *Chemical Week* said of his appointment:

> And speaking as an outdoorsman and father himself—two boys, two girls, ages six to 14—Garrison says, "I don't want my kids exposed to harmful chemicals anymore than the next man." But his response to outcries against chlorinated hydrocarbon insecticides is "to push for governmental approval and development of finite tolerances for these products in their important agricultural uses. Studies show they leave little if any residue to contaminate food or environment."[23]

Garrison had to achieve such changes in policy because his firm's profits continued to be more and more essential to Heineman's strategy of building a conglomerate sufficient to sustain his railroad.

WHAT'S PHILADELPHIA AND READING?

In 1968, Heineman's transition to full conglomerate operation took a giant and fateful step when he discovered that Philadelphia and Reading, itself already a conglomerate, was interested in a merger. Many years previously, the company had operated the Reading Railroad as part of its massive mining operations in the anthracite coalfields in northeast Pennsylvania. In 1902 its owner, George Baer, angered Teddy Roosevelt with his plutocratic arrogance when he said, "The rights and interests of the laboring man will be protected and cared for—not by labor agitators, but by the Christian men to whom God in His infinite wisdom has given the control of the property interests of the country, and upon the successful management of which so much depends."[24]

Baer was long gone in 1952 when a new generation of plutocratic investment bankers from New York borrowed funds to take over the Philadelphia and Reading and make it the shell of a conglomerate. In 1955 they made two more strategic decisions, turning over leadership to Howard "Mickey" Newman and acquiring Jacob Goldfarb's Union Underwear.[25] The next year they bought Acme Boot, of Clarksville, Tennessee. Later they added Imperial Reading, a supplier of "private label" clothing to chain stores, and Universal Manufacturing, producer of fluorescent-light hardware. Union, Acme, and Universal were the dominant firms in their industries. Over the next decade the conglomerate added Fruit of the Loom, which had maintained its finishing and licensing operations in Rhode Island near Bill Farley's home. Philadelphia and Reading closed the Rhode Island operation and began to blend Fruit into its former licensee, Union Underwear.

This diversification strategy seemed to work perfectly for Philadelphia and Reading. In an era when anthracite coal tonnage fell from 8.4 million in 1951 to 3.8 million in 1957, sales climbed modestly and profits soared. The new operations had far lower overhead costs, making the net profits greater. In 1951 the company had operating income from sales of $81,090. In 1957 it had $10,771,339. By 1967 operating income from sales had soared to $47 million. Buried in all this good news was the rise in "notes payable," from $1 million to $90 million.[26] Those obligations would become less noticeable if they were merged into a larger company. This is where the Chicago and North Western Railway came into play.

The word was out that Ben Heineman wanted to expand the Chicago and North Western beyond Velsicol and Michigan Chemical. Mickey Newman decided to act. He called Heineman at his summer home in Wisconsin. Heineman said his first response was, "Who or what is Philadelphia and Reading?" They worked out a deal where Heineman would be chairman and Newman president. Like so many future overleveraged conglomerate leaders, Newman was so desperate for the deal that he accepted this demotion to get at Heineman's assets.

The reorganized Chicago and North Western, which Heineman renamed Northwest Industries, now owned the old Chicago and North Western Railway, Velsicol, Lone Start Steel, Acme Boot, Universal Manufacturing, Imperial Reading, Union Underwear, and Fruit of the Loom. From a railroad with about 16,000 employees in 1960, it had become a complex firm with 36,000 workers. Its sales had climbed to $701 million from $138 in 1960. Net earnings were $61.8 million, up from $1.3 million at the start of the decade. However, it had a long-term debt of $402 million and annual interest payments of $19.3 million. By 1971 Heineman admitted, "It became clear to me how disadvantageous a railroad is. The returns were so high from the Philadelphia and Reading companies that I became familiar with the advantages of manufacturing businesses."[27] Accordingly, he offered to sell the railroad to its employees. The sale price was exceptionally low: $19 million to be paid over a twenty-year period. However, the catch was that the employees had to assume $401 million of the c&nw's debt.[28]

With his new awareness, Heineman became an outspoken promoter of the values coming to shareholders from production of goods. At a time when the most innovative members of the business community were beginning to move offshore, Heineman would resist the trend and build new plants, much as he resisted the full abandonment of the railroad. The consistency of his outlook was reflected in the opening lines of the first *Northwest Industries Annual Report* in 1968: "Economic growth is, after all, really a product of people at work."[29] How could anything go wrong?[30]

NEW CHEMICALS AND REGULATIONS

Throughout the late 1960s and early 1970s, environmental problems at Northwest's chemical subsidiaries regularly surfaced. Northwest reported that "Significant sums are . . . earmarked for air and water pollution control equipment [at Michigan Chemical]. Michigan continues to cooperate with state air and water pollution control authorities to improve the environment in the vicinity of plants." Meanwhile at Velsicol, Northwest noted that "sales and earnings in 1970 continued to suffer from the pesticide controversy." Taking the usual positive approach in reports to shareholders, Northwest announced, "Velsicol has launched a vigorous domestic program aimed at increasing sales of Chlordane. While Chlordane belongs to the family of chlorinated hydrocarbons, it is less toxic in application and less persistent in the environment than others."[31] As was so often the case with the company's boasts, the next year they had to retract their announcement. Now they said the "Velsicol Chemical Corporation is gradually reducing its dependence on its chlorinated hydrocarbon insecticides which are threatened by adverse regulatory pressure and restrictions."[32] Buried in these reports, however, were clues to a variety of problems. First, chemical sales had been in significant decline for several years. In 1968, they had totaled $71,800,000. They fell to $64.2 million in 1969, $62.8 in 1970, and recovered only to $67.4 in 1971. There must have been intense pressure on John Kirk, president at Michigan Chemical, to raise sales by any means possible. Second, any marketing success grew from introduction of new products that other firms elected not to produce: Phosvel at Velsicol, and Fire-master or BP-6, Michigan Chemical's fire retardant. Ironically the company's annual report tried to show the importance of the chemical subsidiaries to conglomerate profits with a photo of a child wearing pajamas laced with Michigan Chemical's fire retardants. The company also announced in mid-1974 that Velsicol was moving ahead on its biggest herbicide plant in Beaumont, Texas, while Phosvel production soared at the nearby Bayport plant. Between 1973 and 1974, chemical sales shot up 60 percent from $105.2 million to $167.8 million.[33]

As a reward, Heineman promoted Kirk to be chairman and CEO of

Michigan Chemical, and Theodore Girard, another sales genius and formerly vice president of marketing, assumed the presidency. Since Heineman needed to maximize sales to keep the cash flowing up through the conglomerate, Girard (who said, "I'm a born salesman") fit his needs perfectly. Like other recent presidents of the company, Girard lived in Chicago, not in the community with the original and biggest plant, St. Louis, Michigan. He liked the fact that Northwest gave him much freedom in running the company, only reviewing new applications for capital spending.[34] If Girard or Heineman had paid more attention to the communities where their facilities operated, they might have guessed long before 1974 that they were sitting on a powder keg.

Specifically in St. Louis, everyone from state natural-resource staff to local residents complained about the egregious dumping from the plant. Inside the plant, there were problems ranging from general housekeeping to specific equipment failures. A company inspector noted in 1973, "Age of facilities along with spilled-leaked salt accumulations in various areas present a poor picture. . . . There was also a considerable amount of miscellaneous trash-litter laying around."[35] After a review of the plant in 1977, the National Institute for Occupational Safety and Health reported filter and exhaust improvements and installation of lids on various equipment that had allowed workers to ingest various carcinogens. In addition the report called for labeling all piping with the chemicals inside.[36] A local resident who was a retiree from Dow Chemical, Ralph Boyles, felt that the plant was "worn out and the railroad company that owns it has bled all the money it could out."[37]

Externally, the plant's impact was egregious. Young and old residents noted the problems. Joe Scholtz, in the 1970s a young gas-station manager who heard the plant might close, observed, "Without that plant, the paint on a lot of houses will stop peeling. . . . [S]omething yellow [released from the plant] a few years back, killed all the gardens around—like frost in midsummer. . . . And we can't go fishing on the Pine River. All you get is carp."[38] Jim Kelly, a policeman, noted that the plant dumped so much waste into the 250-foot-wide river that "I could hardly paddle down the Pine River. The bottom of my canoe was covered with a yellowish-white mass. That's sickening." Two high school girls agreed that "it smells around here all the time and gives you a headache." The father of one, Richard Wrath, added, "Fumes from the plant about turn your stomach." Before the plant was built, the area had been a picnic area and campground; perhaps exaggerating, Wrath noted, "[It was the] most beautiful picnic ground in the world. Now look at that mess."[39]

In 1955 the Michigan Water Resources Commission had conducted "A Biological Survey" of the Pine River that flowed past the Michigan Chemical plant. The commission's researchers collected samples at six locations, two above Michigan Chemical, one at the plant, and three downstream. On the two sides of the river next to the plant, no animals were found. The report found:

> On the right side of the river below the Michigan Chemical Company the bottom was covered with DDT crystals at sampling station 7. There were no bottom animals present at this station. In mid river the bottom was of black very oily mud with a refinery odor. The debris in the sample was saturated with oil. . . . Severely polluted conditions are indicated at the stations [below the dam that barred migration].[40]

Later reports found pollution only getting worse.

In 1967, the state not only found significant flows of contaminants still coming from the plant, but noted, "Large sludge deposits have accumulated in the Pine River." The Water Resources Commission staff sent to measure the emissions noted a fundamental problem when "[The] sharp acidity disintegrated the ¹⁄₆₄ inch thick aluminum flashing which was used to line the crest of the weir. This disintegration was noticed approximately 48 hours after the weir was installed."[41] Two years later, the state Department of Natural Resources (DNR) reported the company's dumping of wastewater on land owned northeast of town. The state geologist urged seeking cooperation of the company, noting, "We are vitally interested in preventing this type [of] contamination, even before it has a chance to get as far as it has in the present case."[42] Despite these findings, Michigan's agencies could take action only in crises, or to facilitate uncontested cleanups.[43]

The events of the era were to become an excellent test of theories of voluntary compliance with good natural-resource practice. Despite repeated requests from state officials, the cash-strapped company countered with proposals to reclassify the river to "commercial use." In July 1970, the state urged the company to dredge up to 15,000 tons of solids from the river, after finding that sodium concentrations below Michigan Chemical had increased significantly between 1967 and 1970.[44] As in 1955, they found that "The impoundment bottom below the Michigan Chemical Company was coated with a hard, white precipitate."[45] In a memo accompanying the larger report, the team summarized the results found below Michigan Chemical as follows: "All quantitative dredge samples contained large amounts of oil and were devoid of animal life. The bottom muds consisted of black oily ooze overlain with a thin layer of hardened calcium carbonate. . . . The sampling equipment was fouled by oil that re-suspended when the shoreline was disturbed."[46]

Simultaneously, the company contaminated a host of other sites in the region with wastes removed from the plant site. Across the river from the plant, the company owned thirty-seven acres, which it used for waste. At least some of the waste was pumped under the river from the plant. The company used this site from 1936 until 1970, when it sold the land to golf-course developers. While no records exist stating which chemicals and how much was pumped to the site, the company admitted in 1980 that it had probably sent DDT, 2, 3,-dibromopropyl (TRIS), and heavy metals there. In the 1980s the golf-course area that had been used for waste disposal became one of the community's three Superfund sites. The county landfill, which all local residents used for trash, became a second Superfund site because the company had dumped hazardous wastes there along with office trash. Finally, several miles northeast of town, the company had its radioactive-waste disposal site. In 1980, a review of company records showed that 303,174 pounds of wastes were buried there between 1967 and 1970.[47]

PRELUDE TO ACCIDENT

Given this background, what happened sometime in 1973, probably late May, was surprising neither to local residents nor to state natural-resource officials. It only came as a shock to the distant Chicago management and federal regulators. Neighbors and workers at the plant knew that ramshackle buildings, trash, and poor safety equipment marked the St. Louis facility. As the company squeezed every possible cent from the operations, labor relations continued to deteriorate into a late 1973 strike. But fire retardants seemed a way out of the problems, just as DDT and rare earths had been earlier. As with DDT production during and after World War II, federal government policy played a key role in the success of the company.[48] A new federal fire-safety concern created a potential replacement product for rare earths.

Unfortunately for the public, the very compounds that had remarkable fire-retarding properties had disturbing characteristics when in contact with organisms. By the early 1970s, several researchers concluded that the compounds were linked to cancer and genetic damage.[49] Since halogenated hydrocarbons, such as PBB, tended not to break down and pass out of an organism, an animal low in the food chain would retain all the PBB it had consumed and pass it to predators. At the top of the food chain, people eating river fish, living near the plant, or working in it would retain all PBB inhaled or otherwise ingested. Because of such negative properties, several companies elected not to produce PBB.[50]

The experience of Michigan Chemical over the next few years with PBB illustrated the weaknesses in a largely voluntary environmental and epidemiological system, especially when a firm faced pressures for high profits. For example, Michigan Chemical commissioned a low-cost study of PBB dangers from a firm in Cincinnati, Hilltop Labs. Two flaws undermined the value of this "independent" analysis. Hilltop largely ignored the developing literature on PBB toxicity, reporting back merely that while nontoxic, BP-6 should not come in contact with food.[51] Not only did Hilltop underestimate the compound's toxicity, Michigan Chemical exacerbated the problem by underestimating the relevance of the study.[52]

As with the early DDT research showing toxicity problems, Michigan Chemical chose to throw caution to the wind and begin full production of PBB in 1971.[53] Continuing past practice, Michigan Chemical expected to profit handsomely from this risky behavior, assuming it would bear none of the externalities. It continued to have low-cost use of the county landfill and river, where it dumped tons of PBB wastes. Under the market system in 1973, the St. Louis plant passed on the external PBB costs to the workers, the community, and the wider population of the upper Midwest.

In addition to corporate carelessness, a major PBB problem became more likely as a result of the limitations on government regulators. At the time PBB production began, Michigan officials believed the state was among the nation's leaders in the control of industrial pollution. Since the Progressive reformers of the early twentieth century, such as Governor Chase Osborn, the state had pioneered the development of public environmental policies.[54] In the 1920s it created an early state Natural Resource Commission and Stream Control Commission to promote conservation and protect water quality.[55] Beginning immediately after World War II, these commissions gave special attention to pollution in the Saginaw Bay Watershed of Lake Huron, in which St. Louis was located.[56] Two generations of state policymakers had refined law and administration to regulate and prevent environmental degradation. However, with limited staff until the 1960s, they gave little attention to relatively small firms such as Michigan Chemical. More obvious sources of hazards, especially the larger and older Dow Chemical plant downstream in Midland, overshadowed it.[57]

When assuming office in 1969, the new governor, William Milliken, even more than his predecessors, emphasized environmental protection. In his first speech to the legislature, he said, "Concern for our natural environment must occupy a high place on our state agenda. Those who prosper from our resources must help to preserve them."[58] During the next four years, he reorganized state environmental administration and sought additional legislative authorization for environmental regulations. At about the same time, new federal legislation, such as the Clean Water Act of 1972, reinforced state pollution control.[59] Two weeks before a catastrophic PBB accident, the state launched a new Wastewater Discharge Permit Program to be administered by a reorganized Water Resources Commission. However, given major urban pollution sources, the preliminary ranking of wastewater threats rated Michigan Chemical as a secondary polluter, subject to a delayed permitting process. Without staff to monitor what were considered secondary pollution sources, the commission did not follow up on the Pine River's pollution until after massive food-chain contamination focused concern on the plant.[60]

THE INEVITABLE MISTAKE

The PBB disaster began when Michigan Chemical Company accidentally shipped some PB-6, trade name "Fire-master," to a Michigan Farm Bureau Services feed-supply yard in place of a cattle-feed supplement, trade-named "Nutri-master." Because of the similarity of the names Fire-master and Nutri-master, compounded by the inability of livestock to detect the difference, the PB-6 not only was shipped but also consumed by animals on farms around the state from 1973 into early 1974. The first indications of the mistake surfaced on individual farms as livestock, especially cattle, lost appetite, developed deformities, and bore deformed or dead offspring.[61]

Initially, the farmers with problems received advice from the agricultural extension system. However, when individual farmers linked the problems to their feed supply from the powerful Farm Bureau, the help from public agricultural extension officials at Michigan State University and the state Department of Agriculture faded. Inaction by those responsible for the public welfare allowed a small problem to grow throughout the region, and for animal products to pass into human consumption for months after the first complaints surfaced. Agencies and institutions supported by the general public acted to protect their clients more than the general citizenry. Many of their clients, especially small dairy farmers, were even allowed to get into deeper trouble than necessary by lack of attention to the public welfare. By the time firm action was taken to halt food-chain contamination, many of the eight million people in Michigan's Lower Peninsula had absorbed PBB into body fat.

Of course, any assessment of responsibility for this accident must begin with Velsicol. They had cut corners for so many years to milk Michigan Chemical of cash. Later investigations found products such as BP-6 stored in the same places as the food products, of which Michigan Chemical made many. Compounding cost-cutting, which left the factory dirty and poorly organized, was the fixation on marketing. The company decided that the word "master" tacked onto their chemicals helped "brand" their work. Yet, the similarity in names made the mixing of adjacent bags even more likely. A shortage of paper bags in early 1973 caused temporary packaging of Fire-master in old Nutri-master bags, with mere restenciling of Fire-master on the bags. Even after company records reported bags of BP-6 missing in June 1973, the employee-strapped company did not launch an emergency search for their destination.[62]

The Farm Bureau, however, was not purely a victim in this accident. Its behavior only complicated problems and harmed its members. The first farmers to note problems in their herds included one persistent man with chemistry training, Rick Halbert. He pursued a cause for his problems relentlessly and fairly quickly traced the problem to the feed he had bought from the Farm Bureau supply center in Climax, Michigan. By fall of 1973, complaints from him and others had been sufficiently noticed by the Farm Bureau that staff meetings discussed what could be wrong. As with so many institutions, the Farm Bureau quickly shifted from helping its member clients into protecting itself—it did not tell individuals that others were complaining. It recalled contaminated feed without telling members why. It did not tell members that in February 1974, it had initiated calls to Michigan Chemical about the quality of its feed additive. After it became aware of the nature of the accident in May, the Farm Bureau shifted fully into a mode of protecting itself.[63]

The state regulatory structure for agriculture focused on helping the Farm Bureau, not all citizens. The Michigan Department of Agriculture (MDA) saw the PBB accident in economic, not environmental-health terms. With a director appointed by the "nonpartisan" Agricultural Commission controlled by the Farm Bureau, the department could not be forced to work with public health or natural-resource officials. The governor's special aide for PBB, Kathy Starika, lamented that her boss could not fire the department's director, B. Dale Ball.[64] MDA's goal was to protect farmers and the state's agribusiness interests generally from the adverse consequences of the accident.[65]

However, that approach allowed the health consequences of the accident to expand, and the financial consequences for some farmers to worsen. A factor in MDA behavior was its decision to include only special interests in initial meetings to resolve the crisis. After the accident was exposed in a *Wall Street Journal* article on May 8, 1974, the MDA held an emergency meeting with the Farm Bureau and Michigan Chemical, including the company's attorneys, but allowed only a low-level public-health official to attend. While state natural-resource staff who had long been familiar with the company rushed to St. Louis, MDA kept control of statewide response for as long as possible. MDA managed to convince the governor to keep epidemiologists at bay for several months.[66]

Complicating the problem of government failure to act, even the land-grant university, Michigan State, had its behavior compromised by traditional links to the Farm Bureau and the habitual research behavior of its scientists. Generally, the university's agricultural programs had close ties to the Farm Bureau, and as soon as the bureau's possible legal liability became clear, the university backed away from helping farmers such as Halbert. More insidious was the process of grant support for research. Generally, faculty only did funded research, and none showed interest in a new, important issue for which they were not paid. Since the only early funding available for PBB work came from the Farm Bureau, any help for farmers from the university was unlikely. Further, since the Farm Bureau virtually controlled the university's county agricultural agents, these front-line experts turned a deaf ear to the farmers and the general public. When tests were needed, veterinarians working for individual farmers turned to internationally recognized agricultural universities outside the state, especially Purdue in Indiana and the University of Guelph in Ontario.[67]

REGULATING AN ACCIDENT

Despite the intense scrutiny of the cattle-feed problems, only limited lessons were learned directly from the cattle-feed mistake. It was purely an accident, not the goal of routine manufacturing.[68] The PBB crisis provided an excellent illustration of recurring environmental-policy problems ranging from chronic market failure to overconfidence in the effectiveness of formal regulations. It showed the limitations inherent in administrative remedies applied without widespread public participation. Particularly when studied from the local perspective, the inequity and inadequacy of PBB policy illustrated the problems of a process that discounted knowledge of the public interest, curtailed judicial protection of individuals, and restrained effective public management by tolerating capture of administrative agencies by regulated interests.[69]

The failure of the environmental-policy process to protect the local environment provides vitally important context from which to evaluate the environmental-policy debates in the late twentieth century. In the twenty years after the PBB crisis, that debate focused primarily on the choice between preemptive regulation and reliance on market forces. Pointing to recurring environmental crises, proponents of federal government regulation succeeded in getting many specific federal laws passed regulating air, water, solid wastes, and in 1980 highly contaminated sites under Superfund.[70] Many states passed similar legislation, sometimes before federal action and often in response to it, allowing the states to play a significant role in collaboration with the national government. This seemingly impressive new field of policymaking, symbolized by the creation of US EPA in 1970, came in the context of interest-group liberalism and, with each passing year, the newer version called neoliberalism.[71]

During the 1970s, neoliberal supporters of deregulation launched a generally successful critique of preemptive regulation. They emphasized the absence of sufficient knowledge to make rational regulatory decisions. They believed the market could respond more quickly and precisely to control threats to public health or the environment, with administrative adjudication resolving disputes. They also backed reliance on private-sector consultants rather than public administrators for proper implementation of existing regulatory laws.[72] Yet, the reality was that neither approach worked to control the PBB problem, and that failure should have alerted policy scholars of the error inherent in the terms of the debate, guided by the shared liberal and neoliberal assumptions about "creative destruction," the role of elite leaders, respect for technical and professional experts, and the contrasting contempt for public officials.

The PBB crisis provided one of the best tests of both the liberal environmental reforms of the 1960s and 1970s and the neoliberal counter-reforms starting in the 1970s. The PBB crisis had its origins in corporate behaviors developed before most late twentieth-century environmental laws took effect. Clearly, it demonstrated the failure of loosely regulated market mechanisms to control pollution and protect public health. Yet, the crisis continued into the period after the implementation of the new body of regulatory law.[73] Supposedly tough regulatory laws implemented in a pluralist policy context dominated by articulate and affluent interests would not achieve good results for citizens. Efforts at containment of the PBB crisis lasted long enough that it continued to be wrestled with in the era of "deregulation" that followed 1980. Since failure took place in all eras, the crisis revealed how the neoliberal approach, regardless of policy, period, failed to serve the long-term public interest. In all eras, the failure to address the crisis showed how a subgovernment of key state and company officials set policy advantageous to themselves, and how their reluctance to be limited by popular concerns or administrative expertise resulted in failure to restore either the local environment or the economy.[74] The responsiveness of local legislators and civil servants from outside the subgovernment made this failure particularly notable and indicated the value of both vigorous public debate and expert public administration as solutions to environmental crises.

Despite years of state environmental-policy experience, the state was unprepared when the pollution originating from Michigan Chemical came to public attention. Michigan had not implemented regulations to control pollution emanating from such firms, nor had it performed comprehensive risk assessments of their impact on the environment.[75] The market for the company's products, the cost of extracting raw materials (primarily bromine from beneath St. Louis), and wage rates provided the only behavioral incentives. Since there was no formal mechanism to collect information on toxins before the crisis, neither the government nor workers knew they should make special environmental or health demands on the company. There was no effort by government to assess the company for the costs of stream pollution, and the union did not demand compensation to offset potential health risks. Of course, advocates of deregulation rationalized how the earlier failure proved that no regulation was best. Writing in 1983, one proponent of deregulation said, "The most compelling point is that the absolutist approach to regulation, with rigid standards intended to eliminate risks, has yet to produce significant improvements in workers' well-being but has nevertheless inflicted substantial burdens on firms."[76]

WHO PAYS FOR ACCIDENTS?

While the environmental impact of Michigan Chemical's operations was largely unregulated before 1974, once the cattle-feed accident focused attention upon the company, demands for regulatory changes intensified. In response, Michigan Chemical and the Farm Bureau went into damage control. They simultaneously blamed each other, while searching for support from both their insurance companies and each other. Observing the two, Kenneth Frankland, the governor's legal counsel, said, "Unbelievable is perhaps the best adjective to describe the final negotiations to bring this matter to conclusion. . . . The workings of the human mind may be explainable to Sigmund Freud, but not to me."[77] They also made symbolic gestures after dragging their feet. For example, the chemical company kept refusing to stop PBB production and then suddenly announced a halt on November 20, 1974. Although still profitable, Velsicol no longer focused primarily on profit margins at Michigan Chemical, but rather on avoiding as many of the costs of PBB cleanup as possible. They even prepared to strike a deal with the government to minimize penalties by discontinuing operations. Unfortunately for the workers and the community, regulatory changes since the early 1970s had not improved their position in the policy process. No regulations mandated that production or closure decisions weigh the costs to the community or the workers. In fact, tax loss rules made cleanup and closure relatively painless for the parent company.[78]

As would be evident in later cases with other Fruit of the Loom subsidiaries, the Oil, Chemical and Atomic Workers Union was not prepared to confront the threat of closure. Customarily, union concerns included only wage rates, basic work rules, and conventional occupational safety issues.[79] As the PBB disaster unfolded, the workers wrestled with the dilemma of assigning priorities to health protection, compensation for illness, and job preservation. With the county unemployment rate averaging at least 10 percent in all but one year in the 1970s, the workers feared losing the jobs that made them part of the county elite. As one news story noted, "While angry at the company for not warning them of the dangers of PBB, . . . they would help to keep the plant open."[80] As so many other workers in areas of high unemployment, they were not in a position to absorb the costs of policy decisions.

Likewise, the city was not prepared to assume the costs. The need for stable businesses in St. Louis led residents to accept environmental degradation as the price of prosperity. PBB production was not the first time city residents made such a bargain. Throughout the first century of white settlement, the region accepted profound ecological change as the cost of development. In the nineteenth century, St. Louis profited from deforestation and became a regional resort for visitors seeking the medicinal powers of the mineral waters pumped from under the city. It was only one more step in an established tradition when the city welcomed the boom that followed conversion of its mineral waters into raw materials for Michigan Chemical.[81] Some residents complained about the obvious pollution resulting from Michigan

Chemical's operation, but many hoped "to solve the problems [without driving] . . . out the one large employer in St. Louis."[82] Certainly the city was not prepared to bear a major cost of any plant closure, particularly given the contamination by the plant of sites such as the local golf course, deep wells, and the county landfill.[83]

While the initial responses to the PBB crisis did not threaten the worker and community cost-benefit tradeoff, this changed by the mid-1970s. After two years of negotiations, the state and Velsicol announced in 1976 a settlement that included a token fine and closure of the plant. Given the scale of environmental damage known since the 1930s, the outcome of the negotiations bordered upon the absurd. The firm agreed to pay $20,000 to the state Water Resources Commission to restore fish to the river, and in exchange received a blanket release from future claims from the state to any damages to "the waters of the state."[84] Much too late, Stewart Freeman, the state attorney who had negotiated the settlement, appeared before the new Michigan Environmental Review Board (MERB) to defend what his boss called a "fair and balanced" settlement. The press described the meeting as a "fiery session." Rodger Young, an attorney member of the MERB who represented "the public," argued, "I don't understand why for $20,000 prior to any investigation you release these people from any liability."[85]

One of the reasons for mistakes such as those made by the Water Resources Commission was that the company and its experts effectively used scientific uncertainty to their advantage.[86] For example, when the governor's office sought to learn what was known of the dangers of PBB, the special interests concerned with liability for the mistake used uncertainty to call for further action.[87] Meanwhile the press raised expectations of prompt, decisive action by public officials. There were risks to elected officials in acting too quickly or too slowly, or in admitting ignorance of PBB and policy alternatives. Only the company had a clear sense of the risks of its options. Closure would cut its losses, especially if public officials continued to exempt it from responsibility for complete cleanup.[88]

Elected officials faced two other barriers to good policymaking: the coordination of a variety of government agencies, and the lack of responsiveness of issue networks centered in specific agencies.[89] A variety of public officials had a role in addressing the PBB problem. The Michigan departments of Agriculture, Natural Resources, and Public Health had key roles in protecting the public from PBB. Because the PBB crisis came to public attention through the contamination of cattle and then entered the food chain, primary responsibility for risk assessment and control moved across departmental lines. Once the public became aware of the potential PBB problem, the Department of Natural Resources needed to follow up on its long-term evaluations of environmental problems in St. Louis and other PBB contamination sites. Coordination became as important as the capacity of each department to implement policy.[90]

Complicating dependence on a variety of agencies, the governor faced an additional barrier to control of two of the departments. The directors of both the Michigan Department of Agriculture and the Department of Natural Resources reported to commissions, not to Milliken.[91] This structure resulted from Progressive Era efforts to insulate administrators from "inefficient and partisan" control and to subject them to oversight by concerned special interests.[92] It allowed such special interests to achieve their policy goals primarily by building a cozy relationship with the commissioners of the agency, which served their needs. Departmental directors worried about the goals of these narrow constituencies, ignoring elected officials and their concern with the public interest.

In agriculture, a classic cozy triangle developed with the Commission of Agriculture dominated by the Farm Bureau. The Farm Bureau and commission also had special links to legislators with key agricultural policymaking roles.[93] The triangle may have served the interests of most farmers before the PBB crisis, when they and the Farm Bureau shared many policy goals. The weaknesses of cozy triangles became abundantly clear once blame was assigned for the PBB accident. The PBBs had been mixed with cattle feed at a Michigan Farm Bureau Services–owned operation in Battle Creek. Soon the Farm Bureau as an institution was liable for losses suffered by members. When some farmers began asking for public investigation of the sources of their problems, the Farm Bureau sought delay. For a year the agricultural triangle backed the Farm Bureau, not its members.[94]

As government failed to respond to the crisis because of lack of information, poor interagency cooperation, and conflicts of interest, attention shifted from the private-sector blunder of an irresponsible company to questions of public leadership and management. The 1977 British TV documentary "The Poisoning of Michigan" criticized the state departments of both Agriculture and Public Health, as well as the governor, while devoting little attention to Velsicol.[95] Clearly, officials had underestimated the extent of exposure to PBB before and immediately after the cattle-feed mix-up, but they had not caused the PBB crisis; Michigan Chemical bore that responsibility. By 1976, there were legislated responses to PBB-type problems, including the federal Toxic Substances Control Act, which established a testing and record-keeping procedure for all chemicals.[96] Despite this policy response, the press confused the timing of public policy failure and policy responsiveness. Media coverage of the PBB accident became more intensive and dramatic, and politicians faced embarrassing portrayals of their roles. The British documentary was followed by a planned fall 1978 dramatization on the popular CBS-TV *Lou Grant Show*.[97]

Because of the career concerns of elected officials, election schedules began to coincide with PBB decisions. In 1978, it seemed some political careers would be helped and others terminated by the disaster. William Fitzgerald, the Democratic challenger to Governor Milliken, repeatedly linked the governor to mistakes in handling the crisis. In the congressional district that included St. Louis, the challenger, state legislator Don Albosta, built his campaign around his responsiveness to the PBB problems of farmers and the region's residents.[98] To avoid an election disaster, the governor took several steps. He added to his staff Frederick Halbert, the first farmer to bring his herd's contamination to media attention.[99] Meanwhile, Republicans pressured CBS into delaying the "Lou Grant" episode until after the elections.[100] Finally, the governor and key state administrators effectively removed Velsicol from Michigan several weeks before the election by concluding the closure of the St. Louis plant on September 1, 1978.[101] Of course the closure shifted costs for the PBB accident from the company to the employees, St. Louis residents, and neighbors on the Pine River.[102]

Of all of the lessons of the PBB incident, the most important arise from the decision making leading to the departure of Velsicol from the state. Repeatedly from 1976 to 1978, both the company and groups of employees offered to continue to operate the plant and clean up the environment.[103] That solution, if enforced, would have been the better one. The community would have retained jobs and its tax base. The river would have been cleaned at no cost to the public. Prime riverfront land would have been usable. By contrast, plant closure and toxic containment forced community residents and former employees to pay most of the costs of the disaster. They were not consulted until it was too late. The most inclusive congressional hearings, which gave former employees, residents, and local officials an opportunity to be heard, came on November 19, 1979, fourteen months after the plant ceased operation. After the elections in 1978, it was too late to correct the major policy decisions related to plant closure and limited corporate liability.[104] A policy subgovernment dominated the earlier stages of policymaking, including statewide and federal elected officials, key appointees, private environmental experts, and company officials.[105]

The closure decision illustrated the policy mistakes inherent in a process that excluded some of the key stakeholders. It also made clear that confusion about risks worked to the benefit of those with most control over choices.

Despite widespread skepticism about the possibility of simple cost-benefit analysis of risks, Velsicol had a very clear assessment of potential costs and benefits from policy options. The closure assured it would assume proportionally little of the costs. Within three years it again was profitable and continued to operate outside of Michigan.

THE ELITE PROBLEM

The failure of the process in St. Louis raised fundamental questions about two extremes in the limited environmental choices usually offered to U.S. citizens. First, the egregious market failure of Velsicol exposed the fundamental flaw in free-market environmentalism.[106] However, it also showed that the free-market advocates were right to rebel against those who look to an elite-controlled bureaucratic state imposing regulations for the good of the citizens.[107] Rather, it pointed to the value of an alternative approach advocated many years ago by John Passmore, who based his hope for good policy on American traditions, frustrated intentionally in St. Louis. He praised the fact that "in the United States, particularly, the habit of local action, the capacity of individuals to initiate legal proceedings, the tradition of public discourse, are powerful weapons in the fight against ecological destruction."[108]

The problems evident in the responses to the crisis by state officials, leaders of the Farm Bureau, and researchers in higher education reveal a flaw in the assumptions that would entrust control either to private market forces or to elite leadership. The elite leadership of the Farm Bureau, the Michigan Department of Agriculture, and relevant departments at Michigan State University followed customary procedures when the crisis began. They met with Michigan Chemical officials and trusted the company for too long, allowing widespread food-chain contamination. Consultation with impacted farmers, lower-level bureaucrats in state agencies such as the Department of Natural Resources, or residents in places like St. Louis would have revealed fundamental problems with the company, and should have resulted in prompt action that would have halted most food-chain contamination.

While it is inevitable and understandable that leaders in an industry such as the Farm Bureau, or in government, talk primarily to leaders of a company with which they have an issue, it is not necessary that they ignore other sources of knowledge when they see a problem linked to that company.

There can be processes established that require broader information gathering to assure the flow of warning signs up to those with responsibility for community well-being. A retrospective review of what went wrong in St. Louis related to Velsicol is challenging, yet there are intriguing indications of how the system failed.

The best evidence of the fundamental failure of the PBB policy process can be found in the stories produced by a team of three reporters who descended on St. Louis in the aftermath of the PBB accident. The stories the reporters gathered quoted fifteen people, based on interviews with a number of local leaders and average citizens. At the time, the company had just announced it would close the plant in two years. These extended interviews reveal an interesting pattern of thought. While limited in number, the reporters' sampling produced a stratified random set of responses that hints at the contrasting perspectives of leaders and average citizens.

Those interviewed included six current community leaders; three business owners, including a former mayor; and six "average" citizens, including two teenagers. The leaders interviewed included the school superintendent, president of the school board, head of the chamber of commerce, city manager, city clerk, and the head of the county economic-development agency. The business owners included the manager of a bank, the owner of a restaurant, and a furniture storeowner who was a former mayor. The county economic-development chief did not make comments related to corporate behavior, and therefore cannot be included further in this analysis. The "average" citizens included a policeman, an auto worker, a gas-station manager, a retiree from Dow Chemical (based thirty miles downriver in Midland, Michigan), and two high school students—the only women interviewed.

Of the five current community leaders, the comments were uniformly sympathetic to the company. Ignoring years of documented reports of company contamination of the environment, the city manager said, "[The charges against the company] simply aren't true . . . I think Michigan Chemical, given the time, would have abated the problems within their financial framework, but it just wasn't economically feasible for them at this time."[109] The city clerk excused the company's PBB mistake, saying, "This mistake thing is just human error, but I guess you can't be too careful."[110] The head of the chamber of commerce took the opportunity of the interview to attack state business regulation, not the company: "This is a terrible place to do business and people are going to other states where the laws are more accommodating."[111] The president of the school board, a local physician, turned his attacks on company critics: "The ecologists have gone overboard. I predict in the immediate future those ecologists will cause the

American people a lot more trouble."[112] Finally, the school superintendent added, "We think of Michigan Chemical as a friend."[113]

Two of the business leaders shared these opinions, while the former mayor, the only popularly elected person interviewed, took a somewhat different tack. The bank vice president and manager admitted that "environmentalists have a point." However, almost as if he were powerless, he added, "But the question is how can the two, jobs versus the environment, best work together."[114] The local restaurant owner echoed the theme of the chamber of commerce, calling for an improved business climate: "We'll have to start making a better effort if we hope to survive. . . . If we don't and the state doesn't, all the good businesses will be moving south."[115] The former mayor and furniture-store owner, joined by his wife for the interview, was the only leader to voice some skepticism about the company. "He thought the cost of the pollution control may only have been an excuse of the parent holding company (North Western) [*sic*] to 'bail out' on a low-earning unit." But even he, using graphic images, expressed the shared powerlessness of the elite. "It's like an atomic bomb. Epicenter is the chemical plant. But direct and indirect explosion and damage will be miles in every direction. Terror and damage are there."[116]

Average citizens, in contrast to the community leaders, took a more critical perspective, focusing on the company's irresponsible environmental behavior. Perhaps most interesting was the response of the young women in the high school. Unlike their school superintendent and the head of the school board, they did not see Michigan Chemical as a friend. The sophomore noted that with the company's impending closure, "the town will look better and some of the sickness might go away," including the asthma of her younger sister.[117] Her freshman friend did more sociological analysis than the adult leaders, adding, "A lot of men will be on the streets and in bars. And the place won't stink as much."[118] Unlike the school superintendent, who saw the company as a friend, the service-station manager's friends were neighbors who suffered from the plant's behavior. Ironically, considering that the company made chemicals for color picture tubes, he noted: "And a friend of mine who lives across the street from the plant says the stuff corrodes television antennas, resulting in poor color reception."[119] The auto worker, instead of attacking state regulators, said that the closing of the plant "is the best move this state ever made." He assessed blame on only one institution: "It's the plants [*sic*] fault this is happening."[120] The reporters added that the policeman likewise noted, "The plant has caused air and water pollution problems in the area since before he was born (1942). . . . the real reason the plant was folding was its trouble with the PBB."[121] Finally, the Dow retiree and volunteer fireman lamented the loss of jobs and tax base, but placed blame on Northwest Industries for taking out assets and failing to clean up the pollution they had created. "It's a helluva note. . . . I used to catch pike and bass in that river when I was young. Kids today don't believe me."[122]

The most fascinating question raised by the contrasting responses of elites and average citizens is why the more powerful members of the community became defensive, while the least powerful were more assertive. Perhaps the responses in this case show that the local elite feared that admitting errors of the leading company in their community would only prove that the local elite had little power. To make the company innocent and powerless protected the local elite's claims to leadership. In contrast, the average citizens made no pretense of being powerful and could see clearly the manipulations at the top, as when the Dow retiree correctly described Northwest Industries' milking of assets from the company, or when the policeman noted that everyone knew of the problems, yet management did not act quickly to correct them.

Clearly local leaders bring to issues of this type a broader perspective than many residents. The resident may think only of odors or deposits in their garden; the leader must worry about maintaining a tax base or attracting future business development. Both are valid responses but may put people into fundamental conflict with the leaders who seem indifferent to daily hardships. If handled poorly, such problems can lead to cynicism about leadership and to various forms of resistance. If the statements captured in the news reports are accurate reflections of broader attitudes of school leaders and students, is there any wonder that teenagers would rebel against the school routine? It must be presumed that all participants in the interviews eagerly purchased the paper and read the comments of not just themselves but the others. It also must have been quite humorous for the two students to read the superintendent's and school-board president's remarks, conveying such misplaced admiration for the "stinky company." Likewise the other "average" citizens must have chuckled at the blindness or naiveté of their "leadership."

The most important question revealed by this case is not why the leadership tended to be blind to company faults, but how to avoid such blindness in other circumstances. Given the ultimate fate of Michigan Chemical's operations, and even the ultimate consequence for its parent company and the nation's taxpayers, methods or processes should be found to reduce the gap in perceptions. To label this process, a way should be found to improve "the politics of local knowledge." Ultimately, the mistakes of Michigan

Chemical and its local backers did not protect the company or save local jobs or the environment. They only discredited the leaders involved and allowed potentially minor problems to become unsolvable. The entire PBB crisis could have been avoided if the most basic reforms had been made in production of chemicals at the plant. Clearly Jim Kelly, the policeman, had a better assessment of the consequences of decisions than did the leadership. The state counted on the closure of Michigan Chemical to solve its embarrassing PBB crisis. Kelly warned, "[The end of the plant] will not end the problem. The mess just won't fold up and go away. I bet it'll cost a bundle."[123] Obviously, he should have been involved in the leadership in the 1970s. The point of considering the politics of local knowledge is to learn processes that would build insights, such as the policeman's, into the decision-making process of the leaders. Likewise at the schools, the ideal situation would be for the leadership to learn from the wisdom of their students. Perhaps such a restructuring would avoid some of the dysfunctional nature of our schools, which seem more committed to turning out obedient followers than a wise public, and which are too much involved in teaching the correct answers rather than how to ask good questions.[124]

Ultimately, the basic institutional corrective for these problems worked—that is, democratic elections. Jim Kelly eventually became a St. Louis City Council member and has participated in the process of advising the EPA at the end of the century on its massive cleanup of the Pine River. Similarly, Congressman Elbert Cederburg, who had risen to a leadership position within Congress while ignoring his district, went down to defeat at the hands of an angry dairy farmer. The problem with awaiting this electoral solution is that most policy, and all policy implementation, is made below the electoral level. Velsicol and Northwest Industries were not going to allow the forced closure in St. Louis to be a defeat. In the tradition of the interest-group system, they would try to escape responsibility with the help of a regulatory process that they would dominate.

This process in turn would taint the name of Ben Heineman, as it had earlier tainted the philanthropy of Joe Regenstein, at least in the eyes of those informed about both the local and national impacts of their corporate management. The transportation innovator and urban-policy leader had been reduced to desperate entrepreneur whose search for cash at all costs brought on an accident. The subsequent effort to escape the consequences of the accident would further tarnish the reputation built on earlier leadership and integrity, would undermine respect for the regulatory policies his party favored, would transfer costs to future generations, and for decades leave contaminants in place to harm human health and the natural world.

CHAPTER 3

Disempowering Communities

IF VELSICOL'S SOLE PROBLEM IN THE 1970S HAD BEEN THE ONE IN St. Louis, Michigan, the firm might be seen as the unfortunate victim of an accident. If the deals that shut the St. Louis plant were the only ones signed by the firm, Northwest Industries might have been regarded as a model of corporate propriety with one blemish on its record. As an individual political actor, Heineman continued to support leaders in environmental protection.[1] Yet, a cursory review of the era's news revealed Velsicol as a repeat offender, driven by corporate policy not just to cut costs on environmental disposal, but to vigorously resist product sales restrictions and occupational safety rules, seek public bailouts for private liabilities, and oppose restrictions on foreign sales of products determined harmful in the United States. In addition to many human-health and environmental consequences, Velsicol's behavior was abominable because it undermined public confidence in civic institutions.

In the 1970s, besides the St. Louis damage, the company faced major environmental pollution cases at sites in Arkansas, New Jersey, Texas, and at three different locations in Tennessee. Its overseas sales, backed by the U.S. Agency for International Development, were linked to contamination in a number of places in the developing world, especially in Central America, Egypt, Indonesia, and Vietnam. Periodically during the decade, the company's alleged behavior brought it into the news as a litigant, with suits initiated by federal and state governments, neighbors of sites, and workers. Losses in many of these cases demonstrated that a number of the allegations possessed merit.

The experiences of the 1970s sowed the seeds for the most egregious corporate corruption of the policy process in the early 1980s. With a new administration celebrating the benefits of the deregulation Northwest sought, the company received regulatory and legislative favors saving it more than $100 million in obligations. These benefits empowered Ben Heineman to extricate the conglomerate from enough liabilities to find a willing buyer and exit his life of corporate innovation with a fortune. As with the closure of Michigan Chemical in the late 1970s, however, someone had to bear the costs for irresponsible behavior. That expense was spread among the individuals and ecosystems exposed to the company's products, the manufacturing communities, several prominent public officials blamed for indiscretions in assisting the firm, and all the taxpayers of the United States, who for generations would pay the costs to clean up the messes abandoned by Velsicol.

A review of the contamination incidents in the 1970s not only provides a perspective on the firm's behavior but also on the U.S. regulatory process. The general convention in environmental policy studies is that the 1970s marked the peak of liberal environmental regulation, which was eclipsed by the 1980s rise of neoliberal deregulation. Yet, in reality the experiences of the 1970s showed the relative weakness of the liberal approach, especially when corporate interests had perfected legal impediments to environmental laws.[2] Perceiving the weakness of the 1970s laws does not minimize the Reagan administration's impact on Velsicol. Its celebration of deregulation and overt attacks on government competence, including the appointment of EPA officials hostile to federal environmental regulation, enabled clear interest-group corruption of the policy process.[3]

NEW JERSEY AND TEXAS PROBLEMS

Berry's Creek in Woodridge, New Jersey, was a prime example. Wood Ridge Chemical processed mercury and, as with other Velsicol operations, released large quantities of the product into the creek and onto an adjacent dumpsite. Even after the sale of the factory, Velsicol continued mercury-dumping at the 33-acre site. From the beginning of operations through the plant's shutdown in 1974, six years after Velsicol sold it, 268 tons of mercury were discarded at the site, "the highest found in fresh-water sediments in the world."[4]

Velsicol acquired the plant in 1960, known as F. W. Berk and Company. Four years earlier, the New Jersey Department of Health pointed out that the settlement tanks that discharged into the creek contained "virgin mercury." Whatever Velsicol knew of these reports, as soon as it purchased Berk in 1960, the local sewerage authority advised it that it "would not process the company's effluent." The company responded that it would construct its own treatment facility, but did not do so.[5] Again, a state industrial toxicologist reported "excessive levels of mercury contamination" in the creek in 1966.[6] However, as the court record stated, "State officials insisted that remedial action be taken, but their warnings resulted only in a prolonged course of evasive action by the operating companies . . . the discharges of pollutants thereafter occurred intentionally."[7] The appellate court concluded, "Velsicol was 'responsible' because it knowingly permitted Wood Ridge to illegally discharge hazardous substances when it had the power, both as a landowner and as a corporate parent, to stop the polluting conduct."[8] However, the firm prevailed and evaded most costs at the site when in the early 1980s, during the era of Reagan deregulation, the dredging of the creek was abandoned. Consequently, citizen confidence in both government and business was undermined. A crabber, James Amato, interviewed by the creek in 1997, expressed the cynicism that results from this process: "Let's face it, the whole world is contaminated."[9]

PHOSVEL: BAYPORT AND THE WORLD

The most serious Velsicol contamination of the 1970s outside St. Louis and Berry's Creek occurred in Bayport, Texas. In 1971 at its plant near Houston, Velsicol had begun producing an organophosphate pesticide, leptophos, for use on cotton and various tropical crops. Trade-named Phosvel, the company never received permission for sale in the United States, but nevertheless supplied it to a large global market. In fact, as with other Velsicol products, the U.S. government assisted the company with global marketing, buying almost 14 million pounds under foreign aid programs for shipment to fifty countries.

Reports of harm within the Bayport factory and from users around the world surfaced within two years of the start of production. For almost another two years, these claims were ignored, output increased, and sales pushed into new markets. When workers recounted "bizarre neurological" problems in 1973, the company noticed sufficiently to provide safety equipment, but it neither reduced production nor stressed the use of protective gear.[10] Velsicol officials claimed, "As soon as we became aware of the serious potential occupational health problem we retained independent medical experts to review the health of our employees. We have reported fully on this matter to the appropriate government agencies."[11] As in the case of PBB, the differences between natural resource, health, and agricultural agencies delayed a response. Finally, the National Institute for Occupational Safety and Health (NIOSH) launched the largest study. As often with such studies, the report found significant numbers of health problems, but no conclusive link to Phosvel. For example, NIOSH discovered that 7 of the 300 employees displayed symptoms of multiple sclerosis (MS), compared to an expected rate of 0.041 per 300 people.[12] Yet, even with a rate of MS more than 2,000 times likely than expected, NIOSH could not prove conclusively that Phosvel caused the disorder. Although NIOSH was as definitive as a health study could be, concluding that the rates of neurotoxic symptoms were excessive and likely linked to the plant, it had to qualify the causal explanation. That qualification became the loophole allowing the company to explain away its responsibility. Similarly, when the EPA granted the firm a one-year permit pending a final study of Phosvel, the company treated the permit as proof that the chemical was safe.[13] After stopping production in early 1976, the firm still shipped most of its inventory overseas to Syria and other cotton-producing countries. Reports found that Phosvel use in Egypt was implicated in the death of peasants and the poisoning of as many as a thousand water buffalos.[14] Although these accounts probably were known by the company at that time, the firm accepted government aid to promote sales and misused government documents to convince foreign users of safety.

FIRE RETARDANTS AND TERMITES

As if the Phosvel controversy at Velsicol was not sufficient to demonstrate a pattern of irresponsible behavior, more discord surfaced simultaneously, related to the fire retardant TRIS. As with most of Velsicol's other environmental disasters, government support for product sales, coupled with the weak response of regulators in other government agencies and poor safety research, resulted in a potential crisis for the company. However, as had happened before, skillful manipulation of the policy process rescued the firm from its dilemma.

Government support for TRIS emerged in the late 1960s. In an effort to reduce the high rates of fire deaths, especially among children, Congress amended the Flammable Standards Act to allow federal agencies to impose flammability standards on consumer products. The law and regulations did not mandate a method to make products fire-resistant, but Michigan Chemical-Velsicol jumped into the new market with TRIS.[15] In expanding

TRIS production, Michigan Chemical followed its usual procedure. It skipped thorough toxicological testing and went straight to a marketing study.[16] Simultaneously with the Phosvel debate, evidence surfaced that TRIS was a carcinogen.

The TRIS debacle was unique among Velsicol's disasters. TRIS was not more dangerous than Phosvel or even PBB in cattle feed, but this time a Velsicol product injured other businesses in the children's sleepwear industry. Consequently, well-funded plaintiffs, not Texas workers or Michigan farmers, threatened the company.[17] The TRIS situation clearly demanded creative action on the part of the company. Over the next few months, Velsicol transformed from adversary to lobbying partner of the sleepwear industry. That Michigan Chemical had neglected full toxicity testing in the rush to fill a market void was overlooked in the campaign to blame misdirected regulations.[18]

Velsicol and its lobbying allies formed a coalition with textile mill worker unions to seek indemnification from the government. In 1978, Congress passed the first compensation bill, but President Carter vetoed it.[19] The veto demonstrated how Carter was the "odd man out" in the liberal policy era, with his perception of necessary environmental restraint to achieve sustainable outcomes.[20] After Carter's defeat, however, success materialized with the "free-market" Reagan administration. With the sponsorship of Senator Strom Thurmond, the bill passed and was signed by Reagan in December 1982. It provided $56 million to the sleepwear industry and freed Velsicol from future lawsuits.[21]

Velsicol's recurring problems attracted the attention of the Justice Department in 1975. At the request of the EPA, Justice impaneled a grand jury to investigate the firm's suppression of information related to the toxicity of two termiticides: heptachlor and chlordane. Allegedly, the company failed to share a study with regulators showing that mice exposed to the chemicals developed tumors.[22] On December 12, 1977, the Justice Department handed down an eleven-count indictment. According to the indictment, six employees, including the general counsel and another corporate attorney, collaborated in suppressing information.[23] It was the first time the agency had pursued criminal penalties for concealing research results.[24]

As with its other failings, the company again managed to escape seemingly certain punishment. Velsicol retained the best counsel money could buy: Williams, Connolly, and Califano, the Washington firm that included Joseph Califano among its principals.[25] Califano left the firm by the end of the case to serve as President Carter's secretary of Health, Education, and Welfare, where he employed the son of Ben Heineman. In April 1978, a federal judge threw out the indictment on the grounds of prosecutorial misconduct. In 1980, the government settled out of court, ordering each employee to pay civil fines of $600 and the company $1,000. Anyone witnessing this process, such as EPA associate general counsel Richard Denney Jr., had to be impressed that Velsicol prevailed. Soon, Denney became one of the key EPA officials to switch sides.[26]

TENNESSEE DUMPS

This pattern was interrupted in Hardeman County, Tennessee, fifty miles northeast of Memphis. After the Mississippi fish kills were linked to endrin-dumping in the city's sewers, Velsicol bought a 1,000-acre dump in the county near the town of Toone. Rather than refrain from putting hazardous chemicals into the environment, the firm simply transferred its dumping from the city sewers of Memphis to a rural field surrounded by woods.[27] Between 1965 and 1972, Velsicol buried 350,000 fifty-five-gallon drums of wastes from its Memphis pesticide plant. When sinks in Toone turned brown in houses using well water, residents prepared to fight. Velsicol, however, was not ready for angry citizens who had no stake in good relations with the company.[28]

The U.S. Geological Survey had already reported that dumped hydrocarbons were leaching through the subsoil and had contaminated some of the surface. It also noted that the groundwater was subject to pollution from the site.[29] Despite the report, Velsicol continued to ditch waste at Toone until in 1972 the state ordered a shutdown. Yet, the existing drums remained and continued to deteriorate over the years. In 1976, the Geological Survey finally alerted the state to the site's impact on wells, but Velsicol did not respond.[30] Forty-two plaintiffs filed a suit seeking $1.5 billion in compensatory damages and $1 billion in punitive damages.[31] Sensing its inability to apply any of its usual approaches to avoid responsibility, Velsicol attempted to buy off the litigants. Twenty-seven of the original forty-two took an out-of-court settlement, but the remaining fifteen found additional neighbors and carried on the fight, winning a large settlement in 1988.

Nonetheless, the contaminants remained buried at Toone in vain hopes that a pumping system would prevent future water-table contamination.

Velsicol was more successful back in Memphis fending off criticisms from the largely African American population living near the Hollywood Dump and Cypress Creek north of its pesticide plant. Repeatedly, the citizens complained of health problems they linked to the contamination. Even after state officials found dieldrin in the yards of all ninety-five residents along the creek, Velsicol avoided change or liability.[32] As with other company responsibilities, those throughout Tennessee were deferred sufficiently that cleanup became the responsibility of the bankruptcy trustee who inherited the contamination sites in 2002. A model for neoliberal concepts of entrepreneurship, Velsicol only bore a fraction of the total cost for its operations, while realizing all the profits.[33]

REORGANIZATION?

These difficulties within the chemical subsidiaries of Northwest Industries brought change from above after the *Wall Street Journal* ran a front-page story about the company's problems.[34] Heineman began by hiring Ray Ver Hoeve as the president at Velsicol and folded Michigan Chemical into the company. Ver Hoeve said the firm now was "dedicated to being the best, and doing the best." In the summer of 1978, together with Heineman, he arranged a meeting in Chicago of the top two hundred managers from around the country to introduce corporate improvements. Called the "Challenge and the Choice," the 1978 conference launched what management called the "new" Velsicol. In hindsight, it proved to be a blend of public relations and leadership techniques with little substance. They brought in a recent special assistant to the EPA administrator to serve as corporate ombudsmen, who said bluntly, "Government doesn't like you right now. A good part of that, whether by bad judgment or bad luck, you richly deserve." Heineman told the managers, "Had we anticipated it, we probably would not have bought Velsicol. . . . Our first priority in the allocation of capital is not for . . . expansion but to make the plants and products environmentally secure and safe." Most of the meeting proceeded to focus on public relations. Heineman hired Rick Blewitt away from FMC, the large pesticide company, to head public affairs. He also brought to the meeting a public-relations consultant from Monsanto.

The firm's counsel announced that Velsicol was modifying its approach to the EPA. It planned to soften the hard line of the past "to make reasonable settlements . . . with the goal of coming to a peace, or at least a truce, with the United States Environmental Protection Agency."[35] However, the meeting did not bring substantive change. In Michigan, when a team from the state Department of Natural Resources arrived at the old Michigan Chemical plant on the morning of December 3, 1979, to do water sampling, the facility manager, Ed Smith, refused to grant entry. The team required a search warrant to come onto the plant grounds, even though the company had not had any production at the factory in fourteen months.[36] Equally disturbing, three months before the incident in St. Louis, Ray Ver Hoeve, hired to create the "new" Velsicol, resigned.[37]

Heineman picked the next leader from among his most trusted lieutenants. In January 1980, W. Howard Beasley III rose from assistant to Heineman to CEO at the chemical company. As with many of the "new" Velsicol moves, the appointment of Beasley included elements both of hope and of concern for anyone impacted by the company. Fitting a new pattern of officials, Beasley had financial training and significant experience in government. An economics major at Duke, he held a Ph.D. in finance from the University of Texas. Immediately out of graduate school, he managed to get a job as special assistant to Treasury secretary John Connally.[38]

Beasley's appointment at Velsicol was either an excellent break with the past or a sinister new effort to prevent responsibility from being assessed. *Chemical Week* called Beasley's endeavors "a race with the past." But instead of expressing concern with the past, Beasley stated, "To be honest, I haven't given that much thought to the history. . . . I'm chiefly concerned with what we're doing today so that we don't have embarrassing situations now or in the future. Our record is fairly clean for the recent past." He then proceeded to be personally involved in resolving problems of the past. He traveled to Hardeman County and agreed to pay residents "for their trouble," including buying some houses at "fair market value" and paying for connections to a public water supply for families drinking from contaminated wells. Terry Cothron of the Tennessee Water Quality Division noted, "Velsicol had to be prodded, poked and sued to get them to work on cleaning up problems. [Now] they're making an extra effort to turn over a new leaf and do things before they get dragged into them by regulatory agencies." However, Ruffin Harris of the Environmental Defense Fund criticized the size of payments to families, which totaled $700 to $1,500 and were only given if people signed medical release forms.[39]

Beasley's career path from well-connected government service into an intensely regulated industry also raised questions. More worrisome was his conscious defense of the practice of recruitment from regulatory agencies. After Beasley joined Velsicol, the company filled a number of key positions with leading EPA or state environmental officials. Most disturbing was the hiring of John Rademacher away from the EPA to be the vice president for Environmental, Health and Regulatory Affairs. Rademacher, a civil engineer who began his career working in state and local government, joined the

Public Health Service in 1956 in its new water-pollution control division, which became part of the EPA after 1970. There he rose to be one of ten regional administrators in the agency.[40] He accumulated twenty-three years of federal service, seemingly not the ideal background for private-sector leadership. Beasley unapologetically justified the hiring strategy by saying there should be no "Chinese Wall" between business and government. He criticized federal regulations that were implemented after Rademacher's departure, designed to prevent movement of government officials into jobs at industries they formerly regulated. The new rules, he said, "[are] going to deprive government in the future of some of the best talent . . . and it will freeze in the government some very well-qualified people we desperately need in industry."[41]

If Rademacher had been the only official hired, Beasley's defense might be considered the reaction of someone wrongly accused, singling out an isolated case to claim existence of a pattern of behavior. However, Rademacher was not the only or even the most important official recruited.[42] In the early '80s, as negotiations moved forward on the settlement in St. Louis with the EPA and the State of Michigan, Velsicol replaced its general counsel with Richard J. Denney Jr. A Stanford graduate with a J.D. from Harvard, Denney was one of the pioneer attorneys at the EPA. From 1971 to 1976, he served as assistant general counsel with the Air, Noise, and Radiation Division. He gained direct experience with Velsicol as associate general counsel for the Pesticide and Toxic Substances Division for the next two years, and from 1978 to 1980 functioned as associate general counsel in the separate Toxic Substance Division.[43] Denney left Velsicol shortly after it achieved its agreement with the EPA in St. Louis.

The ultimate coup for Velsicol was the replacement of its CEO. Beasley switched over to become president and CEO of Northwest Industries before the EPA finalized the St. Louis settlement. In his place, Velsicol positioned Michael Moskow. Like Beasley, Moskow was not a chemist, but had economics training, starting at Lafayette College and concluding with a Ph.D. from the University of Pennsylvania. In 1969 the Nixon administration invited him to act as senior staff member of the Council of Economic Advisors, becoming director of the Council on Wage and Price Stability and undersecretary of Labor in the Ford administration. *Chemical Week* summarized his appointment as president of Velsicol in the summer of 1982 as: "A seasoned bureaucrat takes over at Velsicol." With a Republican administration managing the EPA, what could be better for the company than to have men like Moskow, Rademacher, and Denney negotiating with the adversaries of the company

Beasley explicitly defended why Moskow, with no experience in the chemical industry, was qualified to head Velsicol. "One major contribution that Moskow makes to Velsicol is his knowledge of how government agencies work. This is particularly important, given that chemical companies must constantly fence with regulatory bodies." Old friends in Washington apparently inquired as to why he would join such a notorious company.[44] They asked the same thing when Rademacher quit the EPA; one official there said the announcement from Velsicol "provoked a few smiles."[45] Denny's recruitment represented an identical situation, which *Chemical Week* put in perspective, announcing, "[Velsicol] has raided the Environmental Protection Agency to get top in-house counsel on environmental questions."[46]

The transition from regulator to regulated was effortless and uncomplicated for them since none expected to remain at Velsicol. For example, Moskow resigned only two years after joining the company, and Denney soon went to California to practice law. They were hired to do a job, and they had done it. It was also comfortable for some, like Moskow, since they supported deregulation in principle. Moskow eventually served as president of the Federal Reserve Bank of Chicago, where he regularly defended the wonders of the free-market economy.[47]

With Heineman's empire continually in debt, options to improve the company came with limitations. Even when launching the "new Velsicol," Heineman mentioned being dissatisfied with corporate profits.[48] When he handpicked Beasley to manage Velsicol, he reminded him that the boss was not giving him a blank check. "Clean it up, spend whatever money is necessary, but don't gold plate it" was the directive.[49] "Reform" meant better public relations, not real change.

MICHIGAN HEALTH AND SAFETY

Its responsibility for the health and safety problems of its St. Louis facility obsessed the company in 1982. These were not potential or minor difficulties. The St. Louis plant's production introduced pollutants into the bodies of millions of people in the upper Midwest through air and water emissions, solid-waste dumping, drum burial, and the PBB food-chain contamination. Numerous peer-reviewed health studies described the potential consequences. These studies served as concrete refutation of the myth in anti-regulation circles that the environmental dangers linked to firms such as Velsicol were exaggerations. The evidence that exposure to substances such as PBB and DDT correlates with a variety of fundamental health problems, ranging from various cancers to thyroid dysfunction, must be included in any assessment of the net cost and benefit of these substances. Certain applications of these chemicals save lives. The question remains, are the total lives saved greater than the overall negative impact on human health?[50] Clearly, fears of immediate death from cancer from slight exposures to substances like PBB were exaggerated. However, exaggerated alarm often arose because of the indifference of company experts to health concerns from intense exposures during production.

One problem with any evaluation of health impacts from exposure is the caution required of scientists assessing data from large samples. Obviously, in individual cases, a variety of life choices and exposures not related to the chemical of concern can impact health consequences. Yet, that research glitch is controlled by large sample sizes and techniques, such as age adjustment of rates of illness or death. Given all of the circumspection necessary for responsible assessment in epidemiological studies, the problems linked to Velsicol's emissions in St. Louis are as close to conclusive as can be expected.

Most health studies of Michigan Chemical focused on PBB. These, rather than DDT, were the contaminants of concern in the 1970s and early 1980s. The massive amounts of DDT in the regional environment were ignored until later, although they were deposited first. Since attention centered on PBB while the product was still manufactured, epidemiological studies used cohorts of exposed individuals who were identified easily. Thus, exceptional amounts of information are known about the work and life experiences of exposed populations. Additionally, several other studies of regional populations, less directly connected to PBB sampling, help assess exposure impacts.

Because of the fear and outrage among farmers exposed to PBB, a massive study of their families' health has continued for over a quarter century. In the early years of the crisis, employees of Michigan Chemical were included. With almost four thousand individuals in the cohort, the PBB study "was the largest and longest running exposure registry in the United States."[51] As a whole, and in subparts, it provided an excellent test of the impact of known toxins on people in the general population. In addition to the overall cohort, smaller samples were used to obtain emergency assessments in the 1970s that provided details of some individuals in the larger study. One early study looked at blood serum levels of the two major contaminants driving cleanup of the Michigan chemical site: DDT and PBB.[52] It separated subjects into six major groups: Michigan farmers from farms with high PBB levels in farm animals (called quarantined farms), nonquarantined farmers, consumers of products from quarantined farms, Michigan consumers of products from nonquarantined farms, Michigan Chemical employees, and the control group, Wisconsin farmers. Several more detailed, peer-reviewed studies used this classification method, showing the higher exposures of the Michigan Chemical workers. Table 1 shows these results.

Inexplicably, in 1990 the public-health officials dropped the workers from the long-term study. Investigators stopped surveying the former employees, because they had higher exposures to PBB and high levels of other contaminants, primarily DDT and its derivatives. Consequently, the researchers did not provide extensive follow-up of the people with the greatest likelihood of health repercussions. Nonetheless, the studies that did review the protracted health of the exposed populations suggest that the results for the highly exposed farmers and farm-product consumers applied to the Michigan Chemical workers and community residents.

Of course not all workers or farmers and consumers from quarantined farms bore similar levels of serum PBB. Table 2 shows the range of levels:

Subsequent investigations, consequently, have divided analysis between those with higher and lower blood levels. The data are somewhat different due to distribution, depending upon which part of the four thousand in the overall cohort was included.

Four sets of studies from the PBB group are especially important. The first set was completed in the last years of the company's operation and reviewed contrasts in organ function and child development. This probe

Table 1. Serum PBB and DDT Medians

	PBB	*DDE*
Quarantined farmers	3.9 ppb	9.2 ppb
Non-quarantined farmers	1.4 ppb	10.1 ppb
Consumers from quarantined farms	4.2 ppb	11.0 ppb
Consumers from other Michigan farms	2.2 ppb	8.1 ppb
Michigan Chemical workers	9.3 ppb	14.5 ppb
Wisconsin farmers	0 ppb	9.7 ppb

Source: From tables 1 and 2 in Henry A. Anderson et al., "Investigation of the Health Status of Michigan Chemical Corporation Employees," Environmental Sciences Laboratory, Mt. Sinai School of Medicine, New York, 29 November 1977; also in Mary S. Wolff et al., "Correlation of DDE and PBB Serum levels in Farm Residents, Consumers, and Michigan Chemical Corporation Employees," *Environmental Health Perspectives* 23 (1978): 177–181. Quarantined farms were those with herds with pbb levels that prevented sale of products.

Table 2. Frequency Distribution of Serum PBB

	Michigan Chemical Workers	Quarantined Farmers	Quarantined Farm Consumers
<0.2 ppb	0	9	0
>0.2–10	28	206	28
10.1–50	17	45	6
50.1–500.9	6	19	4
>500.0	4	4	2
n	55	283	40

Source: Mary S. Wolff et al., "Correlation of DDE and PBB Serum levels in Farm Residents, Consumers, and Michigan Chemical Corporation Employees," *Environmental Health Perspectives* 23 (1978): 178.

was conducted by a team from New York's Mt. Sinai School of Medicine under Irving Selikoff, who published results in 1978 revealing impaired liver function among workers and families from quarantined farms.[53] In the early 1980s, an investigation examined the developmental abilities of children exposed in utero or in infancy to PBB. This study, conducted in cooperation with the Michigan Department of Public Health, discovered impaired developmental abilities correlated with higher PBB body levels.[54] However, other scholars revealed flaws in the statistical methods utilized in the research.[55] While questioning whether the author had found the problems identified, the critics noted that the analysis of the PBB cohort raised issues about perceptual/motor weaknesses among the exposed children. The *American Journal of Public Health* recognized that the potential problems with PBB warranted further examination of the state of PBB knowledge.[56]

A new series of reports appeared in the late 1990s identifying more specific and disturbing conclusions. Following 1,925 women from the cohort, one study learned that "Women with higher serum PBB levels had an increased risk for developing breast cancer compared with women with lower serum PBB levels."[57] The authors qualified their findings by noting that the higher rates fell within the confidence interval, meaning the results could be explained by chance. Four years later, a larger investigation focused on all 3,899 people in the cohort, searching for evidence of any cancer. Two types of the disease correlated significantly with serum levels of PBB. The first was digestive-system cancer, which included liver, stomach, and similar cancers. Animal studies and human observation research showed that higher-than-expected rates of liver cancer existed among those exposed to

PBB, supporting the results from the PBB cohort study. The second correlation, found in 1999, was with lymphomas. While such a connection was not recognized previously with PBB, some have noted correlations between PCB exposures and lymphomas.[58]

The third set of PBB studies involved one of the more worrisome health effects associated with Velsicol: a cancer cluster in the town of Breckenridge, a few miles east and downwind from St. Louis. In 1975 Caroline Coulson, a twenty-five-year-old woman from Breckenridge, suffered from Hodgkin's disease. While receiving treatment at the University of Michigan Hospital in Ann Arbor, she indicated that she knew ten other people in town with the same illness. Dr. Joseph Silva immediately followed up on this information, since that number, if true, in a town of 1,300 people was many times the expected rate.[59] Dr. Silva and colleagues, Richard Schwartz and Jeffrey Callen from the Department of Internal Medicine, initiated a study of the community in cooperation with the state Department of Public Health.

Their investigation revealed that Breckenridge, a place of little population migration, contained ten cases of Hodgkin's disease and four of non-Hodgkin's lymphoma, for rates of 41.3 and 22.5 per 100,000, in the twenty-year period from 1954 until 1973. All the victims had been born in town and had remained there. Various national studies of Hodgkin's disease show rates ranging from 2.6 for females to a general age-adjusted rate of 4.1 for the general population. The expected rate in Breckenridge in a twenty-year period was less than one case. The actual rate was 12 times the expected rate, and the death rate 22 times the expected rate.[60]

Silva and his colleagues first presented their findings at the American Federation for Clinical Research conference in January 1977, hypothesizing that the explanation had to be environmental. From their visits to the town, they concluded that the only possible cause in Breckenridge, which had no industry, was a large grain elevator that towers like a medieval cathedral over the area. They suspected that the dust from navy beans stored in the elevator exposed residents to "phytohemagglutinin (PHA), a mitogen which selectively stimulates thymus-derived cells eventually bringing on Hodgkin's or related lymphomas."[61]

Having received positive feedback in New Orleans, a revised paper appeared in 1978 still hypothesizing on the navy-bean theory. The amended article concluded with an addendum noting the addition of an eleventh case in 1975.[62] The following year, joined by biochemists from the University of Michigan, the authors tried to ascertain how the navy beans caused the cancer cluster.[63] This was medical science at its best, tied to a clinical response. Dr. Silva and his colleagues uncovered the Breckenridge problem by listening to local knowledge from a nonacademic young women, Caroline Coulson. They traveled to the community to investigate, bringing in specialists in possible causes. They shared their results with the experts through the scholarly press and welcomed a critique.

Educated criticism by a peer reviewer surfaced from a draft of the first published version of the Silva study. Matthew Zack from the Centers for Disease Control, who collaborated on the project, worked with colleagues from the state Department of Public Health and studied other communities with navy-bean elevators. Concluding that the Silva study held an interesting hypothesis, Zack did not find supporting evidence for navy-bean links to Hodgkin's disease. In all the communities with navy-bean elevators, the observed rate of Hodgkin's disease was 4.0 per 100,000, below the 4.6 expected rate.[64] A larger unpublished paper by Zack documented the methods and samples. He collected data in fourteen communities in the state with a mean population of 2,321 and compared them to a group of communities of the same size without navy-bean elevators.[65]

While Zack raised basic questions about the navy-bean theory, he did not dismiss the assumption that environmental factors induced Hodgkin's cancer. Even people in town doubted the navy-bean hypothesis. Since the grain elevator had operated since 1909, why were there no problems before the 1950s? The residents looked toward St. Louis to the west, although *Newsweek*'s story of the cancer cluster only referred to a refinery in nearby Alma as another suspected cause.[66] While state officials denied any possible link to Velsicol, a Freedom of Information Act request filed in 1999 with the state Department of Environmental Quality uncovered a "Breckenridge Cancer Cluster" file in the company's records. It was natural to link the two. Only the shutdown at Velsicol coincided with the end of the problem. The Alma refinery and the Breckenridge elevator continued to operate long after the Hodgkin's cluster stopped. Production at Velsicol, however, corresponded in years to the start and end of the matter.

The last and most recent indication of health problems related to PBB originated from a follow-up of girls exposed to PBB in the womb. The study obtained the cooperation of 327 girls, with serum measurements of PBB in them, their mothers, and in their mother's breast milk. The average exposure in the womb was 17.3 parts per billion. Breast-fed girls exposed to high levels of PBB (over seven parts per billion) reached puberty a year earlier than those with low exposure.[67] This one-year difference raised troubling concerns about PBB impact on children and their endocrine systems.[68]

While Velsicol had no knowledge of this last study, the company accumulated ample evidence that PBB generated significant health consequences.

The Hill Top Lab's report of May 1970 documented disturbing effects on rats and mice exposed to the compound, such as "extreme growth suppression" and "small kidneys."[69] Congressional hearings after the PBB accident uncovered a 1972 DuPont study of brominated biphenyls that found exposure resulted in changes in the liver and increased birth defects in rats. The last paragraph of the report summarized the import of the findings:

> In conclusion, we should like to mention that because of the toxicity picture presented here (and in the subsequent paper) and the likelihood that it would accumulate in the environment the same way as PCB, OBB [DuPont called the compound octabromobiphenyl] was abandoned by DuPont as a candidate flame retardant for synthetic fibers.[70]

Complicating Velsicol's potential defense related to PBB exposure were its decisions to ignore grievances from workers about conditions in the PB-6 department, even before the cattle-feed accident became known. On February 21, 1974, the union filed a grievance asking for coveralls and shower time because the substances "are both irritating to the skin and very destructive of clothing."[71] The company's own study of worker exposure, not completed until 1975, confirmed the troubles with the ubiquitous PBB dust.[72] Only after the publicity about the PBB accident did the firm act to reduce exposures, and then after more than six months. In November 1974, it reported that officials ordered coveralls, overshoes, gloves, and masks, and were granting shower time while forbidding eating in the PB-6 production area.[73]

Other disturbing reports existed of exposures to the general public. A month after the accident became known, a company study discovered trace levels of PBB in a variety of its other products. For example, bulk salt contained 10.6 ppb in May 1974, which was only reduced to 7.7 ppb in June. BP-6 dust was so widespread in the factory that the firm leased laboratory space at Alma College in order to test without "atmospheric contamination" of equipment.[74] An agricultural chemist who visited Michigan Chemical described conditions at the plant as pathetic, with dust from products everywhere. Whether the PB-6 detected in consumer salts was actually present or a consequence of dirty laboratories cannot be known. Documents on the conditions at Michigan Chemical became public during the various congressional hearings that centered specifically on the company and PBB.[75]

All this publicity about Velsicol's contamination ignored the DDT pollution. The consequences of the PBB accident served to distract regulators from concern with earlier dumping. Although reports noted higher levels of DDT than PBB, the narratives focused on the latter. For example, the 1993 Public

Health Service "Site Review and Update" revealed DDT concentrations in carp of 35.5 ppm compared with PBB of only 0.425 ppm. Nonetheless, the document used the next two paragraphs to examine the details of PBB exposure, without returning to do an analysis of DDT.[76] Such a response appeared natural given the outcry about the PBB accident; but the company understood what really lay underwater in the Pine River, hundreds of thousands of tons of DDT, and each study confirmed it was increasingly present, not diminishing in fish and other organisms with which people came in contact.

DDT had been the subject of controversy for a decade between the publication of *Silent Spring* and the federal ban on all but essential uses. Regularly, news of hearings and decisions about the chemical communicated a growing sense of threat to the public. Michigan remained at the center of much of this contention due to the DDT spraying in Detroit and East Lansing critiqued in *Silent Spring*, and the dispute about fruit spraying on the shores of Lake Michigan. The Environmental Defense Fund went to court to block the latter, but was opposed by B. Dale Ball, the head of the state agriculture department. Ball became infamous after the PBB accident as a defender of the Farm Bureau.[77] The state finally established DDT regulations after the FDA seized 22,000 pounds of DDT-tainted Coho salmon.[78]

DDT statistics continued to accumulate in reports about St. Louis. The company had to suspect that time might defame its reputation and legacy. The team from the Mt. Sinai School of Medicine found higher levels of DDT than PBB in Michigan Chemical workers in 1977, although the substance had not been manufactured at the plant since the 1960s.[79] Later studies disclosed that DDT "possibly because of its estrogenic properties" interfered with mothers' ability to lactate.[80] Other reports showed correlations between the chemical and pancreatic cancer "under circumstances of heavy and prolonged exposure."[81] There were hotly debated connections between "breast cancer and serum levels of DDE"[82] and links to testicular cancer.[83] Researchers discovered DDE, a metabolite of DDT, associated with preterm and premature babies.[84] While many of these analogies might remain unproven, it seemed likely that the company would face massive costs to clean up the regional environment as well as pay compensation to victims. In the early 1990s, Gratiot County, Michigan, the location of the plant, continued to have the fifth highest age-adjusted breast cancer rate in the state, 68.6:100,000—significantly above the state mean of 55:100,000.[85]

Not only did Velsicol contend with the serious health effects from its products and their manufacturing, but it also encountered responsibility for environmental degradation. As John Heese of the state Department of Natural Resources reported in 1974, "The recent accidental contamination of animal feed . . . stimulated an independent evaluation of the potential for water contamination from production, product formulation and use of this class of compounds."[86] When he and others investigated, they found PBB in the water, and rapid uptake in minnows left in cages.[87] As a result of these early studies, the state imposed a "no-consumption" fishing advisory on the river in November, the highest-level warning used, and it remains a quarter century later. Demonstrating the company's liability, members of the Pine River Superfund Task Force won a multi-million-dollar claim in federal court for lost fishing rights in 2005.[88]

Over the next several years, the state and the EPA gave regular attention to conditions in the watershed, focusing upon Velsicol's impact. They summarized their findings in a comprehensive report on the river issued in the summer of 1979. This document might have alarmed Velsicol as it contemplated its exit from responsibility in St. Louis. It listed exceedingly high levels of contaminants at and far below St. Louis:

> In 1974 sediments from the St. Louis impoundment below Michigan Chemical showed high levels of total DDT (293 mg/kg), phthalates (19.5 mg/kg), PBB (9.0 mg/kg) and oils (19,000 mg/kg). Levels of arsenic (19 mg/kg), copper (1000 mg/kg), cadmium (17 mg/kg), chromium (3000 mg/kg), zinc (2800 mg/kg), nickel (740 mg/kg) and lead (1900 mg/kg) were also significantly elevated in St. Louis impoundment sediments. Total DDT, PBB and copper concentrations remained elevated downstream of the impoundment in Pine River sediments throughout the study reach [33 miles]. Lead remained elevated in sediments 12 km downstream.[89]

Conditions continued largely unchanged in 1979, making the fishing ban permanent, with the DNR adding a warning about the dam that formed the lake by the old factory.

> The St. Louis dam should be maintained in sound condition and precautions taken in any extensive drawdown to reduce the possibility of flushing polluted sediment downstream in the Pine River. Dredging in the St. Louis impoundment and the Pine River directly below the impoundment should not be permitted without approved disposal of contaminated dredge spoils and precautions to prevent contaminant flushing downstream.[90]

Reports on the fish sampling confirmed the seriousness of contamination. Conversations in 1992 with John Heese, the lead state aquatic biologist in the studies, still brought back vivid memories of unprecedented toxicity in the impoundment. In May 1974, some river water left in a new galvanized pan at the state's lab trailer was found the following week with a hole eaten through it. The written report depicted the extent of toxins, elaborating at one point:

> Conditions of severity were observed when fish were collected from the St. Louis impoundment below Michigan Chemical Corporation. A carp, netted in this vicinity during two 15-minute midday sets, was turning white due to slime coagulation when captured. Carp and bullheads caught in an overnight set at this location were partially dissolved. Many of the head bones of the netted fish were free from the fish skulls and remained in the net when the fish were removed.[91]

The research detected PBB uptake in other wildlife consumed by the public, including ducks.[92] Later studies noted it in a number of different species, including deer and cattle, that grazed by the river but had not eaten contaminated feed.

The Pine River was not the only place in the region containing pollution from Velsicol. There were four specific locations of intense concern: the plant itself, part of a golf course across the river, the county landfill, and a radioactive-waste disposal about five miles northeast close to Breckenridge, Michigan. All but the nuclear-waste dump became Superfund sites after the passage of CERCLA in 1980.[93] The plant site, called the Velsicol Chemical Site (Michigan) by the EPA to distinguish it from the other Velsicol locations, was 54 acres in size and contained a variety of contaminants heavily concentrated in a lagoon area, where the company collected dredged sediments from the river (to keep its pipes clear), and a drum-storage area. The three-acre Gratiot County Golf Course received liquid wastes from 1956 until 1970, including several thousand gallons of DDT refuse. The 40-acre Gratiot County Landfill, one mile south of St. Louis, possessed at least 269,000 lbs. of PBB. The Breckenridge dump held over 300,000 lbs. of low-level radioactive wastes.[94]

COMPROMISING THE POLICY PROCESS

Of all the troubles facing Velsicol at the end of the 1970s, potentially the most devastating for company profits was the complex situation in St. Louis. Closing the plant at the end of 1978 did not solve the old problems, but only stopped creating new ones. Three categories of litigants sued the company in Michigan: farmers who lost dairy herds, workers exposed to PBB on the job, and federal and state environmental regulators. While Velsicol had already settled about seven hundred claims from farmers for the PBB mix-up, it had about a tenth that number still disputed. In addition, there were four hundred former workers who, once they lost their jobs, were quite willing to seek a settlement for potential health problems. Finally, the state and federal governments wanted the company to pay for containment and possibly cleanup at multiple sites in the community. The settlements with farmers, while expensive, were shared with the Farm Bureau and insurers.[95] The lawsuit brought by the Oil, Chemical and Atomic Workers Union sought $250 million for the 318 people who labored at the plant during PBB production. Velsicol managed to delay the end of that trial from 1978 until 1991, when a compromise was reached that gave 306 former workers $500 each.[96] Clearly the conclusion was unfair. If the workers did not suffer any consequences, they deserved nothing. If they had serious health consequences, $500 would be of little help.

Worse than affecting the farmers and workers, the food-chain contamination meant that many citizens were attentive to how the government regulators resolved their claims against the company. The national and international media focused attention on the site on three occasions. Thames Television's *This Week* series produced a documentary on the "Poisoning of Michigan," with graphic footage and narrative about the plant.[97] A book and "made for TV" movie called *Bitter Harvest* described the plight of the PBB-exposed farmers."[98] The broadcast of a dramatization on CBS's *Lou Grant Show* became a drama itself when Republican Party operatives pressured the network to delay its airing until after the 1978 elections.[99] The company obtained the support it needed to restrict public participation in the process and allow passions to cool. In a remarkable letter to a county official, the governor's special assistant for the environment urged the county commissioner to help prevent formation of a citizen advisory committee: "We have learned from past experience that it's awfully easy to frighten the public to a degree which may be unwarranted about toxic chemical problems."[100] The governor's office received that assistance. Despite a Democratic president and Congress, and eight years of regulatory experience at the EPA, the "special" interests had learned how to control the liberal policy process.

Unfortunately, the public had reason to be frightened, not only about the pollution but also about the secret negotiations in which its officials were engaged. They paid dearly in lost revenue from the settlement that emerged a few years later. Velsicol's strategy of hiring former EPA and federal officials resulted in agreements that protected the interests of both the company and elected leaders, but sacrificed the welfare of the local community and the general good of the American taxpayer.

The public learned the consequences of these meetings, if they paid attention to the news, through a well-orchestrated press conference on November 18, 1982. That day in Lansing, John Rademacher of Velsicol, the state attorney general, Congressman Don Albosta, EPA regional administrator Valdus Adamkus, and U.S. attorneys celebrated what they called "the biggest financial settlement ever reached with a company for a cleanup."[101] The announcement of the settlement stimulated new controversy. However, Velsicol did not suffer any consequences from the debate and discord, as did some public officials responsible for it.

The day before the press conference, the Michigan attorney general, Frank Kelly, criticized the Reagan administration's efforts to force the settlement before the November elections. Kelly charged that they had tried to rush the agreement to help the cause of Republican candidates in the state, especially Rep. Jim Dunn of Lansing.[102] He alleged that "After months of not caring, suddenly they're in a big hurry—it was very obvious what they wanted."[103] Yet these were the mild criticisms coming from the settlement. On the day of the announcement, Philip Shabecoff, the *New York Times*'s perceptive environmental correspondent, reported the claim of EPA administrator Anne Gorsuch that the settlement "clearly demonstrates that our initial enforcement approach of dealing with the regulatory community with a presumption of good faith can bring about the prompt cleanup of hazardous waste." After quoting Velsicol's John Rademacher, who said the company was trying "to put past problems behind us," Shabecoff reminded readers that the firm's vice president had been an EPA official.[104]

The press conference presented the settlement as a victory. It ordered Velsicol to pay $38.5 million in damages to the government—the largest compensation in such a case. Anyone with knowledge of Velsicol's past and its hiring of EPA officials to negotiate with the EPA had to question this rosy scenario. Shabecoff searched for another opinion. He interviewed Khristine Hall, an Environmental Defense Fund attorney, who volunteered, "It makes me nervous to think they are putting that stuff in clay so close to the river. It is very chancy." Then he asked Hugh Kaufman his perception. Assistant to the director of the EPA's Division of Hazardous Site Control, Kaufman had worked on the establishment of Superfund. Shabecoff's summary of his comments probably infuriated its director, Rita Lavelle. "The action today sent out a 'clear signal' that companies that improperly dispose of hazardous wastes could negotiate with the agency at the last minute and wind up paying only administrative costs."[105]

Kaufman knowingly rejected the assumption that the settlement represented a victory for the agency. First, the advertised sum of $38.5 million was smaller than the $120 million that the state alone requested. Second, most of the $38.5 million existed only on paper as estimates of the expenses remunerated by Velsicol in completing the requirements under the consent judgment. In addition, it included many costs already incurred at the site. The actual amount transferred from Velsicol to the federal and state governments totaled $14 million, not $38.5 million, and was specifically designated as not a fine, keeping it tax-deductible.[106] Of greater value to the company was its acceptance to contain contaminants at the plant site and thereby be freed of liability at the others. The state assumed responsibility for the county landfill, with Velsicol supplying, at no cost to the government, clay and access to a deep injection well for contaminated water disposal. The pollution from the golf course supposedly had been removed and trucked to the plant site. Importantly, Velsicol paid no penalties and survived, released from responsibility for river remediation.

Once Kaufman denounced the settlement, events unfolded quickly. Because Gorsuch and Lavelle already disapproved of his remarks on other Superfund matters, the leadership of the EPA commenced a private investigation, hoping to uncover information to force Kaufman's resignation. In early December, he turned to the U.S. Department of Labor for protection from agency harassment since the department had statutory authority under Superfund to investigate and enforce employee security. In its opinion, Labor held, "EPA's actions were found to be an apparent attempt to discredit Mr. Kaufman's expertise and silence the communication of his ideas."[107] In one incident in Kaufman's case, an investigator from Lavelle trailed him to Pittsburgh in the summer of 1982 and photographed him exiting a motel with a woman. Confronting him with the evidence, Kaufman enjoyed the last laugh. The woman was his wife, and the photo proved that his boss was inappropriately trying to build a pretext for his dismissal.

On December 16, Rita Lavelle was summoned before the House Science and Technology Subcommittee and under oath lied about ordering the harassment of Kaufman. After Christmas, the committee launched an investigation of the facts, which resulted in perjury charges.[108] For several weeks, President Ronald Reagan continued to support Lavelle, who had worked for him when he was governor of California, but by February, he was concerned about the bad publicity and fired her. Eventually, EPA administrator Gorsuch and several other key EPA officials resigned as well. The perjury conviction landed Lavelle in jail for several months.[109]

BURYING THE PAST

For the next few years, work progressed in St. Louis to bury the old plant under a clay cap surrounded by a "slurry wall." While Lavelle and Gorsuch suffered minimally from the settlement, the primary victims were St. Louis and former plant workers. The EPA and state environmental officials celebrated their firm response to the crisis while Velsicol repeatedly boasted of making things right, but none from outside the community fathomed the local hardship and affliction. The symbol of the gap between the community and the policymakers was the tombstone on the old plant site. Government and company attorneys agreed in the 1982 consent judgment to position a granite marker to warn future generations of the dangers buried there. The last local economic benefit from the settlement was the purchase of the tombstone from the town's monument maker. The consent decree even described the message to be engraved:

WARNING DO NOT ENTER.

THIS FENCED AREA WAS THE SITE OF A CHEMICAL PLANT. THE GROUND CONTAINS CHEMICALS WHICH MAY BE TOXIC AND HAZARDOUS AND ALSO CONTAINS LOW LEVEL RADIOACTIVE WASTE. THE AREA HAS BEEN CAPPED AND SECURED. TRESPASSING STRICTLY PROHIBITED.

When government officials beamed with pride at their accomplishment in front of the media, the town's mayor, Jim Ayers, reputedly lost his temper, exclaiming, "How can you be proud of this? You have marked us forever as a contaminated town." Actually, the regulators were not thinking of "forever," but they did figure three hundred years. That was the length of time they boasted the slurry wall would last.[110] By the end of the 1990s, the slurry wall collapsed but the tombstone remained, although pushed unceremoniously behind an EPA trailer when the agency returned in 1998 to perform an emergency cleanup, correcting after fifteen years the "300-year solution."

The failure to understand cleanup technology and community concerns was an important legacy of the era's environmental-policy process. This laxity taught lessons about expert arrogance and overconfidence, as well as corporate manipulation of the system. After construction ended on October 12, 1984, at the Velsicol Superfund site, the EPA quickly claimed total victory. John Perrecone from the federal government assured everyone: "It means there is no longer a public health threat or environmental threat at the site."[111] The EPA reiterated the same verdict as it completed the first phase of containment at the Gratiot Landfill outside of town.[112] They discussed removing all the city's sites from the National Priorities List as they did with the Gratiot Golf Course.[113] After a second phase of water reduction at the landfill occurred, they declared certainty that the sites would be forever contained.[114]

Less than a decade later, the EPA was back in town working on early studies to address increased contamination at the plant site. Nearly simultaneously, the landfill containment failed and the state began emergency pumping of polluted water flowing from it. By the start of the new millennium, the cost of plant-site remediation rose to fifth place on the list of EPA-funded cleanups, and renewed study centered on buried contaminants at the "delisted" golf course. In fact, remediation contractors for the state admitted in 2006 that the boundaries of the golf-course site did not correspond to the contamination.[115]

All the while, residents and former workers stressed over possible and probable health impacts. Officials ignored investigation of links between Velsicol's emissions and the cancer cluster downwind in Breckenridge. Over a hundred residents nearest to the plant volunteered for blood tests and completed an eight-page survey.[116] They only learned that they had PBB in their bodies.[117] In the lower peninsula of Michigan, the health department estimated that 96 percent of nursing mothers in the 1970s had it in their milk.[118] Some responded like Mayor Ayers, believing, "The sad thing is that it's never been shown that PBB was toxic and that it hurt anyone."[119] Not until the 1990s did the studies appear that correlated exposures with some cancers; but even then, the results were maddeningly suggestive, not conclusive. Exemplifying expert indifference to the concerns of average citizens, health officials who had taken blood samples from residents did not return to report on the existence of new research, or to help residents and local health professionals respond to it.

In 1991, the workers who had filed a class-action suit in the 1970s for health damages against the company finally received their token settlement offer of $500 each in exchange for waiving rights to sue. The meeting with attorneys at the Czecho-Slovak Association Hall in St. Louis on a cold winter day provided one of the rare bits of certainty regarding health.[120] Clearly, the workers who would suffer no consequences from exposure deserved nothing, and $500 was surely not enough for those who would.

Occasionally, residents mocked the government that helped cause the town's notoriety. When the city ran into trouble with the Federal Energy Regulatory Commission over the relicensing of the municipal hydroelectric dam, locals laughed at the "bureaucracy" that condemned the fish in the river and then demanded a plan to protect them from being killed in the turbines. As Mayor Ayers observed, "No one fishes downstream anyhow."[121] Residents could not understand why the elimination of contaminated fish represented a problem rather than a health benefit.

For the next fifteen years, the only visible sign of Velsicol in St. Louis was the grass-cutting at the old plant site, and increasingly frequent visits of its in-house environmental consultants from Memphis Environmental. By the mid-1990s, the company and state officials began to notice worrisome indications that the cap and slurry wall had failed. However, locals hardly perceived this extra activity until the end of the decade. They also remained unaware of the corporate changes happening at Velsicol and its parent firms. The transformed Velsicol that confronted people at the end of the century was different from the old partner of Michigan Chemical. It also caused great confusion in later years when Fruit of the Loom owned primary responsibility for a site where there never had been a factory manufacturing its products. That Fruit of the Loom's NWI Land Management paid the taxes on the property and not Velsicol suggested that the corporate irresponsibility witnessed by St. Louis was part of a bigger pattern.

CONCLUSION

Regardless of the question of ownership of the Velsicol site in St. Louis or responsibility for the Hardeman or Hollywood dumps or Berry's Creek, the firm's survival beyond the early 1980s demonstrated the combination of strategy and luck that protected Velsicol's leadership from its irresponsibility. The company repeatedly took advantage of six features of policymaking in the United States:

1. Experts, particularly from universities and health research centers, generally failed to report back to the laypeople whom they studied to make the results of research useful.
2. Empirical standards of proving causality, especially with emphasis on cost-benefit analysis of risks, were so difficult to meet outside of controlled experiments that scientific caution served the interests of polluters.
3. Communities in modest-income regions of the Midwest and South were so concerned with maintaining jobs and tax base that they tended to respond weakly to longer-term threats to human health and the environment.
4. The use of former government employees and the capture of bureaucrats working in service agencies, such as agriculture and commerce departments,

often brought overt government support for production and methods of business structure that proved adverse to the safety or general well-being of average citizens.

5. Regulatory indecision and caution often led regulators into inaction or delay, which made the bureaucrats the primary subject of public scorn for problems created by regulated industries.

6. And finally, the consequence of the above failures and inadequacies was to undermine public expectations for good policy and thereby the capacity of civic leaders and institutions to deliver such policy.

Corrective action requires both changed expectations about personal behavior and the identification of community priorities. The brief era of anxiety and precaution that followed the publication of *Silent Spring* was terminated by a relentless campaign to prefer and enjoy the benefits, such as short-term jobs, rather than the long-term costs to human health and the environment. Likewise, a momentary period in the 1970s of concern for public integrity, following Watergate and a failed war policy in Vietnam, ended quickly with the rise of an unprecedented arrogance on the part of experts and officials who shifted back and forth from "public service" to profit maximization. Neither the liberal policies of the 1970s nor those of the neoliberals in the 1980s served the public interest. Is there a possible solution to these problems?

Sally Mueller, a city commissioner in neighboring Alma, confirmed that the locals possessed perceptive and innovative observations. In a meeting in Washington with EPA officials before the 1982 consent decree, she endorsed the concept that was later called a community advisory group or CAG. She prophetically declared, "Toxic waste sites need local advocates to prompt clean-up procedures."[122] It was a decade and a half before the EPA heard the message. When it implemented a policy to encourage community advisory groups (CAGs), the agency opened the door to a process that subjected firms to an adversary that asked more questions, demanded answers, and did not relent in efforts to correct mistakes of the past. Likewise, the CAG possessed the energy and resolve to hold bureaucrats to their responsibilities while demanding that experts return to communities after their field research to inform citizens of findings. In the 1970s and 1980s, that was in the future.

CHAPTER 4

New Management

It was not surprising that Ben Heineman paid little attention to Velsicol except after crises that had festered for years. Throughout
Heineman's ownership, Northwest repeatedly faced the challenge of stay-
ing ahead of the creditors. Since many of its subsidiaries resembled Velsicol
in their liabilities, the preferred way to pay debts was to buy new firms,
refinance the conglomerate, and use short-term profits to diminish the
mounting losses. However, although Heineman pioneered the conglomerate style, he did so with moderation, always keeping debt within bounds
so could serve with honor on presidential advisory panels and various
boards, such as that of the University of Chicago.

Heineman's successor, Bill Farley, came from a new generation of conglomerate leaders. He planned to build a larger empire on Heineman's
foundation and do so in the spirit of the times. He used junk bonds—
nearly worthless pieces of paper acquired by mortgaging one's future, but
that allowed anyone with daring to become the owner of a huge enterprise.
This new approach enabled Farley to toy with a level of public involvement
never envisioned by Heineman—a run for the presidency—while cursing
the companies he touched with massive indebtedness. The only way out
of these obligations was an occasional "fire sale." In one of those, Velsicol was "spun off" but remained hopelessly entangled with its parent as a
result of the financial manipulations that created the empire. Eventually
time revealed the disastrous consequences this type of growth spawned for
other parts of the conglomerate. While Heineman retained and created
jobs, except in St. Louis, and only left a negative environmental legacy, the
new style of management cost communities and even investors much more.

CHANGING NORTHWEST INDUSTRIES

After Ben Heineman sold the Chicago and North Western Railway to its employees, Northwest Industries included subsidiaries such as Union Underwear-Fruit of the Loom, Acme Boot, Universal Manufacturing, and Velsicol-Michigan Chemical that operated fifty major production facilities. During the 1970s, the number of facilities rose slightly, as shown in table 3. Acme Boot produced a variety of Western and fashion boots at four factories in Tennessee, one in El Paso, and one in Massachusetts. Velsicol-Michigan Chemical owned two factories in Texas and Tennessee; one each in Illinois, Arkansas, Arizona, and Mexico; and one in Michigan until closure there. Union Underwear had production only in the South, with five plants in Alabama, three each in Kentucky and North Carolina, two each in Mississippi and Louisiana, one in Oklahoma, and one in South Carolina. Universal Machine, which made fluorescent-light ballasts, had three in Mississippi, two in New Jersey, one each in Connecticut and Arkansas. Several new subsidiaries owned the other factories. Fruit of the Loom was the odd firm in the group.

As the *New York Times* said, "There are millions of dollars worth of apparel sold under the name of Fruit of the Loom, Inc., but the company

Table 3. Production Facilities of Northwest Industries, by Subsidiary

	# Production Facilities	
	1974	*1978*
Acme Boot	6	7
Buckingham Corp.	0	0
Fruit of the Loom	0	–
General Battery	12	17
Lone Star Steel	2	2
Michigan Chemical	2	–
Microdot	–	30
Union Underwear	16	18
Universal Machine	7	7
Velsicol	5	8

Source: Northwest Industries, *Annual Reports*; using the *Annual Report* information, Microdot was not purchased until 1976. In 1976 Michigan Chemical merged with Velsicol and one plant was retained, and Fruit of Loom fully merged with Union Underwear.

that owns the name doesn't produce a thing."[1] In 1974, Fruit negotiated twenty-eight contracts with others to fabricate its clothing, including Union Underwear in the United States and companies in Japan, Canada, and South America. Until that year, all the foreign licensees only produced garments for their home markets.[2]

In 1974 Union Underwear and Fruit of the Loom underwent a significant transformation. First, when Union founder and CEO Jacob Goldfarb retired from both companies, the two firms merged. Second and more ominously, Fruit contracted a new style of license, its twenty-ninth. It went to CBS Apparel, a subsidiary of Warnaco, which manufactured 80 percent of its products in Asia for importation into the United States. With this license, the company pioneered the shifting of manufacturing outside the U.S. South to low-wage labor markets in war-torn Southeast Asia.[3] Despite the seeming growth in the early 1970s, the reality was that Heineman could not pay for all of his acquisitions. With Velsicol's settlement costs soaring, Northwest sold off subsidiaries.[4] At the end of 1984, the corporation had 25,000 employees, down from 41,000 three years earlier.[5]

However, Heineman did not choose merely to own a smaller, leaner conglomerate. He resolved to retire while Northwest was regarded as a valuable enterprise. Throughout the early 1980s, as the crises at Velsicol loomed, Heineman increasingly adopted financial strategies to disguise the problems. In the fall of 1981, the company announced the first of several stock buybacks. Reducing the number of shares tended to augment the value per share of the corporation. At the time of the buyback, the company offered $55 per share.[6] There were two alternative interpretations of these buybacks, Heineman's and that of "Wall Street analysts." The business press quoted Heineman saying, "The fact is, we think our stock is a better buy than the market thinks it is. We're greatly undervalued. It's a better buy than any other company." While some claimed he had no plan of what to do with Northwest's money, most stock analysts were hoodwinked by Heineman's classy and bland assurances. Sidney Heller of Lehman Brothers acknowledged the absence of a strategy, but asserted, "Mr. Heineman's last two moves indicate that he doesn't need a master plan."[7]

Eugene W. Lerner, speaking from the independence of higher education at Northwestern University's Kellogg School of Management, judged more clearly than the stock analysts. He labeled Heineman's behavior "cannibalization." Lerner only erred in this assessment with his follow-up summary: "It's a marvelous act of folly." Folly depended upon who you were. After the February 1982 buyback, Heineman started to unload high-priced Northwest shares. By May 1983, he had reduced his holdings from 647,000 to

150,000.[8] He was approaching retirement with quite a nest egg, regardless of the decimated remains of his empire.

At the same time as he downsized his holdings, he went public with his wish to retire. In May 1983, he announced that any assets of the conglomerate "are for sale at the right price."[9] However, none of the recent strategies worked to enhance Northwest's value. Stock prices fell into the thirties in 1984, and profits stayed slim at most divisions. Heineman decided to remain on "past retirement" to help the firm through the rough times. After one buyout failed in early fall, speculation grew about the potential liabilities at Velsicol, particularly with the litigation in Hardeman County. In addition, the Bhopal disaster soured the company's future, since Velsicol had hoped to market a new pesticide based on methyl isocyanate, the poison that leaked from Union Carbide's plant in India.[10] Heineman grew desperate, but found deliverance in an unlikely person, William F. Farley, owner of Farley Metals, a small privately held conglomerate headquartered in Chicago.

BILL FARLEY

Farley's background resembled Heineman's: a lawyer turned entrepreneur who mostly "succeeded" with borrowed cash. He also shared Heineman's interest in politics, but he differed in many ways from the Northwest founder. Perhaps he revealed the tragic consequences that had resulted from Heineman's worst tendencies, not just his noble public-policy commitments to civil rights, poverty amelioration, and mass transportation. Although Heineman presided over Velsicol's disasters and enhanced his stock's value above reality, he did have a positive impact on public life. It was not as clear that his successor possessed such goals. Even with charity, Farley preferred the institutions of the privileged, especially in private education.[11]

Farley was the son of a letter carrier in Rhode Island whose athletic prowess at St. Raphael's High School in Pawtucket won him admission to Bowdoin College. After college, he briefly sold *Collier's Encyclopedias*. Then, he attended Boston College Law School and worked in corporate planning at National Lead Industries (NL). While at NL, he picked up an MBA at New York University and developed an interest in business strategic planning. He gained an appreciation of good conglomerate management and the way to success. After moving on to Lehman Brothers as a corporate analyst, he fixated on one small subsidiary of NL, Anaheim Citrus. While controlling 40 percent of the pectin market, it was losing money. He realized that any firm with significant market share should be profitable, since management could raise prices sufficiently without fear of much lost business.[12]

With this insight, he decided to buy Anaheim Citrus from his former bosses at NL. While they wanted to sell, he only had $25,000 in cash, and the company was worth $1.4 million. With persistence, he found bankers who would risk a loan to a guy with a dream. He was hooked on leveraged buyouts, the technique of buying what you cannot afford and paying for it by either milking profits or selling assets. To sell assets and survive as a conglomerate, Farley needed subsidiaries. Each time he assumed possession of a new company, he had to restructure the conglomerate's debt, gaining time to reshuffle the subsidiaries, sell some assets to raise cash, and prepare for the next addition. Essential to this strategy was a continual flow of good publicity.

The year after Farley bought Anaheim Citrus, he added Baumfolder, another underperforming subsidiary of a major conglomerate, Bell and Howell.[13] In the Baumfolder deal, he put up $800,000 taken out of his Anaheim assets and borrowed $6.5 million. A few years later, he relinquished Baumfolder to its management for $10 million, earning profits to help pay off creditors and generate cash for more expansion.[14] That happened when his old friends at NL sold him their "Metals Group" for $125 million.[15] As a result of the purchase, he acquired a railroad-equipment producer named Magnus and an aluminum-casting company, Doehler-Jarvis. In just a half-dozen years he borrowed ten times the amount he had pursued to buy his first company. The encyclopedia salesman succeeded in assembling an empire worth a fortune with only $25,000.

In 1985 came the grand opportunity to climb into big business. Heineman was desperate, and he owned the classic Farley-type business—not Velsicol or Lone Star Steel, but Union Underwear. It controlled large market share, which an owner could milk for profits to fund other grandiose ventures. By this time Farley had attracted the attention of Leon Black, a director at Drexel Burnham Lambert, one of the new junk-bond investment firms.[16] Nine years after Farley procured a loan to purchase Anaheim for $1.4 million, he bought Northwest Industries for $1.4 billion.

At Northwest, the management team prepared to work for Bill Farley. Ellyn Spragins, a reporter for *Business Week*, noted a slight agitation when Heineman introduced the new owner at the headquarters in the Sears Tower. All two hundred or so Northwest managers appeared worried about their future. However, Farley only talked about himself. Spragins noted, "Farley's self-absorption is almost a religion. Boyish, handsome, and affable, he sees himself as a living testament to American enterprise, and he has been eager to see his bootstrap story splashed across the general and financial press." A public-relations executive retained by Farley to promote his image summarized the case: "Like any 42 year-old, Farley needs to be constantly legitimized."[17]

Eighteen years later, in a seminar on corporate failure at Dartmouth's Tuck School of Business, Professor Sydney Finkelstein appropriated Farley as an ideal case study. Farley had destroyed most of Northwest Industries and driven a host of other acquisitions into bankruptcy. His approach to management cost thousands of jobs, left many former workers without health and pension plans, and created abandoned Superfund sites and worthless stock. Finkelstein placed Farley under his spin master's classification as the personification of the problems arising from "believing all problems are public relations issues."[18] For most of the period between his introduction as the new owner of Northwest and Finkelstein's lecture, Farley dazzled business leaders (especially bankers and financiers), political leaders in both parties, educators, government officials, and many of his employees with his persona, nurtured with the ingenious talents of public-relations experts.

While Farley told Spragins, "I sleep very well," she noted a number of inadequacies in his lack of both real technical skill and independent expertise. By contrast, Heineman had been a technical innovator from his days at the railroad through the end of his tenure at Northwest. Farley distinguished himself by employing only the rhetoric of efficiency. He surrounded himself with "groupies" who liked his style, ranging from aerobics to his anti-smoking obsession and his love of herbal tea. Spragins quoted a former Northwest official who said after the takeover, "Right now it is amateur hour over there." By contrast, a vice president of General Electric Credit, Michael Dabney, who helped provide some of Farley's huge credit line, thought Farley "smart, tough, and honest."[19]

How did Farley fool a vice president of General Electric and experts from Bankers Trust into supplying massive loans? First, endless positive public-relations accounts bragged about his hands-on management and rumored that he was a people person who valued workers. GE's Dabney asserted, "He buys right and manages well." Kenneth Labich, writing in *Fortune*, exemplified the mindless adulation of Farley in the business press. He commended Farley for making few enemies despite his fast rise, and claimed that associates had "nothing but praise for his character and business acumen." In his 1985 story on Farley's "dream machine," Labich briefly raised the issue of the large debt Farley had incurred, but then gave the entrepreneur's reasoning more prominence. In masterful double talk, he summarized Farley's position by saying, "He argues stoutly that a manager who can accelerate cash flow of his operations should be willing to carry a lot of debt for a while." He then quoted Farley: "In this country we have become arbitrarily too conservative in our capital structure."[20]

The good publicity paid off handsomely for Farley. *U.S. News* said, "His management style is straightforward and simple: Long hours, constant attention to detail, cutting fat in management and candor with bankers."[21] In an interview with *Industry Week*, he contrasted his leveraged buyouts with those of unethical buyers, who start to raid a firm in order to get a counteroffer to purchase back the shares they have acquired. Farley made sure journalists thought his approach different. "Bill Farley, in contrast, has bought the companies for which he's bid. And he very much runs them." They quoted him: "I focus my activities. I get involved with our businesses as much as I need to." At his automotive parts die-casting business, Doehler-Jarvis, they reported, "Mr. Farley absorbs himself in the nitty-gritty details of quality improvement and capacity planning." "I like to get a worm's-eye view of what's going on, rather than reading reports from a distance." *Industry Week* also related that Farley possessed a "people-oriented approach to management" and an "emphasis on values" at the company. They let Farley add, "Every employee deserves dignity and respect. . . . That's fundamental."[22] Considering that within a decade of this story, Doehler-Jarvis closed its factory after losing contracts because of obsolete technology and failure to own sufficient capacity—and its workers subsequently lost their pensions—Farley's principled and sanctimonious words did not resemble his management style.

Articles also recounted his entrepreneurial skills. The 1980s represented one of many periods in U.S. history when entrepreneurship was equated with sainthood by the mass media, and as brilliance by business academics. The press published Farley's stories about his rise to success from obscurity. *U.S. News* narrated his explanation of his first leveraged buyout of Anaheim Fruit, specifying that "Farley told himself, 'If I don't do this deal, I'll be like everyone else who just dreams.'"[23] The *Washington Post* identified him as someone who "has made a fortune out of buying and selling the mundane, the unwanted and the unglamorous."[24] Another tale described how while he was out running on a golf course in Colorado, a groundskeeper told him about the poor quality of Fruit of the Loom elastic. The story claimed that his response proved his attention to detail. Farley was quoted: "So I went back to our guys and said, 'What's the problem?' They agreed we did have a problem. Anyway, to make a long story short, we agreed at that meeting to increase the strength."[25]

The public-relations effort of Farley bore fruit in the honors he received, which in turn provided more public-relations fodder. In 1984, the Boston College Law School Alumni Association selected him as the Outstanding Businessman among the graduates.[26] A more prestigious award followed in 1986 from the Horatio Alger Association. The citation reviewed his rise from newsboy through Anaheim Citrus to Fruit of the Loom. It applauded his belief that "management and owners sink or swim together." It also reiterated Farley's emphasis upon "[treating] workers as individuals and [involving] them in decision making when possible," concluding with the upbeat quotation "Working hard and seizing opportunities that come your way are the keys to making positive things happen."[27] A few years later, when he seemed out of options in the midst of another hostile takeover, *U.S. News* concluded, "But this real-life Horatio Alger character has beaten the odds before and could surprise the doomsayers with another death defying escape."[28]

Using his public persona with amazing skill, Farley capitalized on the greed of those in the investment-banking system to convert tiny amounts of real assets into a massive bulk of borrowed capital. Joining with Leon Black and others at the ultimate junk-bond firm of the 1980s, Drexel Burnham Lambert, Inc., Farley lured previously respected banks and regular investors into entrusting vast sums to his care. Table 4 shows the growth of Farley's leveraged empire.

Table 4. Farley's Leveraged Experience, 1976–1985

Year	Company	Annual Sales	Purchase Price	Farley's Equity
		(in millions of dollars)		
1976	Anaheim Citrus	$16.0	$1.7	$0.025
1977	Baumfolder	31.0	6.5	0.4
1982	Farley Metals	430.0	150.0	4.0
1984	Condec	250.0	131.0	8.0
1985	Northwest Industries	1,400.0	1,400.0	70.0

Source: Ellyn Spragins, "Northwest Industries: The Acid Test for Bill Farley's Offbeat Style," *Business Week*, 9 September 1985, 68.

The problem inherent in the $70 million he used to "buy" Northwest was that much of it was borrowed against the presumed value of his other subsidiaries. While Velsicol made $13.1 million, all the Northwest companies only generated $68.3 million. Yet, the interest on Farley's purchase totaled $79.4 million, and he owed $21 million in severance pay to the managers he had dumped to cut costs. Finally, he needed to compensate his junk-bond bankers $50 million in fees.[29] Fortunately, the Doehler-Jarvis subsidiary of Farley Metals provided a small cash cow with a solid reputation and market share.

Table 4 also identifies another problem with the Northwest purchase, and in fact, with each subsequent addition as Farley enlarged his conglomerate. In contrast to the price for Anaheim Citrus, where sales equaled almost ten times what he paid or owed, at each subsequent step, sales were a smaller fraction of the price, until in 1985, when buying Northwest, he spent as much as annual sales. Furthermore, his subsidiaries added fundamental problems to the empire. Farley's scheme intended to combine Fruit of the Loom/Union Underwear, Acme Boot, and Doehler-Jarvis into the core of a new conglomerate while dumping the rest for sufficient cash to keep the creditors at bay. Public relations became crucial.

Farley's dealings with the media were planned to conceal his existing financial challenges and to shift blame for any problems onto what he had inherited from Heineman. Barbara Schwartz at Drexel fed the line to the press about their joint leveraging venture with Farley. Referring to the creative financing in the months after the acquisition, she said, "That's definitely no problem. . . . We knew in the beginning there'd be a cash crunch." Another financier, Guy Dove, joined in the deception, asserting, "That was the structure of the deal." Farley agreed that the massive cash outflows at the start of Farley-Northwest were "just what we expected." Independent observers, while less sanguine about the future, nonetheless held their concerns in check.[30]

Farley predicted in 1985 that his second-half performance would be better than the first. In reality, things were getting much worse. In 1985, Northwest's companies lost $72.9 million, and he still had to pay off Drexel's $500 million loan, which accrued interest at 14 to 15.75 percent. He owed another half billion to other banks. As one solution, Farley resolved to speed up sales of subsidiaries. In January 1986, Farley reached agreement with MagneTek to sell Universal Manufacturing for $72.5 million.[31] Universal Manufacturing then passed through MagneTek to become an independent Universal Lighting Technologies, with at least two of its former factories remaining.[32]

THE VELSICOL DEAL

Velsicol Chemical required more careful handling than any of the other subsidiaries. As soon as Farley took over, he ordered his managers to find a way to sell the firm.[33] In 1985 the EPA was pressuring the company to cease all production of its profitable anti-termite chlordane and heptachlor. In addition, there were potentially massive settlements with former workers exposed to PBB in Michigan, with neighbors of its Hardeman County dump, and with the state of New Jersey for pollution in Berry's Creek. In addition, the company experienced a continual management crisis in the mid-1980s. Arthur Sigel, who headed the pesticide division of Velsicol, replaced Michael Moskow.

When Sigel sought a buyer for Velsicol, he found little interest in the whole company and its "good will." Instead, he received offers only for lucrative parts, such as functioning factories and products. Of course, fundamental challenges existed in selling a company with a host of current and potential Superfund sites. The Swiss rescued Farley, as they did on several occasions over the next few years. In the spring of 1986, Sandoz offered to buy the "agrichemical business" of Velsicol, but only if they did not assume liability for any of Velsicol's contamination.[34]

With its debt reduced from $500 million to $100 million by this transaction, the company's rump still included four aging factories in Illinois, Tennessee, and Texas; a host of Superfund sites and other environmentally damaged areas; and potentially injured residents and workers. Sigel and four other principal managers agreed to a management buyout. The difficulty, however, was that Farley needed more cash than either he or the managers had, and the latter preferred not to assume the excessive environmental liabilities of Velsicol.

Never concerned about environmental or financial sustainability, Farley and the group struck a deal. To give Farley more assets, they virtually doubled the book value of Velsicol and handed him, free of charge, 49.9999 percent of the stock. He had an asset, and the managers had control. To make this work, they forged a holding company to direct Velsicol called True Specialty, in which Farley had 49.9999 percent ownership. Sigel and the other managers possessed 50.0001 percent. As for the environmental liabilities, they developed a two-step solution. Fruit of the Loom formed a subsidiary, NWI Land Management, and then granted it title to the seven worst sites, including the old Velsicol plant in St. Louis, and the Breckenridge, Michigan, radioactive-waste dump. Since such a business was not exactly a high-profit operation, NWI also received title to two useful properties, the Velsicol company headquarters on prime land in Chicago's near north side and 7,000 shares in the golf course in St. Louis, Michigan, which was created on the old company dump. Supposedly, all the contaminants from the golf course had been hauled off and redeposited, without treatment, on the abandoned plant site in St. Louis.

The Assumption and Indemnity (A&I) Agreement, signed on December 12, 1986, between Farley and Velsicol, establishing NWI and assigning property ownership, also included a sliding scale of responsibility for soil and groundwater cleanup costs for any claims at the four plants Velsicol bought from Farley Industries. After twenty-one years, Velsicol finally assumed full and unlimited liability for soil and groundwater problems. Farley agreed to assume responsibility for any injuries to workers or users of products from the four operating facilities that had occurred before December 12, 1986. Repeated references to insurance payments implied that Farley promised to maintain insurance adequate to pay for liabilities at sites. The problem for communities near these sites and for public regulators generally arose from the inadequacy of the insurance and other company resources when remediation was required or compensation when health problems surfaced.

Given Farley's precarious financial situation in late 1986, this agreement bought valuable time. It completely removed Velsicol's liabilities from the books of Farley-Northwest. Since the A&I Agreement was treated as "highly confidential proprietary information of Velsicol Chemical Corporation not for public disclosure," the two companies pledged to only use it in the case of lost litigation.[35] When, for example, a farm family that had leased a well to Michigan Chemical fought to block NWI's use of it for disposing of contaminated water from the St. Louis plant site, the agreement was never disclosed to them even though they had sued the wrong firm, Velsicol.[36]

The negligence of Farley and Velsicol in 1986 resulted because no regulatory oversight existed to assure that sufficient resources were set aside by the companies to pay for future remediation of sites or for indemnification of workers and residents. The general public absorbed the vast majority of costs as one last massive set of subsidies, much like the earlier appropriation for TRIS liability and for promotion of the sale of Phosvel. Farley, on the other hand, received a mythical 49.9999 percent of the overvalued Velsicol from the separation of the two companies, which he used to balance his books.

ESTABLISHING FRUIT

In order to raise cash in 1986, Farley decided to make Fruit of the Loom public. To prepare the markets, in mid-September 1986 he announced the payment of "a regular quarterly dividend" to shareholders in Farley-Northwest, the only publicly held part of Farley Industries.[37] The business press and "independent" analysts, who previously helped finance Farley, praised the move. Michael Dabney, who had transferred from GE Capital to Bear Stearns, noted, "All entrepreneurs like to remain private. . . . But a hot stock market is hard to turn down."[38]

Even after selling Velsicol, Farley had $823 million in short-term debt in early 1987, and total obligations of about a billion.[39] Worse, there were other cash problems.

His junk-bond friends, such as Leon Black, drove hard bargains for supporting his survival efforts. They demanded almost a quarter of the equity in Farley Industries and loaned him new bonds at 15.1 percent interest. Farley desperately needed cash from Northwest's two cash cows, Fruit of the Loom and Acme Boot, but to obtain funds for the Northwest purchase, he had promised not to tap their cash flows.[40] He decided to sell Acme for $38.4 million to his privately controlled Farley Industries. Then he could raid Acme's assets without public oversight.[41] In the short run, the restructuring saved his schemes, but the consequences for Acme's workers and home communities proved disastrous.

Since Farley owed more than $25 million in annual interest expenses, the meager profits from Farley Industries supplied little help. Doehler-Jarvis, and now Acme generated only $11 million in earnings. While some criticized Farley's tactics, he was quoted at the time with his "Horatio Alger-esque" bravado that the media loved: "I enjoy buying companies. No one ever told Americans that they have to stop at a certain point."[42] In hindsight, Farley's survival in the late 1980s was absurd. However, he used three techniques as well as his general advertising and public-relations skills to deceive and persist. Internally, he built a completely loyal business structure, particularly in his public companies, by creating boards of directors fully compromised by their links to his schemes. Second, he toyed with the highest levels of the political system to keep government critics off guard. Finally, he continued to enjoy the life of the successful tycoon, intimidating critics in a period of American history in which the most audacious robber baron could receive the adulation of a saint. Collectively, this strategy obscured criticism arising from any likely quarter, at least any quarter from inside the power structure.

COMPROMISED CRITICS

Like others who manipulate corporate finances, Farley's goal was to reduce the potential for oversight either by private or public regulatory mechanisms. He relied on control of his boards of directors to gain political influence, and an aggressive public-relations effort designed to celebrate him as a leader.[43] The board of Fruit of the Loom exemplified Farley's technique to avoid potential trouble from a group that might otherwise second-guess its CEO. He stacked the board with himself; his lobbyist, John Albertine; William Hall, the president of the company; Richard Cion, chief financial officer; Kenneth Greenbaum, chief counsel; Robert Meier, controller; Leon Black, the junk-bond backer from Drexel; and Douglas Kinney, an associate of Black's. Most amazing, perhaps, was Hall, who had taught business administration at the University of Michigan and Harvard and should have understood the independence essential to an effective board.[44]

Albertine was an especially compromised figure on the board, repeatedly serving as an expert witness to defend the wisdom of Farley's actions. In the spring of 1987, he pointed out for the business press that Fruit's projected growth in market share was "a done deal," as if he would say otherwise.[45] A few months later, the press publicized Albertine's proclamation that "Bill [Farley] is a visionary."[46] In another story in the *New York Times*, he suggested the reporter get a second opinion from an analyst at Goldman Sachs. Not surprisingly, the second opinion from a Farley supporter concluded, "Mr. Farley is one of the great proponents of the leveraged buyout. . . . He's done them very successfully."[47]

Farley's political involvement also contrasted with Heineman's. The latter volunteered his time to take on difficult advisory roles. The former's political involvement was decidedly more personal and menacing. During his financial crises, Farley spoke openly and repeatedly of adding to his activities a run for high office. He threatened to seek an elected position and talked about running as a Republican against Alan Dixon for a U.S. Senate seat from Illinois in 1986. He even announced that he had hired Ronald Reagan's pollster, Richard Wirthin, to "gauge his chances."[48] In 1987, between elections, he spent four million dollars from his nearly bankrupt empire to tell Americans how the country again could be number one in manufacturing.[49] Later in the year, he publicized his chairmanship of the New Age Coalition, an organization modestly described as "a group of politicians and businessmen who plan to tell Congress and America how to keep the economy strong and industry competitive."[50] Unlike Heineman, who directed task forces at the request of a public official, the New Age

Coalition was an entrepreneurial activity of a type unknown before Farley launched it.

As the next election cycle came closer, Albertine told *Industry Week*, after a glowing assessment of Farley, that someday he would be called "Governor Farley." The reporter who heard this remark possessed the good sense to record also that Farley had "no previous background" in running the businesses he headed.[51] However, he had no intention of starting his career with the menial job of state governor. Instead, in 1988, he set his sights on the presidency. By now he was a Democrat, and he actually launched a small Iowa caucus campaign, hiring a former manager of the recently discredited Senator Gary Hart. He dreamed of being the Democrats' "first 'entrepreneurial' candidate."[52] After Farley abandoned his effort early in the year, his "political adviser" affirmed, "I think he feels qualified to live in the White House."[53]

In addition to his overt political activities, Farley advertised that he lived a very special life. He eventually bought the old Auchincloss Estate in Newport, Rhode Island—Hammersmith Farms—where "Camelot had come to America" with the marriage of Jackie and John Kennedy. To be bipartisan, he owned a house in Kennebunkport, which he claimed provided him with a view into the Bush summer home. In keeping with his political pretensions, he served as director of the U.S.-U.S.S.R. Trade and Economic Council and joined the Council on Foreign Relations.[54] Having married twice before his rise to prominence, he used his third wife and later the fourth, an heir to the MacArthur fortune, to further his ambitions. A slight embarrassment surfaced when wife number two, Susan Stanford, a psychologist whom he divorced in 1976, authored a book with derogatory characterizations of a person named Frank, a possible depiction of Farley.[55] The divorce of number three, Jackie Farley, followed in the late 1980s, after she served as the fitness chief of the Farley empire. His fourth wife best accommodated the lifestyle of a tycoon with political aspirations as she hosted lavish parties for political leaders.

BLIND POOL

Just when things seemed to be looking up, at least for Fruit of the Loom, Black Monday hit the stock market in October 1987, and share prices fell from $9.75 to $5.75. A subsequent announcement that Farley planned to buy back a million shares helped a little.[56] However, clearly he needed a major initiative to rescue him from his debt trap. Breaking with financial reality, Farley sought a half billion with no clearly defined purpose, again turning to the junk-bond market. Over the winter of 1987–88, Farley's friends at Drexel, which itself was about to fall into bankruptcy, worked to help him raise $500 million in a "blind pool" fund, which he would use to make some undefined acquisition. When he finally decided on a new firm to purchase, he mistakenly chose a community and a company that would fight his assault, eventually sending him reeling back to the protection of his underwear empire. However, those who resisted him were never the same after the experience, and many suffered job loss, social disruption, and a decline unimagined before they learned Bill Farley's name.

Farley sought ownership of a firm headquartered in West Point, Georgia, and with factories in a group of small cities in the Chattahoochee River Valley along the state border with Alabama. West Point was a fitting site for the Farley assault, as it had been the location of one of the last futile battles in the Civil War, when a small group of wounded and convalescing Confederate soldiers tried to hold off a Union army at the Chattahoochee Bridge on Easter 1865. They clearly were outgunned by the Yankees, and that scenario seemed to be unfolding again when the Horatio Alger of Pawtucket attacked the town's only large employer.

Globalization appeared in West Point when the founding Lanier family engineered the merger with Pepperell Manufacturing of New England in 1965.[57] Pepperell was formed late in America's first industrial revolution in Lewiston, Maine. By mid-century, Pepperell was itself moving south, eventually retaining only one northern mill. When it merged with West Point in 1965, the combined West Point-Pepperell had nearly 20,000 employees. From the 1960s through the 1980s, West Point-Pepperell picked off one mill after another throughout the South. The character of these additions changed in the mid-1980s as Lanier took the company into the merger mania of the Reagan decade. Now, debt did not matter, only size.

In 1985 West Point-Pepperell bought Cluett Peabody, the New York manufacturer of Arrow Shirts, for $383 million. Cluett was one of the northern apparel firms that was facing stiff competition from globalization. Later, it developed and patented a process of fabric treatment called Sanforizing that made Arrow the easiest to care for shirts in the world.[58] Resisting the option to move offshore, Cluett perfected a global network by licensing manufacturers in each country to supply its garments. Thus, the Mexican licensee produced shirts for the Mexican market, but not for export back to the United States.[59] Cluett ignored centering production in the South. Its large factory in Atlanta produced regionally, but most North American production came from northern factories and Canada. Unfortunately, the company mistakenly assumed that innovation could check the global trends in textiles.

As if the absorption of Cluett were not enough of a challenge, on May 7, 1988, West Point-Pepperell acquired J.P. Stevens for $1.2 billion. Stevens had been the largest U.S. textile manufacturer, with over eighty plants and almost 50,000 employees in the 1970s.[60] In the 1980s, Whitney Stevens of the founding family still controlled the declining firm, but was looking for a buyer and apparently held conversations with Farley as well as West Point. The discussions with Stevens and his subsequent merger with West Point attracted Farley's insatiable ambition. Desperate to spend his $500 million "blind pool" fund to earn money, Farley gambled that the time was right for his ultimate leveraged buyout. He planned to restructure his operations and debt briefly in order to buy all of West Point-Pepperell, including Cluett and Stevens; organize it as a private enterprise; and then sell its assets to retire his junk-bond debt. In October 1988, he offered $48 per share, expecting to spend $1.4 billion to gain complete control.[61] Workers and the wider "Valley" community rebelled. The owner of the seed and feed store, Nancy Adcock, organized local opposition, urging people to burn their underwear. Scrutinizing Farley's public-relations routine, she labeled Farley "a silver tongued devil."[62] Her response and that of other average citizens in the "Valley" epitomized their homegrown wisdom in contrast to the foolishness of the national business media and various ideologues of entrepreneurship. Terrell Whaley at WCJM Radio displayed some southern humor by re-recording the country music hit "The Devil Went Down to Georgia," replacing Lucifer with Farley.[63] In a graphic interview with CBS News, Whaley observed, "[Farley's] the greatest BS'r of all time. He BS'd the bankers. They believed every word of it."[64]

Generally, most average citizens from the Valley interviewed by the national media displayed the same wisdom as Adcock and Whaley. They discerned that the creative financing that enabled Farley to buy firms like West Point resembled credit fraud more than innovative finance capitalism. Whaley shook his head in amazement when he recalled that bankers "let [Farley] have four billion dollars." Drew Spradlin added, "Mr. Farley took a perfectly viable, profitable company, overloaded it with debt and that's why he can't do the things that were always done before."

The worldviews of "Valley" residents, as those of citizens of St. Louis, Michigan, fundamentally contrasted with the ideology and unsustainable behavior of businessmen like Farley. The latter supported the "creative destruction" version of entrepreneurship celebrated by Joseph Schumpeter.[65]

He and his progeny among economists saw human behavior solely in material terms, and defined success and even goodness as the greater accumulation of wealth. It is hardly coincidental that Schumpeter's family owned a textile mill in Bohemia that grew at the very time the first generation of mills in England declined with their obsolete equipment. Endorsing the materialistic creed, Farley exhibited only indifference to the criticisms in the "Valley." He said, "I'm not bothered that the town is against me, they were against me in Bowling Green, Kentucky, when I bought Fruit of the Loom. Now I'm the second-largest employer in Kentucky. They'll come around in Georgia too."[66] The people in Georgia were too smart to trust his boastful promises about the size of his employment. The 9,800 in Kentucky should have been equally wary.

In November 1988, as Farley pushed his takeover bid, the town rallied around the scion of the founding family, Joe Lanier. Eight thousand showed up at a high school football stadium to express displeasure with the prospective owner. Trying to block shareholders from accepting his offer, Lanier claimed, "We'll fight till hell freezes over and then we'll fight him on the ice."[67] As a result of Lanier's resistance, the price of West Point's stock rose to $58 per share, and Farley needed about $3 billion to complete the acquisition. Those associated with West Point management faced conflicting pressures. Many owned shares from stock ownership plans and could benefit from the sale. On the other hand, their fellow "Valley" residents urged them to refuse to sell. Because Farley's offer of $58 per share exceeded the company's value, most sold and secured a nice profit, all the while complaining about Farley.[68]

When first interviewed about his plans, Farley deceitfully alleged that he wanted the company because it was so well operated. Of course, he coveted the assets and intended to trim management to obtain more profits. One manager, Steve Lyons, captured the contradiction: "It's a very well run company. It's not broken, so why fix it?" Another member of his family, Suzie Lyons, guessed the conclusion: "I'm scared because I don't know from day to day if I'll have a job."

By mid-1989, Farley was trapped. Because the purchase of West Point exceeded the original price, Farley did not acquire complete ownership. He bought 95 percent of the shares, but he needed 100 percent in order to privatize the firm and seize its cash. To get enough money to obtain the remaining shares, he tried to increase productivity, but ran into worker resistance.[69] They grew especially critical as they witnessed his destruction of Cluett Peabody.

Immediately after accumulating most of West Point, Farley sold Cluett at fire-sale prices, and the new owners closed all factories and profited from simply licensing the famous name. The company that had once only licensed production of its shirts for sale within the manufacturer's home country now became a licensee of the right to label Arrow shirts and export them anywhere in the world. Although money came into its headquarters, management lost its ability to develop new technology like its innovation of "Sanforizing." It possessed no garment technicians, only marketers and accountants. The model developed by Jacob Goldfarb and the founders of Fruit of the Loom altered its direction. Goldfarb had used licensing to get many American factories producing a quality product. Now licensing supported offshore production. Junk bonds were resulting in junk management.

Farley next announced he would trim the "fat" at West Point. Since "fat" had long ago been trimmed at what he previously called a well-managed company, his fat was basic production personnel. Much of the "Valley," including its spirit, entered a downturn. Bankruptcies and foreclosures soared. For Sale signs went up everywhere. One young man, Tommy Whitlock, committed suicide. His mother, Mary, lamented, "If Farley would've never taken over, I believe, Tommy would still have a job and still be livin'. I feel he's the cause . . . a big cause of my son being dead today." DuWayne Bridges, owner of a truck stop in town, used an analogy to explain the impact of Farley's cuts. "When you approach a deer in the road . . . they're afraid, they're scared. You see it in their eyes, and that's what's taken place in our community." Farley's response: "I think this area has been very reliant on . . . a paternalistic attitude . . . too reliant. . . . I'm sympathetic to them and I wish it were different. But, that's the way it is."[70]

While Farley advised others to wean themselves from paternalism, he pleaded with his bankers to save him from disaster, as angry workers tried to push him over the financial brink.[71] He managed to raise enough money to refinance his debt.[72] However, at Farley Inc., his private holding company, the situation continued to decline. He only managed to escape bankruptcy by selling Doehler-Jarvis to a new firm owned by himself and a few partners, including a Farley clone, Vincent Naimoli. As with the sale of Velsicol's nonagricultural chemical unit to management, the inflated sale price that appeared on the paperwork added more assets to Farley's empire.

Farley succeeded in landing these deals because his creditors were corrupted by the excessive loans they had offered him. Bank of New York, which extended a $150 million credit line to Farley, and Banker's Trust, another big creditor, apparently advanced much of the money to finance the Doehler sale. While the deal included many overvalued assets, they, as he, needed cover to avoid collapse of the Farley façade.[73] Farley's difficulties originated both from his own actions and from modern business practices that tolerated and blessed risky behavior.

Unfortunately for the communities in which Farley operated, his corporate indebtedness often concealed a variety of accounting gimmicks that hid other massive liabilities. As Farley spun off subsidiaries, whether Velsicol, Doehler-Jarvis, or his apparel and textile operations, he adeptly assumed pension and environmental liabilities to assure the cash exchanges he needed. At Velsicol, he created NWI Land Management, a firm with little assets, to own Superfund sites. When the Doehler sale encountered both pension and environmental questions, Farley offered the unwitting United Auto Workers Farley Industries shares to "cover" the pensions and health obligations.

NEW-STYLE LEADERSHIP

Although the seven deadly sins in their various guises have been around since expulsion from the "Garden," some eras and leaders have perhaps been more nefarious than others. History does detect trends and patterns, as when a "Gilded Age" is followed by a "Progressive Era." Both to contemporaries and later historians, the leaders of the Gilded Age at the end of the nineteenth century included more scoundrels, while the reformers of the Progressive Era from 1900 to 1920 sought to curb abuses.[74] Business historians also note a change in the era after World War II. As John Kenneth Galbraith observed, major corporations succeeded so well that their shareholders entertained no thought of controlling them. It was an era of prosperity, so few cared.[75] By 1970, conditions changed, perhaps due to the recovery of the war-torn world; however, U.S. business leaders blamed complacent management, not global conditions.[76] Management then seized leadership in the development of the conglomerate merger boom as Heineman created Northwest Industries.[77] According to this account, Farley represented the investors' return to control corporate strategy.[78]

Without oversimplifying a complex story, the shift in power from Regenstein to Heineman to Farley reflected some general changes in the business practices of its leaders and popular expectations of their proper behavior. Despite the egregious impacts on the environment and occasional labor hostility, Regenstein and Heineman accepted their roles as community leaders. Heineman genuinely served the public as the director of various commissions and boards, and Joseph Regenstein Jr. modestly donated large sums of money without seeking credit. The former's public activities were particularly noble, given the unpopularity of civil rights among many and the national hostility to welfare. Regenstein, in a different and more quiet way, was a major benefactor of Chicago, setting a high standard of generous support for the arts, health, education, and the natural world. However ironic, considering Velsicol's fish kills and river poisoning beyond the Windy City, generosity was his legacy at home.

Farley and his compatriots seem like different animals. While occasionally dabbling in public causes, Farley appeared guided by an awareness of the profits available from strategic civic involvement. Heineman served on commissions from the local to the national level that subjected him to verbal abuse and taxed his skills as a mediator. He struggled to resolve the open-housing demonstrations in Chicago and to mediate between the mayor and angry blacks. He confronted radicals over the Vietnam War while leading President Johnson's civil rights panel. Heineman did not gain support for his acquisitions by taking on these thankless tasks. Likewise, his decision to live in Hyde Park was a genuine commitment to the city, not comparable to Farley's buying the old Auchincloss Estate in Newport in order to boast of its links to the Kennedy family.

However, publicity also centered on his life as an executive. Farley nurtured an image of himself as corporate leader that was decidedly different from that of Heineman, who could boast of really making trains run on time or hauling chemical managers over the coals. The business leaders of the late 1980s and the 1990s like Farley engaged in politics either to gain policy breaks, exert pressure on regulators, or distract investors. Certainly, Heineman had received major breaks from government for buying Phosvel and shipping it to unwitting peasants in developing countries, and again for the TRIS bailout. But, his years of service tangibly benefited the wider society. Of course, Farley boasted that his activities profited his communities. He told the hapless victims in West Point, Georgia, about the employment he had brought to Kentucky. Unfortunately, Farley's goal was not job creation; he needed to generate cash and would terminate workers to get it. His generation perfected a model for the perversion of the entire U.S. political economy in the 1980s and beyond. He applied it to West Point; others soon applied it across the world.

A two-step process assisted Farley's survival and prosperity. First, he mastered modern communication. Handsome, articulate, and able to vocalize untruths without blinking, he demonstrated a style that permitted him to succeed even when the facts proved him wrong. A typical example appeared early in the West Point takeover when Suzie Lyons challenged him about her concern for her job. Farley deceitfully responded to her passionate fear with a smile and lied: "I say relax. It's OK. It's going to be great."[79] Good communication skills worked because of Farley's second step. He continuously preached a message about the value of entrepreneurial freedom that had become almost a national faith. Not just Wall Street, but those in charge of towns, schools, and even churches believed in the gospel of entrepreneurial leadership. Like any creed, its advocates had difficulty assessing blame when an enterprise failed; then, the culprit was fate. This occurred in St. Louis when Velsicol closed. No institutional leader voiced a determination to fight the system, only a gas-station attendant, a police officer, and some high school girls. Similar behavior appeared in West Point as Mayor H. S. Steele said of Farley, "He was a victim of circumstances." Farley added, "The world changed in the middle of this deal. The world became very different."[80]

The absence of an articulated alternative faith made these problems difficult for citizens in places like West Point. The belief in entrepreneurship had roots in the secularized Puritanism described a century earlier by Max Weber. He had seen Ben Franklin as the personification of the "spirit of capitalism."[81] After two centuries of independence, many accepted Franklin's materialistic Christian heresy as a quasi-religious sanction, reflected in the mid-twentieth century by Norman Vincent Peale and more recently in the "prosperity gospel."[82] It consisted of three components. First, success and progress were measured by material accumulation. Second, the uncertainties willingly accepted by the great entrepreneurs justified the highest possible return from their achievements. Third, constant destruction of old methods and organizations by innovators was necessary and had to be welcomed. Faith in these elements separated the advocates of entrepreneurship, who led American institutions and business, from many average citizens, who focused on security, family, and the pursuit of happiness.[83]

The latter, largely abandoned by their leaders, resorted to orthodox Judeo-Christian beliefs to explain their plight. Thus Nancy Adcock, the owner of West Point's feed store and a real entrepreneur, compared Farley to the devil rather than to herself. Hollis Kelley, who lost his job when Farley closed mills in the Valley, expressed his criticism in the language of the prophets: "I'm glad I hadn't got to judge [Farley]. He's got to go before the judgment bar of God one day and he's going to see all the faces of all those people whose lives he ruined."[84] Many community leaders labeled prophets such as Kelley or Adcock as heretics, symbolizing another tragedy of modern American culture.

Citizens did find some support. A host of popular writers joined in the critique of the new entrepreneurial leadership. A number of books appeared simultaneously with Farley's takeover battle at West Point, capturing the evils of the new finance system. Connie Bruck wrote about the annual "Predators' Ball," where Farley connected with other corporate raiders to plot strategy and meet their prostitutes. Three years later, James Stewart, an editor at the *Wall Street Journal*, attacked insider trading and other manipulations of the stock market perfected by people like Leon Black. Farley made a cameo appearance in both volumes. A number of other books, without reference to Fruit of the Loom, depicted the trends that extended well beyond Farley and his associates.[85]

These books and the general public's distaste for the new capitalists posed a threat that was matched by the vigor of the reaction that followed. The attacks revealed how American conservatism had abandoned its roots. George Baker and George Smith conferred scholarly approval on junk-bond methodology in a work published in cooperation with one of the largest new financial firms, Kohlberg Kravis Roberts. They explicitly denounced Bruck and Stewart, mocking their books and linking them as "two of the genre's typically sinister titles."[86] Probably more indicative of the conservative transformation into neoliberals was the response of the *National Review*. In 1992, it rushed to defend the decaying reputation of the 1980s. It condemned four "myths" that the *Review* claimed were used to criticize the entrepreneurship of the era. It specifically blamed the first one upon Jim Stewart's *Den of Thieves*, which said that "debt destroyed jobs." They listed the other three myths: "Junk debt financed take-overs. . . . Debt destroyed value, [and] . . . Innovative debt instruments were wampum, the market was a 'Ponzi scheme.'"[87] Having abandoned financial prudence, the new "conservative" ideologues of entrepreneurship concealed their shift to neoliberalism and irresponsibility by calling critics, even those from the *Wall Street Journal*, "sinister" authors of "myths." Farley's actions and the consequences for communities in the Chattahoochee Valley were insufficient proof that the four "myths" were true.

That a formerly respected conservative periodical such as the *National Review* promoted an ideology that in fact terrorized communities and workers, destroyed assets built over many years, and burdened those that survived with long-term debt, demonstrated that it had forgotten its founding principles. As Hollis Kelley lamented, "[Farley] raped this valley. He raped this community."[88] Esteeming entrepreneurial freedom as the highest value and priority, the neoliberals justified community rape along with debt as just another market phenomenon. In the 1990s the Farley empire represented and substantiated this unethical ideology.

CHAPTER 5

Creating Junk

THE FUNDAMENTALS OF FARLEY'S METHODS OF OPERATION WERE SO typical of the behavior of a large class of American entrepreneurs that few business leaders would criticize him, until it was too late. By contrast, the people at West Point, Georgia, noted two dangerous features of his methods as soon as they encountered him. One was his excessive debt, which would be a hallmark of both private and public finance at the start of the new millennium.[1] The second was his willingness to abandon established manufacturing facilities and experienced work forces in order to gain short-term financial benefit. The tragedy of the widespread acceptance of his style of leadership would be that his financial problems, rather than being condemned, became a model for other overexposed business speculators. When the inevitable collapse of their speculative ventures neared, the only solution was repayment of loans with cents on the dollar and vast numbers of job losses. From a contemporary entrepreneurial perspective, this was a brilliant strategy. Farley enriched himself and lived the life of a tycoon. Workers, communities, and investors lost. The catastrophe first fell on his non-Fruit subsidiaries.

Two of the major subsidiaries that abandoned U.S. manufacturing in the era became subjects of special studies: Universal Manufacturing in the book *Mollie's Job*, and Acme Boot, in the film *Booted Out: Exporting Jobs from the U.S.*[2] Such attention might be seen as proving Farley was unique, yet neither account claimed that. Both saw him as representative of a type of irresponsible leader. That is the perspective taken here. The massive job losses in manufacturing after 1990 were not the work solely of Farley, but of a new type of executive, willing to move production anywhere to avoid sharing profits with mere workers.[3] A third case, Doehler-Jarvis, should have received the attention given Universal and Acme, since its destruction was more extreme.

A review of the experiences of the subsidiaries of Fruit of the Loom

133

explains much about the decline of American manufacturing in the era. Even in textiles and apparel, Farley's case shows that industrial decline did not result simply from foreign competition. Rather, his experience raises doubts that such manufacturers inevitably had to move to low-wage labor markets, or to countries, such as China, with undervalued currencies. Clearly low wage rates were fundamental temptations to industry in planning production; however, the financial crises created by junk bonds and related asset manipulation played a fundamental role in the eagerness with which owners of such industries searched for foreign workers to exploit, just as they sought rivers into which to dump wastes. At their worst, executives used unearned corporate profits to lobby for trade policies that would undermine continued U.S. production. As the campaign finance system sank lower into corruption, a generation of American political leaders conspired with those who funded them to sacrifice the heritage of average workers, managers, and communities for a big campaign war chest. Just as the broker state allowed polluters to escape responsibility for their emissions, so it allowed employers to secure trade laws that undermined labor rights.[4]

Both of the case studies in this chapter repeated a similar story, as did others from the Fruit empire. An innovative founder created something out of nothing, providing several generations of workers and their communities with income and social stability. As their founders aged in the 1950s and 1960s, both firms became part of a conglomerate. Even after that change, both survived as productive U.S. manufacturers until the new-style corporate financing of the 1980s destroyed their viability. The point of these stories is to demonstrate that behaviors at Fruit of the Loom and Velsicol were not isolated problems, but part of a pattern of failed business and civic leadership. These cases demonstrate that unsustainability was not only an environmental consequence of a new style of leadership. These new leaders perfected financial, technological, and regulatory manipulation that undermined the viability of firms as much as Velsicol's dumping undermined a sustainable environment.[5]

THE JUNK MODEL

In leading these firms, Farley and his junk-bond allies developed a model for transferring wealth from old entrepreneurs and their communities and workers to the private accounts of corporate raiders. Firms such as Drexel Burnham Lambert marketed high-interest junk bonds, financial instruments sold to speculators to raise cash for a raider. Raiders willingly assumed massive debt and interest payments in order to gain a share of the wealth of the firms they bought, as well as fame and power. At its peak in the mid-1980s, Drexel met with its raiders at an annual Drexel High Yield Bond Conference, colloquially called the "Predators' Ball." Initially, Michael Milken perfected this conference model to match investors with small companies with poor credit ratings. The justification for this process was that the pooling of high risks minimized the worst dangers of failure.[6] Milken would make his money by charging consulting fees. These fees were significant transfers of cash. Bill Farley and other raiders copied this practice.

The sellers and managers of junk bonds were not the only participants in these schemes. The buyers of the new bonds were the ones who made the schemes work, and these buyers were mostly the money managers for institutions—insurance companies, pension funds, college endowments, trusts, and the so-called "thrifts," the formerly cautious savings and loans. The primary problem of the corporate finance system created in the era was the rise of these institutional investors, who chronically focused on short-term return on investments. The institutional investors' money managers had no patience for long-term growth or planning, or interest in the fate of workers or the environment.

The High Yield Bond Conference became the Predators' Ball when Milken, Leon Black, and others at Drexel found a new, more massive use for junk bonds. They created "pools" of capital for high-risk investors, such as Farley, to buy profitable companies and repay the loans with the assets of the company. With utter contempt for cautious old corporate management, which maintained surpluses for reinvestment, research, and development, and as a hedge against bad times, the junk-bond raiders saw conserved assets as resources to repay excessive debt. Likewise, the raiders had no use for old-style bankers who only loaned to worthy borrowers. Worst, Drexel and the other predators found a devious method to hide the absence of cash to complete their schemes: the "highly confident letter" stating that Drexel had faith in the corporate raider's ability to make the acquisition work.[7]

Of course, all of the participants in the predator process wanted to enrich themselves more than their colleagues, and to assure that any risks from their schemes did not impact them personally. There were two major ways to assure financial success in this highly uncertain world. The first was to extract as many funds as possible from the firms in salary, consulting fees, and stock options. The second was to create related but not interlocking corporate structures that could survive intact even if one part of the empire should be lost to bankruptcy.

Much as polluters transferred community assets, such as a fishable river into their private waste dump, they moved investor assets and borrowed money into their private fortunes. The ingenuity of the raiders, like the brilliance of Velsicol's chemists, justified for public regulators and managers of institutional investments acceptance of these transfers. For far too long, usually until it was too late to recover assets to reimburse losses, they accepted private seizure of shared economic and natural resources as the price of growth. Just as the subsidization of Superfund with general revenue after 2000, the bankruptcy process and institutions such as the Pension Benefit Guarantee Corporation helped to shield the raiders from the civic outrage that should have arisen to stop their behavior. The cases of two Fruit subsidiaries will illustrate this process well: those of Acme Boot and Doehler-Jarvis. The story of Universal Manufacturing will be left to those who read *Mollie's Job*.

ACME BOOT

Acme Boot is the simplest case to illustrate this process. As with other Fruit subsidiaries, Acme's founder, Jessel Cohn, was an early twentieth-century entrepreneur. A shoemaker from England, he had come to Chicago at age twenty-one. After many years, he opened a children's shoe manufacturing business with his oldest son, Sidney. In 1929—just before the start of the Great Depression—Clarksville, Tennessee, offered the Cohns financial help to expand if they relocated their factory from Chicago. With the Depression worsening, Cohn not only managed the factory but also served as chief salesman.

The legend around the plant was that his entrepreneurship paid off while on a sales trip in Dallas. He became fascinated with cowboy boots that sold for several hundred dollars. Cohn not only knew how to sell, he knew how to make a shoe or boot, so he bought a boot and cut it apart in his hotel room. He figured out how he could mass-produce boots for a fraction of the cost he had paid for the boot.[8] Back in Clarksville, he and his son shifted production from shoes to boots. Jessel continued to come into the plant as a consultant, even when he was ninety years old. When he died at his home in Clarksville, he and Sidney had made Acme into the largest boot company in the world, with over a thousand employees.

While the residents of Clarksville noticed little change, in 1956 the Cohns sold the company to Howard Newman's Philadelphia and Reading.

Differing in approach from Farley, Newman retained Sidney Cohn as president and paid consulting fees to Jessel. The new conglomerate appreciated the value of continued innovation in its manufacturing facilities. After Sidney Cohn retired, Newman had Robert Turrentine, a graduate of Clarksville's Austin Peay State University replace Cohn. Although only thirty-six years old, Turrentine already had worked for Acme for fourteen years. He remained in charge for twenty-four years. Clearly Acme valued continuity and community. That virtue had helped give Acme the stability that allowed it to become the largest boot company in the world.

Both in the Philadelphia and Reading years, and later under Northwest Industries, the company continued to expand in Tennessee and to innovate in its production. Ben Heineman at Northwest was fascinated with the possibilities of computers in improving efficiency, whether at railroads or factories, and he endorsed Acme adding computer-aided stitching and order-processing. Acme's promotions reflected its industrial dominance, using celebrities such as Johnny Cash and O. J. Simpson.[9] In one famous scene, the governor of Tennessee presented President Lyndon Johnson with a pair of Acme Boots.[10]

Everything changed in 1986 when Bill Farley took over. To raise desperately needed cash from buying Northwest, he sold Acme to himself, took its cash, and burdened it with $70 million in debt.[11] He and Turrentine also hired away Bruce Gescheider, a UCLA MBA, from Beatrice Foods, a new Chicago conglomerate on its way to collapse.[12] Gescheider had only been at Beatrice for two years, so he knew little about food. He knew less about boots. What any boot maker knew at the time was that to be globally competitive, Acme needed continual investments. That was a responsibility the new generation of business leadership disdained. Neither Farley nor Gescheider were able to cut open a boot like Jessel Cohn, but they knew how to cut costs and extract cash. They knew they could sell boots under the Acme name for several years, with no new investment.

Another feature of the new management was rapid turnover of presidents. In 1991, Farley replaced Gescheider with another new-style manager, Mike Vogel, previously the vice president for sales. Farley said, "Mike Vogel was the natural choice to continue the progress and direction of Acme's multiple businesses." Of course, Acme had only one business: making boots. But then Farley added a more disturbing review of Vogel's credentials. "He has built strong relationships with Acme's key customers, suppliers, and most recently has effected dramatic improvement in our manufacturing operations."[13]

Acme and its brands had such a great reputation that the company could license production of boots. At least in the short run, consumers did not understand or care about the difference between a name and a manufacturer. Always a great communicator, Farley won the trust of the union, which assumed he really was trying to remain competitive and assure their jobs. "[The workers] accepted pay cuts . . . in exchange for Farley's promise of secure jobs for them and their children."[14] In his aborted run for the presidency in 1988, Farley said, "The real war out there is an economic war. And that's a war that if we don't win, I'm not gonna have a job and you're not gonna have a job and that's not where we want to see America. . . . No matter how many computer geniuses we produce, we can't remain a world-class economy with a second-class manufacturing base."[15]

Recalling such words, union leaders and loyal managers rationalized the evidence before them. At a conference on the North American Free Trade Agreement (NAFTA), Dennis Ziolkowski, director of product development, said he did not think Acme would shift production to Mexico. Not knowing Mexico's great tradition of leatherwork, he believed that "sourcing in Mexico could endanger product quality."[16] The union trusted Farley and blamed Wal-Mart. Of course, Wal-Mart showed they could deceive as well as Farley. Their director of corporate affairs said, "It's the union and not the company that's misleading the public. Wal-Mart is making things happen positively in American shoe factories across a large part of the nation. . . . Wal-Mart buys shoes from several Tennessee companies."[17] In reality, Wal-Mart no longer made such purchases. The two deceits then joined. Under the cover of the dispute with retailers, Acme announced it would shut all the plants in Tennessee in 1993. As at other Farley facilities, when layoffs increased, Farley switched from the leader who would return manufacturing prowess to the United States, into a recluse. Tennessee Public Television reported, "Bill Farley declined our request for an interview. In a written statement, Farley industries says, 'Bill Farley was *not* involved in the decision to close the Clarksville plant.'"[18]

Acme moved from being just one more closed factory to a symbol of the dangers of globalization on the brink of the congressional vote on NAFTA. Consequently, the official account of what happened in Clarksville became the inability of American workers to compete. *U.S. News* failed to mention the financial and management manipulations of the half-dozen years of Farley ownership. Instead it focused on labor costs, reporting, "For years, Acme Boot Co. dug in its heels in an effort to compete with cheap foreign imports. But last month the Tennessee-based manufacturer finally capitulated to the offshore onslaught and announced that it would shut down its Clarksville plant."[19]

The negative responses to the closings were difficult to contain, however. The national media came to Clarksville and described poignantly the plight of the community. The *New York Times* reported on a union campaign that listed Acme as one of the "Plant Closing Dirty Dozen."[20] Across the core market area for boots, similar stories ran. The *Memphis Commercial Appeal* described one former employee, Janet Manners, crying and saying, "We're all devastated. We all knew it was coming, yet you're never ready to be told that the job you've had for 29½ years is gone."[21] The *Atlanta Constitution* reported in the spring of 1993: "Not only did Acme's move idle more than 1,200 Clarksville workers . . . ; it ended a way of life in which husbands, wives and their children gladly went about the backbreaking work of forging leather into boots."[22]

The U.S. media explained the closures as inevitable consequences of modern trade policy. It took a foreign critic from a nation with a strong labor tradition to get past inevitability. An Australian review of *Booted Out* called it "an important documentary . . . where the few are willing mercilessly to exploit the many and rationalize their behavior in terms of social and economic necessity."[23] Likewise, Mitchell Tucker, head of the local rubber workers' union, rejected economic fate: "This is an immoral, illegal shutdown. . . . We did everything we could to boost productivity and this is the thanks we get."[24] But, of course, there could be no concessions that would have given Farley what he needed out of Acme. He needed the workers to pay some of his debt and not expect a salary.

As would happen at a few other Farley facilities, efforts to save the plant with an employee buyout failed. Farley didn't want competition from his old factories. When Joe Sisk tried to buy the plant in conjunction with the workers, he failed to reach a deal.[25] Sisk said, "Farley has withdrawn from the sale of the facility."[26] Acme's story was not a boring account of comparative trade advantage. It was a Dickensian story. It is a story that begged the question asked by the Australian reviewer of *Booted Out*: "Is growth [in profits] at the expense of social justice and human rights better than no growth?"[27]

Farley's answer was one more scheme to suck money from the shell of the company. In 1994, after he received a tax break for writing down the value of Acme, he split the firm into Acme Boot, Inc., and Acme Boot Company. Neither made boots; one sold them and the other supplied them to the sales operation.[28] Continuous consumption of the seed corn finally meant that Acme was nothing but a label. H.H. Brown, a rival shoe company owned by Warren Buffett, paid $700,000 for Acme's trademarks and inventory.[29] Farley had not only sucked the manufacturing capability and reputation out of Acme, he also had gotten what the firm had promised to its former workers. In 1998, the Pension Benefit Guarantee Corporation had to strike a deal with Fruit of the Loom to protect the pensions of Acme retirees.[30]

The old Acme factory, on which the Cohn's had built the largest boot-making firm in the world, partially burned after the company shutdown. The surviving structure was such a rabbit warren of passages and small rooms that the Clarksville police acquired it for use in SWAT team practice.[31] Ironically, the newest Clarksville facility, on the strip going south from town, became the home of Star Tek, a company that offered firms "outsourcing." This was a fitting fate, given that H.H. Brown, when it bought the Acme trademark, had affirmed that any production would be "globally sourced."

An attentive visit to Clarksville shows the negative impact of the demise of Acme. While the city has had a solid economic base as a result of nearby Fort Campbell, still there is nostalgia for Acme. How many cities have a "world's largest manufacturer" in their midst? Residents also see the irony of SWAT teams practicing in one old factory and Star Tek filling another. Some local citizens explained what happened better than the economists who believe natural laws necessitate exploitation. Economic practices are a result of human decisions, and they can be altered to do right and wrong. Jessel Cohn made the right decisions for himself, his family, and Clarksville. Farley made wrong ones. Cohn's decisions brought prosperity to a corner of his adopted home, a place where few people shared his deepest faith. Herman Doehler would act similarly in Toledo, and again Farley would undo his work and blame it on fate.

DOEHLER-JARVIS

The benefit of reviewing the Doehler story is that unlike at Acme, there was absolutely no subtext of movement of jobs to low-wage labor markets. The collapse of Doehler arose purely from the constant theft of resources. Its fate shows in the starkest form the consequences, for communities and their workers, of gross mismanagement—excused and blamed on fate by modern business leadership. Its story is one that ends in tragedy because of the irresponsible, self-serving behaviors of contemporary leaders.

Like Acme, Doehler-Jarvis originated in the inventiveness of an immigrant, Herman H. Doehler. Born in Germany, Doehler came to the United States at age nineteen and worked for fifteen years at a print shop in

Brooklyn. It was there that he became interested in die casting and invented a hand-operated die-casting machine that later ended up in the Smithsonian.[32] Long before that, leaders of the new automobile industry had learned about Doehler, whom John Willys had lured to Toledo to die cast his engine bearings.

The contrasts between Herman Doehler and later managers of the company are stark. Four features stand out. First, he began his career as an inventor and continued throughout his life as a technological innovator, perfecting casting light metals including aluminum. This work would yield results far into the future. Second, he understood the technical processes of his company and the good labor-management practices necessary to maintain profits and quality. He even wrote a book on die casting.[33] Third, he cultivated links to the community in which his company was based. While he always liked New York, he settled in Toledo in 1940. Finally, Doehler expected his employees, and especially his managers, to remain with the company for their careers. He hosted an annual Quarter Century Club dinner for all employees with twenty-five years of service.[34] When Doehler left the presidency in 1936, he promoted Frank Koegler, who had worked for him since 1912. When Koegler retired, he had been with the company for forty-eight years.[35]

The war years had seen further expansion and success for Doehler. Near the end of World War II, Doehler merged with the W. B. Jarvis Company. At mid-century, the combined firms had 7,500 employees at its plants in Ohio, Michigan, New York, Pennsylvania, and Illinois. Like Cluett, the company met foreign demand by licensing its processes outside the country to firms in Scotland, Germany, Canada, and ironically, given later practices, Mexico.[36] Also, like so many venerable businesses, it remained rooted by more than just the founders to its home communities. Of thirteen directors, four came from Toledo and all from the states where there were manufacturing plants. As was expected, Frank Koegler served on charitable boards, an ecumenical church committee, and the Toledo industrial development initiative. The hourly employees were part of the United Auto Workers, and all employees had pension benefits.[37] The promotion of Charles Pack, an employee since 1911, as vice president for research and development showed the dual commitments to employee retention and technical innovation.[38]

Doehler-Jarvis would become a leader in a new, more ominous way. The purchase of the firm by National Lead made it the first of the future Fruit of the Loom subsidiaries to lose its independence. At first, the experience was not painful. When Frank Koegler retired in 1962, the company replaced him with Alfred Bauer, who had die-making training from a

German technical institute and had patented a casting process for aluminum engine blocks.[39] The change to conglomerate control only slowly undermined Doehler's old management innovation and focus. In the 1970s, NL Industries (NL) began to move managers among the different divisions. For example, Paul Collier came from the chemical division to management of one of the Toledo casting plants. He replaced L. E. Burkholder, who moved to one of the "petroleum services" subsidiaries headquartered in Houston.[40] Ominously, a few months after the promotion of Collier, J. Robert Hahn became general manager of the Doehler division of NL, after serving as NL's vice president for marketing. In addition to his nontechnical background, he had only been employed for a few years with NL and his two previous employers.

Doehler had transitioned from an era of technological innovation to an era when business prospered by marketing and strategic planning. Shortly before promoting Hahn, the company informed the public not of a new technical breakthrough, but that they were conducting "a 'strategic review' of some of the parent company's non-petroleum businesses." NL wondered "if [the metals group, including Doehler] might prosper better with companies more importantly involved in these businesses."[41] As with so much strategic-planning language in modern America, NL did not mean the words in the press release. The words were part of a marketing strategy to enhance the value of NL shares during the sale.

FARLEY METALS

The company "more importantly involved" in their business turned out to be Farley Industries, a firm that owned a citrus processor, a folding-machine manufacturer, a coal mine, and a health-food distributor. Bill Farley knew about the metals companies only because of his previous employment in the mergers office at NL Industries. Farley did not see NL's effort to sell Doehler as an opportunity to control a stable, innovative firm, symbolic of America's technological prowess. He saw the potential of Doehler-Jarvis and the other metal-fabrication subsidiaries of NL as cash cows to repay loans necessary to allow him to buy an empire on borrowed cash.

As was Farley's style, he did not quietly complete his purchase of helpless Doehler. Instead, he orchestrated the classic marketing event to gain backing from those who might be needed for crucial political support. He invited Ed Weber, the Toledo congressman, to attend the press conference announcing the sale. Farley had secured financial backing from such national sources as General Electric Credit, but used commitments to keeping the headquarters in Toledo and the backing of the congressman to get some financial support from Toledo Trust Co. He did not explain that this local focus did not include a personal move to Toledo. He was too important for Toledo. He would not be a Toledo leader; he was a national leader. He was no Herman Doehler or Alfred Bauer, with patents and even a machine in the Smithsonian. His invention was promotion.

The problem for Doehler was that the company needed active, technically knowledgeable leadership to prosper. Doehler was a solid company in 1982, with a history of innovation and a stable customer base. However, the automobile industry was always changing its demands, and that required a supplier like Doehler to be ahead of the curve. The company didn't need marketing, since the big three auto companies depended on it. What it needed was technical sophistication and inventiveness, so the customers would always trust and value their relationship. In order to remain technologically sophisticated, Doehler needed regular infusions of new cash for research and development and the new technology that followed. This brought the company into fundamental conflict with Farley. He needed cash, and he saw Doehler as one source for the cash.

Always the promoter, Farley occasionally descended upon Toledo to woo employees and civic leaders and blind them to the way he was about to manipulate them. When he needed local backing for buying the company from NL, Farley promised to reverse the trends of the last years of NL ownership. Using athletic metaphors to implant the message that everyone would win with Farley, he said in 1982, "Our game plan is to build employment here, not to reduce it."[42] He easily won over the local civic leadership, desperate for an upbeat message at a time of industrial decline. Not only did George Haigh of Toledo Trust help with financing, he and the local press pushed the positive message about Farley.[43]

Before Farley told the employees what he really wanted, he came to town with his third wife to entertain them with more athletic imagery. Jacqueline Farley introduced the employees—both assembly line and office—to an eight-minute stretch-and-tone exercise program. She told them, "It's the Farley philosophy that this plant should be more than a place to pick-up a paycheck."[44] Testing the gullibility of the United Auto Workers (UAW) leadership, Farley said he got them to join with him in starting a new "sweat shop." The union leadership responded by "praising the company's management techniques." When Farley helped pay for the new ball field behind the union hall, the UAW named it Farley Field. Dave Schneider, the president of UAW Local 1058, said, "It's [William] Farley's philosophy to think of his people, not just the dollars."[45] Like any successful American political leader, Farley knew how to act like the "man of the people" that he was not.

The clue that all was not right should have come when Farley's hand-picked president of Doehler, B. J. Iwarsson, quit suddenly in February 1986, fifteen months after being appointed. While Iwarsson fit the modern management pattern of being selected after a number of short-term leadership positions in different businesses around the country, he seemed to endorse what had been sold as the so-called "Farley philosophy."[46] Commenting on the exercise program nine months after he took over, Iwarsson said Farley did not see the employee health program as a productivity issue. "If you pay attention to your employees and have concern for their health, you'll all live better. We aren't doing studies [on productivity increases], we don't focus on return on investment [in the fitness program]. We've done this because we think it is the right thing to do." In response, Dave Schneider of the UAW observed, "We are getting away from the adversary relationship we had with management."[47]

Without clarifying who had the real company "philosophy," Doehler suddenly announced that Iwarsson had resigned "because of differences in operating philosophies with his superiors." Charles Craig replaced him.[48] In March 1987, Craig revealed the real corporate philosophy, threatening that the old Toledo plant would be closed, and perhaps others, unless Doehler received state and local aid for modernizing facilities, concessions from the local utility, and changes in union relations—a perfect example of modern management's "public-private partnership" dependence. George Haigh, CEO of Toledo Trust and chair of the local development group—the Committee of 100—immediately sought the full cooperation of the State of Ohio, the City of Toledo, the utilities, and the UAW.[49] He did not publicly question why Farley and Craig, fellow business leaders who doubtless included in their shared "philosophy" a commitment to minimal government, would expect state and local financial help in upgrading private facilities. Instead, Haigh sought the deal for Farley.[50] However, Haigh could hardly be criticized for inconsistency or poor thinking; the leaders of the UAW were equally gullible.[51]

Given the reception by workers of the new exercise program, it must have seemed odd that Craig requested a change in "dealings between the union and the company." Two days after news of Craig's request surfaced, the local union vice president, Harrell Lambert, said the members trusted Farley: "He is pretty highly thought of." Lambert pledged to encourage employees to maintain good morale. Farley had given operation of the old plant a "fair try." That's all that could be asked. Fortunately, he noted, Craig had not made the union local aware of any wage concessions that would be sought.[52] Neither Lambert nor Haigh asked why the company with problems meeting production quotas was closing a factory they had claimed to have modernized. At the same time as their other March requests, the company announced it had leased a facility in Sheffield, Alabama. The company needed a creditable threat that it was about to close all operations in the North. At the end of March 1987, the company dropped the other shoe.

Bypassing the local union, Farley announced a news blackout on secret negotiations with the UAW in Detroit to work out a new contract, with significant wage concessions. The local union members were informed that they faced two options. Option one was to reject it, which would mean the immediate closure of the older Toledo plant. Option two would be a cut of $2 per hour, adding $2,000 in employee contributions to the health plan, no cost-of-living raises for five years, and changes in work rules, to allow weekend work at regular pay rates. A Toledo union official summarized the request, saying, "They want a hell of a lot. I'll tell you that. And it's going to be hard to get them."[53] Suddenly it became clear that Farley's "philosophy" did not include commitment to weekends with the worker's family. Sabbaths were not that special when money was at stake. Likewise, neither was collaborative decision making. Jacqueline Farley had said that Farley's philosophy included consultation with the employees about the exercise program. Now negotiations about money were in secret. President Iwarsson had said in 1985, "Here we have a union that is strong enough not to be fearful of [possible] deceit and a company that recognizes the role of the union and which does not go behind the union's back."[54]

Working in secret through the last weekend of March, "top UAW leaders" struck a deal. The company yielded only on the hourly wage cut, reducing their demand from $2 per hour to $1. In exchange, the company pledged to bring new equipment into the newer Toledo factory and build a research center.[55] Finally, the local unions rebelled. Still showing some faith in Farley, local president Dave Schneider said, "We are not trying to block or hurt [corporation President Bill Farley], but he has to listen to the people. . . . It's not us that is [*sic*] the problem, it's mismanagement. If they'd listen to us, they'd make the money they want to make."[56] Since no one seemed to be listening, the two locals rejected the secret deal.

Still trusting Farley, the Toledo elite and Detroit UAW leadership reacted with dismay. They blamed local worker blindness, and renewed calls for more secret negotiations. As if in the era of Herman Doehler and Frank Koegler, when the company's leadership was tied to the community and worked to protect and promote it, the *Blade* editorialized, "This is an issue in which quiet, behind-the-scenes activity from Toledo political and civic leaders might still be helpful."[57] Of course, in 1987, those in control of Doehler's fate had no interest in Toledo. Farley wanted to be a national leader, and Toledo civic officials hardly could matter in his larger plans. Even the union had forgotten to talk with its base. Modern America's strength may have been based, as Tocqueville had noted in the 1830s, on strong local and voluntary institutions.[58] The new generation of American leaders was preparing to milk all the wealth and power from local communities for their own purposes. Worst of all, they were such masters of modern communications that they could fool the locals into thinking the pursuit of the leadership's prosperity and power was synonymous with the pursuit of local well-being.

The company's leadership and their allies now played hardball. Doehler sent all employees a letter announcing dates for the phase-out of the old plant. Meanwhile, the *Blade* criticized "this kind of hard-nosed unionism that surfaces all too often and which continues to hurt this city's image as a place to do business."[59] Amazingly, some good will remained for Farley, the "great communicator" of Farley Industries. An anonymous worker in Toledo observed, "There's a lot of people who like him. But the way he did it, he sent one of his hangmen up here."[60]

Farley came to the factory for the last time and worked his charm on the community. The local union narrowly approved the revised contract. Now they waited for the promised reinvestment in the plants. Included with the contract shown to the workers was a letter from Chuck Craig, the president, promising, "If the company does not live up to its commitment to modernization, eligible employees . . . will be reimbursed for wages lost as a result of this agreement."[61] For two years Craig delayed, and then Toledo civic leaders and workers learned the company was for sale.

Still the promoter, Bill Farley said, "I have strong feelings about Doehler-Jarvis and the future of U.S. basic industry, and have already made a large investment in time, energy, and capital to see it succeed." He pledged, "There will be no sale unless I am certain I have realized the appropriate value for Doehler-Jarvis, and that the best long-run interests of our customers, suppliers, employees and the communities in which we operate are served."[62] Of course, if Herman Doehler had said these words, they would have had meaning.

Needing all possible cash from the company, Farley could not allow funds even to go to routine maintenance. Dave Schneider of the UAW local pointed out that for several months in the summer of 1989, workers were on mandatory overtime to meet the orders from Ford. The overtime was a result of "malfunctioning machines" that could not produce enough castings during the regular week. Schneider concluded, "It is a shame because it is physically exhausting and affects family life." After two explosions from failed bolts on the cap of pressure vessels, one of which killed a worker, Farley visited Toledo, but did not show his face on the shop floor. As sale efforts continued, Chuck Craig quit, "to pursue other career interests."[63] The only serious bidders were another group of Chicago corporate raiders with junk-bond financing. They offered far too little for his only major asset.[64]

DOEHLER-JARVIS LIMITED PARTNERSHIP

Farley resolved this dilemma with his usual ingenuity and with the help of Vincent Naimoli, another award-winning entrepreneur. Farley and Naimoli created a management group or limited partnership to pay Farley Industries for the company. In 1990, the Doehler-Jarvis Limited Partnership transferred cash desperately needed for reinvestment to Farley Industries and to Naimoli's Anchor Industries International.[65] Their new president, George Wells, went before the Toledo-Lucas County Port Authority, with which George Haigh worked, to get tax-exempt bonds to pay for the long-promised research and development center.[66]

Wells, like so many of his recent predecessors at Doehler, knew how to say the right thing, even if reality contradicted it. Doehler had no extra cash for even routine maintenance, let alone for meaningful research. The modern manager, however, learned to say many things that stretched the truth. In an appearance before the Toledo Area Small Business Association in late 1991, Wells talked about the need for American business "to get rid of its short-term mentality."[67] As if to dramatize that the statements had no validity, two weeks after Wells talked about rejecting "short-term thinking," he was fired and Naimoli took direct command of Doehler.[68] Naimoli had nurtured in the business press "a reputation for being an excellent organizer and a man who pays attention to detail."[69] Like Farley, he had an obsession with athletics. Both would come together to play major roles in encouraging major league baseball in Florida. Using scarce cash from their overextended businesses, they put money they had never earned into the Tampa Bay Devil Rays and Florida Marlins. Farley had gained the naming rights to the Marlins' Pro-Player Stadium, and Naimoli would go further on credit, becoming the managing partner for the Devil Rays.

What both knew was that linkage to the national obsession with professional sports would protect them from the adverse publicity possible at any time from their financial manipulations.[70] Of course, they were right. *USA Today* stumbled over itself profiling Naimoli, turning him into the All-American entrepreneur who helped his adopted hometown get a team. How could any guy with a kid's love of baseball be a threat? Of course, the modern yellow journalists stopped their reporting before ever asking such a question.[71] Not surprisingly, in 1995 Naimoli received Florida's Entrepreneur of the Year Award, official recognition of his hero status. Specifically, related to his "turnaround" of Doehler, he was called an innovative manufacturer, with a special niche in aluminum die casting.[72]

The Entrepreneur of the Year Award exemplified the use of public-relations experts to change history and glorify the patron.[73] The role of papers such as *USA Today* showed how the national media had lost the critical perspective to assess the accuracy of the stories fed them by those selling junk as an innovative investment strategy. The award committee added more distortions to Naimoli's record, glorifying his management abilities. "He also looks for companies where manufacturing costs are high, and thus, where operational cost-cutting efforts can have a significant impact."[74] Naimoli reminded investors that Herman Doehler's die-casting machine was at the Smithsonian.[75] He did not mention that none of the current leaders of Doehler could invent a modern equivalent.

In the best public-relations tradition, the promoters of entrepreneurship also did not see the inherent contradiction between their praise of Naimoli's management style and the American distrust of authoritarianism.[76] They noted, "Naimoli specifies that: I have to have complete control to turn things around."[77] Checks and balances or consultation were not what modern entrepreneurs accepted.[78] They needed complete control. When the United Auto Workers demanded that money be set aside to cover the pensions Herman Doehler had pledged to fund indefinitely, Farley used his communication skills to win.[79] He "gave" the pension plan 2.425 million shares in Farley Incorporated, the firm about to file for reorganization once Doehler was under the control of a separate corporate structure.[80] Rather than take out the usual worker's compensation insurance in Ohio, he used an alternative process and purchased a surety bond, with a supposed commitment to pay any future costs. Eventually the state sued, but did not win for a decade.[81] Farley got to use the money saved from worker's compensation as an asset. Ultimately, the citizens of Ohio had to pay for the injuries suffered by his workers.

HARVARD INDUSTRIES

For the next few years, Naimoli rode high in the auto parts business, with the help of Farley. Not only did he chair Doehler-Jarvis after 1991, he used it as a platform to gain control of another public-relations creation, Harvard Industries. The two would use Harvard and its junk bonds to buy out their investment in Doehler-Jarvis Limited Partnership. Their next president, Ron Stewart, and the head of marketing invited the press into the old Toledo plant to touch transmission housings and hear how the new firm would spend $55 million in each of the next two years and then $20 million for the following three to keep the company state-of-the-art. Again the press bought and promoted the deceit: "The cloud of uncertainty that hung over Toledo's Doehler-Jarvis plant five years ago appears to have dissipated."[82] Using the public-private partnership ideology, they won a $3.8 million grant from the federal government to improve aluminum die casting. The irony seemed lost on the media that two Horatio Alger award winners needed a government handout to do research and development. Instead the media celebrated that the grant "underscores Doehler-Jarvis' role as a technology pace setter."[83]

In the midst of this good news, several events threatened to undermine Doehler's image. In early January 1995, word got out that Naimoli had hired a strategic-planning arm of Bankers Trust to investigate "business alternatives" for the company.[84] A week later, Standard and Poor's changed the outlook for Doehler's bonds from "stable" to "developing," saying, "We really don't know what's going to happen."[85] Next, the firm sent letters to 120 salaried retirees telling them the company would no longer cover their health insurance. They advised their former office workers to contact the American Association of Retired Persons for insurance. Simultaneously, the company sent letters to the UAW about the union's retirement benefits. Following usual union secrecy, the UAW said, "We can't divulge anything yet."[86]

On April Fools Day 1995, the public learned that Doehler once again had lost money—$1.7 million in 1994, down from $37.6 million in 1993. How this could take place when the company was operating around the clock begged an explanation. In 1994 alone, sales increased a whopping 14.8 percent. In 1994 Doehler had sales of $233 million, and income before interest payments of $10.8 million. The net income was wiped out by more than $12 million in interest payments. The bad news made Farley and Naimoli determined to act before all was lost at Doehler. They concocted a unique scheme. Harvard Industries, which was just recovering from bankruptcy, would buy Doehler for $104 million and assume $114 million in Doehler's debt backed by global bankers at Credit Suisse/First Boston.[87] To reassure investors, Harvard announced a dividend to preferred stock owners.[88] Harvard Industries' 1995 *Annual Report* became a celebration of the brilliance of the new conglomerate structure. Next to a picture of a production worker at a die-casting machine, readers learned, "Through the years, Doehler-Jarvis has been associated with an array of important 'firsts' in high pressure die casting . . . Doehler-Jarvis has a state-of-the-art research and development (R&D) center which was constructed in 1995 at its Toledo facility."[89]

Yet, a careful reading of the Securities and Exchange Commission filings showed that the $104 million in cash paid by Harvard to buy Doehler was a transfer of Harvard's borrowed corporate funds to the personal accounts of Naimoli and Farley, as shareowners.[90] Furthermore, Farley, his creditors at Bankers Trust, and Naimoli were splitting $314,932 in annual consulting fees.[91] One observer from Bear Stearns found Vince Naimoli's role as head of both the buying and selling company "to be very rare. In my universe of covering the manufacturing sector, I have heard of no such transaction."[92]

In the 1990s, no great skill at investigation was needed to find fundamental problems. As Bernie Thompson, a repairman in the plant, explained, the company even laid off the oilers of machinery to save money. Yet, the orders were flooding in. The Bear Stearns analyst noted, "Their sales were going up, but their earnings were going down. They were running so hard to keep up with all the orders that they never did any maintenance on the machinery."[93] Thompson recalled one incident where he was fixing one of the machines. A supervisor ordered him to stop repairs, saying, "We need to keep it running."[94] The company had so many problems with its poorly maintained equipment that it had to keep operating around the clock, paying overtime, just to meet the orders.

The decade of milking Doehler demanded change just when the conglomerate could face no delay in producing cash. In 1995 the company reached a settlement with the Pension Benefit Guaranty Corporation to immediately transfer $6 million to the fund, and pledged to pay $6 million per year for three years to restore the pension fund's integrity.[95] Then came word that Chrysler would not have Doehler produce its new aluminum engine block.[96] The immensity of Doehler's problems crystallized further when Ford canceled a transmission contract, "citing Doehler-Jarvis' inability to maintain delivery dates." *Automotive News* reported: "Its Toledo plant was running seven days a week trying to fill customer orders for transmission housings. Its machinery had gone without proper maintenance and was breaking down. The inefficiencies continue to run up heavy overtime costs."

Later it added, "The Toledo plant could not meet demand for its aluminum castings, and quality suffered as decrepit machinery broke down."[97] In the arcane language of its annual SEC filing, the company explained, "Doehler lost $21 million in 1996–97 caused primarily by operational inefficiencies offset by curtailment gains with respect to post-retirement obligations."[98] When the company made the inevitable Chapter 11 filing in May 1997, it had to admit, "While Doehler-Jarvis, a pioneer in the castings industry, may be profitable in the long run, Harvard Industries cannot support this subsidiary's current capital requirements."[99]

Before bankruptcy could be resolved, the fatal blow came in November 1997. "GM notified Toledo that GM was resourcing the production of V-6 engine blocks to other parties. The Toledo V-6 program was scheduled to launch in late 1998, and would have constituted approximately 25 percent of the operating revenues of Toledo when in full production." Accordingly, management announced: "[Due to] additional capacity problems and substantial capital expenditures required for its operation generated by its Toledo subsidiary, management has determined to wind down existing operations."[100] Not only could the company not maintain its technological competitiveness, the junk-bond kings—Farley and Naimoli—had raided its pension and health-care plans for cash. In 2002, over five hundred retirees learned that their retirement benefits would be paid not by the company at the more generous rate negotiated by their union, but at the lower rates under the emergency process run by the federal Pension Benefits Guaranty Corporation. The PBGC had to step in because the "hands-on managers"— Farley and Naimoli—had underfunded the pension by $97 million out of a total obligation of $213 million.[101] No doubt their hands had been on some money, but it was money that was not theirs.

LESSONS

The failures of Acme and Doehler teach seven important lessons related to leadership and civic life shaped by the neoliberal ideology of the late twentieth and early twenty-first centuries:[102]

First, the demise of these firms generally highlighted the failure of modern leadership models to expose incompetence. Richard Dawson, a Harvard leader, wrote off blame, saying, "Manufacturing is a funny business. You think you have a problem fixed and a machine breaks down. It's a never ending battle."[103] The assessment reflected the shallowness of Farley's "hands-on" managers. Without romanticizing earlier leadership, it is easy to see that the approach of Herman Doehler contrasted markedly with the rationalizations of Dawson. Herman Doehler, Jessel Cohn, and other leaders of companies such as Doehler and Acme remained models of effective leadership throughout periods of great change. Putting marketing and legal staff in charge of a corporation, unless those tasks are its primary function, is so patently foolish that it seems unnecessary to identify this as the first lesson of the firms' collapse.

Second, the prime reason for the appointment of technically incompetent managers is that the primary focus of leaders such as Farley was to extract assets from the firms they seized. Each of the major business leaders involved in the last years of Doehler or Acme gained control not because they knew more about die casting than any others. Instead, they were the people who convinced financial institutions that, if given the chance, they might repay high-interest loans by gambling with the assets entrusted to them. That everyone involved knew this was foolish is made clear by the repeated public-relations announcements that the new leader had faith in American industry. In fact, the bankers made a pact with gamblers to make high-risk loans on the condition of being repaid handsomely, if necessary from the accumulated assets built by generations of inventors, skilled leaders, and dedicated workers. The behavior of the investment cabal was shameless, as would become evident to the world in 2008 and 2009.[104]

Third, what was new at places like Acme and Doehler was that the institutions that the society trusted to bring fairness to the process and check abuse of power, whether independent bankers or corporate boards of directors, utterly failed. Poor and selfish leaders have been legion throughout human history. Greed, of course, is one of the seven deadly sins. The failure of these independent checks on behavior reflected the wider abandonment in business of any responsibility to the past or future generations or the communities that have provided protection, labor, and financial support that nurtured the business. Here the behavior of Doehler's last leaders again contrasted with that of people like Herman Doehler. The explanation is not that the new generation is more rootless and lack local ties and commitment. Herman Doehler was a German teenager before he set foot in the United States, and he never fully abandoned New York. Yet, he knew Toledo had been good to him. He regularly visited and supported the town, even though he loved to tour the world in his last years. And the same could be said of Jessel Cohn in Clarksville.

These were leaders who knew to whom they owed thanks for their success, and they expressed it in deeds. They also knew that with the privileges of wealth and respect went the responsibility of service and attention to detail. As Doehler-Jarvis was failing in the late 1990s, a banker with Credit Suisse First Boston thought he could explain the collapse by saying "[It's a] pretty tough, old, tired plant. . . . If you closed it, you would actually save money."[105] In fact he was indicting himself, his bank, and the leaders they had funded to take over companies such as Doehler. For years, the new-style leaders lied about investing in new plants and equipment, always in the near future. But, of course, they never had money for the new investments because they were sucking out millions in consulting fees and other schemes. The bankers knew that the sales of Doehler-Jarvis to ICM Industries and then ICM to Harvard were mechanisms for converting borrowed corporate funds into "free money" in the pockets of the leadership. Yet, the creditors did nothing.

Fourth, a primary reason this deception could take place for so long arose from the behavior of other institutions once thought to be independent critics of skullduggery. Business leaders such as Farley and Naimoli regularly received glowing treatment in the business press. Farley was celebrated as a Horatio Alger. The University of Tampa named its "Institute for Business Strategy" for Naimoli. Fairleigh Dickinson University's *FDU Magazine* celebrated his rescue of companies such as Harvard and Doehler-Jarvis, forgetting to mention that Doehler was in liquidation.[106]

Fifth, these honors were given to the new business leaders in spite of the utter contempt they showed for their workers. The contrast with previous leadership could not have been starker. Herman Doehler in a speech in 1950 said, "As you know, the policy of management of our corporation is to be just, to be fair, and to practice the Golden Rule under any and all circumstances. This is the only way I know of to obtain the respect and goodwill of our employees."[107] The modern managers had stripped from their pursuit of gold all ethical rules. Workers now were to be fully exploited and then insulted. For example, in the 2002 bankruptcy settlement at Harvard Industries, 600 retirees from Doehler "received" $800,000 placed in a health-insurance escrow account. Four key executives who had run the company into liquidation were allowed to split $1 million. One of the UAW retirees, John Gilbert, who apparently had better math skills than the bankruptcy judge, concluded, "That's not an equal share."[108] By 2003, the Harvard Industries pensions for 25,485 former employees, of whom 1,156 were from Doehler, had been so underfunded that the workers were transferred to the Pension Benefit Guaranty Corporation.[109] Farley, who had made $7,766,000 in 1996 in the midst of leading multiple companies into bankruptcy, said, "There is no doubt in my mind that the problem had to do with very high wages and fringes that led to an uncompetitive cost structure."[110] Reflecting the national trend, the workers had an average income $\frac{1}{258}$ the size of Farley's salary.[111] Herman Doehler, with an invention in the Smithsonian and a world-renowned company with 7,000 employees, might have expected a salary like Farley's. To pay that much to a person who helped destroy what Doehler had built showed a perfect example of greed.[112] No wonder that a major employment problem discovered in the new millennium was "wage theft."[113]

Sixth, not surprisingly, the contradictions in philosophy of the new business elite extended to government. As they exploited workers, the new entrepreneurs took all possible public resources from their host communities. Repeatedly during Doehler-Jarvis's last decade, Farley and his agents sought subsidies from government in both grants and indirect subsidies, such as tax abatements. As with labor-union wage concessions, the first step in requesting a subsidy was to threaten closure. Of course, using public subsidies to support business development and innovation was not new. A number of the old English colonies had been founded by investors receiving significant backing from government. Such policies continued throughout the nation's history, often leading to new growth that benefited the wider society, not just the investors. The difference now was that the recipients of subsidies not only boasted of being self-made, but also used the subsidies to fund their excessive salaries. In its closing years, Doehler received several tax abatements as rewards for new investment, while allowing the facilities to deteriorate.[114]

Seven, one of the problems that had surfaced as early as the first discussions over contract concessions with Farley in 1987 was the archaic and secretive behavior of the United Auto Workers. Behaving like a turtle on a freeway, the union could not act with speed. Facing catastrophe, the union acted as if business were as usual.[115] The union's failure to move quickly on the Employee Stock Ownership Plan contrasted with their work on a severance package. Ironically, the company was so short of cash that the severance package, retirement, and health benefits on which the union focused would be gone before most received them. Instead of modeling a new-style union involvement in preserving jobs and community assets, the union elected to retreat into old-style "bread and butter" unionism. The consequence would be that the union members would lose much of the "bread and butter" they thought they had saved. Workers had lost meaningful jobs, and a few manipulators of corporate assets had squirreled away a fortune. The fate of other communities and workers with Fruit of the Loom facilities would mirror the Toledo and Clarksville experience if they didn't learn the lessons identified here.

CHAPTER 6

Importing Fruit

SIMILAR TO THE TROUBLES AT ACME AND DOEHLER, AFTER 1990, the problems at Fruit of the Loom devastated communities, workers, and investors. As happened with the other subsidiaries, neither the media nor pubic officials got the story right until too late. After the loss of control of West Point, Farley especially needed to advertise Fruit's economic viability and contribution to communities—but that is an explanation, not an excuse for civic and media leaders who failed to notice ruin as it approached. Their acceptance of neoliberal ideology blinded them to the causes of the abandonment of U.S. manufacturing and massive shifts of production to places where workers could be exploited.[1] Fruit's transformation in the 1990s perfectly illustrated what Charles Maier called the movement of the United States from an "Empire of Production" to an "Empire of Consumption."[2] In the ideology of the promoters of this new "empire," market forces replaced civic deliberation in determining outcomes.[3] In contrast to earlier eras when Fruit of the Loom sought government support for new factory construction, now it lobbied for the freedom to locate the cheapest source of production. To the credit of Bill Farley, he held out as long as possible before fully embracing the "Empire of Consumption." He did it only after competitors and political leaders changed trade policy that made unworkable the old-style subsidized manufacturing. The earlier history of Fruit of the Loom, like that of Jessel Cohn at Acme Boot, focused on winning financial support from communities in need of employment.

Like Cohn, during the Great Depression, Jacob Goldfarb, the owner of Union Underwear, struck deals with Kentucky officials to transfer production to the state from its former center in Indianapolis. Both Goldfarb and Cohn took advantage of the common southern belief that the region's industrial growth needed subsidization. As the mayor of one Tennessee town said, "The little town that wants industry to stop the flow of young people away from its surrounding rural area does what is called 'buying industry' or it doesn't get any. . . . I'm awfully tired of hearing the fine theories of industrial development say to the little town desperate for a payroll, 'Only a sick industry wants a subsidy.'"[4] However, basing growth on low-skill and low-wage industries increased the region's vulnerability to competition from even lower wage areas.[5]

Much like the "banana republics" where Fruit of the Loom eventually moved production in the 1990s, the South offered more than financial subsidies. It also offered cheap labor, low-cost construction, and police and cultural protection from unions.[6] Given the cost of labor and the reduced need for heating compared to the North, the South also offered lower-cost structures, especially when the local community subsidized the expense of construction and ownership. Additionally, the use of police to suppress labor and civil rights lingered longer in the South than the North.[7] Finally, the southern fundamentalist churches frequently attacked union membership as a sin. Lucy Randolph Mason, a famed union organizer and descendant of two of the oldest families in Virginia, described the end of one church service: "The preacher said he had a special message for union members. . . . He first made a vicious attack on the cio, calling it all the bad names he could think of, and finished with a declaration that no cio members could be 'saved' and that the people would have to decide between the church and the union."[8]

Throughout the South, workers faced this hostility from the white churches that had abandoned a prophetic role and confined themselves to condemning individual sins, mostly of the sexual variety.[9] Adopting the material measurement of salvation noted by Max Weber, the churches considered worker organizations evil since they threaten the wealth of entrepreneurs. Clearly, God blessed the rich. Poor workers could claim no such obvious justification.[10] The once traditional South transformed into an ideological bastion of modern individualism and entrepreneurship.[11]

BOWLING GREEN

When Jacob Goldfarb struck the deal that brought the Fruit of the Loom brand name to Union Underwear, he needed to greatly increase production. Enterprising members of the Bowling Green, Kentucky, Chamber of Commerce offered to build a factory for him if he paid 20 percent of the costs.[12] Over the next forty-five years, local, state, and federal governments as well as the Bowling Green chamber worked with Union Underwear to support further expansions. The same sequence of events invariably accompanied each new construction. The company would suggest its hopes and plans to expand while hinting it was considering another city or state. Immediately, the chamber and public officials attempted to satisfy the company's request. However, in exchange for subsidies and tax breaks, the company continued to add jobs, and executives fully participated in civic life.

Both Goldfarb and his long-serving successor, Everett Moore, epitomized the model community leader. Moore served on the board of the local Salvation Army and the local bank, and as a booster of Western Kentucky University. Yet, he was not too important to visit each factory weekly and know all employees by first name. Moore said, "We've told them it's their plant and our doors are always open to aid in solving personal problems which sometimes arise. . . . The big reasons we've been successful are the people who have worked for us and the fine cooperation we have received from everybody in Bowling Green."[13] When he retired as chairman of the board in 1975, after forty-two years, a fourteen-year employee and native of the area, John Holland, succeeded him. Like Moore, he became a widely respected community leader. At the time of Moore's retirement, Ben Heineman owned Union Underwear, and the appointment of Holland symbolized his commitment to continue the practices that made the firm, despite its world leadership in underwear, an active member of the communities in which it operated.[14]

The company experienced a burst of expansion once Northwest Industries took over in 1968. Between then and the mid-1970s, it established two plants in Louisiana and North Carolina, and one each in Mississippi, Oklahoma, and Pennsylvania. Its last expansion occurred in 1980 with the announcement of a large plant in Jamestown, Kentucky.[15] In the early 1980s, Fruit employed over 7,000 people in the state. Outside Kentucky, the company's Louisiana factories had 4,800 and Alabama had 2,075; there were 730 in Mississippi, 900 in North Carolina, 300 at two plants in South Carolina, and 375 at a single Oklahoma facility. The role of Bowling Green increased in the late 1970s with the transfer of the manufacturing planning, accounting, sales, and marketing offices from New York.[16]

FARLEY'S ARRIVAL

Modern subsidized capitalism had become a habit, so none remarked on the contradiction between neoliberal free-market rhetoric and the practice of public financing of business.[17] In 1980, there was no recorded opposition in the state press to the county's issuance of $7.5 million in industrial revenue bonds to fund construction of the architecturally stunning "world headquarters" on the edge of town. Only the county attorney deemed it necessary to qualify the subsidization of the firm, insisting that the company alone assumed risk under the bonds.[18] A few months later, when the company decided to build a new manufacturing facility in Jamestown, about ninety-five miles east of Bowling Green, the U.S. Department of Housing and Urban Development (HUD) provided $2.2 million for water and sewer lines.[19] A $2.9 million grant in 1986 came directly from HUD to pay for plant expansion.[20]

The ideology that justified such subsidies prepared the way for Bill Farley's rise. Farley's administration had mastered the art of extracting subsidies from communities in exchange for jobs. Much as Jacob Goldfarb forty years earlier, Fruit temped Kentucky to offer a subsidy for the Jamestown expansion by announcing it planned a major addition somewhere in the country. Initial publicity reported it as a $25 million facility. Kentucky officials said that "the company would not seek major financial incentives from the state."[21] In reality the state contributed eight to ten million.[22] A few years later, the company succeeded again in receiving state aid for a new "distribution center" in Bowling Green, after Fruit said, "The company was considering building the center in another state."[23] According to the company's account, it needed new office space to handle its new product lines and extensions into Europe and Latin America.[24] It did not reveal that some of the "Latin American expansion" existed at manufacturing facilities that would replace work in Kentucky.

Kentucky officials could not be censured for failing to see the forest for the trees. Despite massive debt, Farley continued to exude confidence and empathy for communities. In 1987, it appeared Farley fully had adopted the old company's style when he reorganized Northwest Industries and renamed it Fruit of the Loom. In a reorganization too complex for most to fathom, the brand name of a subsidiary became the conglomerate's designation. The company said it assumed the new title because it had "restructured its capitalization." Most observers did not notice that the reorganized Fruit board included a group of Bill Farley's friends who did not value, nor would they defend, historic links to Bowling Green.[25]

Over the next few years, while attempting to acquire West Point-Pepperell with resources from the remainder of his empire, Farley would need a cooperative board. However, until the West Point disaster crystallized, he seemed like a savior to Fruit of the Loom. The flamboyant CEO even appeared in underwear on television advertisements. For those worried about junk bonds, Farley warned, "A manager who can accelerate the cash flow of his operations should be willing to carry a lot of debt for a while." He said, "In this country we have become arbitrarily too conservative in our capital structure."[26] Optimistic local officials liked this talk and in the fall of 1987 received Farley as the honored guest of the Bowling Green Chamber of Commerce. They proclaimed him "one of the nation's top industrial leaders." They celebrated the successful completion of his initial public offering of stock in the new Fruit of the Loom.[27] Even when Farley Inc. filed for bankruptcy in 1991 as a result of the West Point deal, the community reported no impact on the former Union Underwear. After an interview with its vice-chair, Joe Medalie, the local paper reported, "'Farley Inc. is totally separate from Fruit of the Loom,' Medalie said today. 'Anything that happens at Farley Inc. is not connected to Fruit of the Loom.'"[28]

Many in the business press persisted in loving Farley's expansion talk. As conditions deteriorated for the company, Farley appeared in Louisiana with another gambler, Governor Edwin Edwards, to talk of opening another plant.[29] The *Financial World* reported with glowing praise: "On a beautiful spring day in Vidalia, La., a tiny Mississippi River community rocked hard by the recession, William Farley's perfect tan has taken on the internal glow that comes from knowing you're saving a town." The article's subheading emphasized, "Like it or not, CEO Bill Farley has helped build Fruit of the Loom into one of the greatest companies of the 1990s."[30]

However, given the recent disaster at West Point and the subsequent bankruptcy of Farley Industries, it is hard to understand how finance writers perceived Fruit of the Loom's greatness. But then, they were not prepared to defy the relentless marketing of Fruit. As soon as Farley distanced himself from the post–West Point bankruptcy, which forced him to sell about half the stock Farley Inc. held in Fruit, he personally acquired another third of Fruit, giving him and his company about half of its stock. Consequently in 1993, he resumed expansion, buying three apparel firms that both positioned the company in "athletic wear" and brought it more financial problems. Through these acquisitions, Farley owned Pro-Player, a brand that he advertised by purchasing the naming rights of the Miami baseball stadium.[31]

Meanwhile, events were spiraling out of control at Fruit and were about to devastate the towns with its facilities. After Fruit announced a sharp decline in earnings in 1995, bond-rating firms warned that they might lower Fruit's bond rating, which stood only one step above speculative or junk status.[32] Since Farley needed to act to avoid increased costs for credit, either he had to retain his debts or his workers. He launched a massive cutback in U.S. production. The company closed thirteen of its domestic factories and laid off eight thousand workers. It quickly acquired manufacturing facilities in Central America by both lease and purchase. The press trusted Farley's explanation that the cuts were an inevitable response to U.S. wage rates: "The going rate in the U.S. for workers is about $10 an hour. In Salvador, where the quality is just as high, it's $1 an hour."[33] Neither the press nor his friends on the board questioned why his salary at the time had to be many times more than a Salvadoran executive's. While his company struggled, he received $12 million in wages and bonuses and charged the firm an additional $10 million in management fees.[34]

When the company announced the 1995 cuts, they blamed the events on everything but current leadership. NAFTA became an easy target. Ronald Sorini, a corporate public-relations representative in Bowling Green, said, "What you are seeing is the cumulative impact of NAFTA and GATT."[35] He did not mention that Farley needed to cut wages to transfer money from employees to service the debt he had assumed in buying companies he could not afford. As in the past, the strategy succeeded in the short run. As Fruit of the Loom reported its cutbacks, it received a gift right before Christmas when its stock price soared by $2.50 per share to $22.63.[36]

In the 1995 *Annual Report*, a smiling Farley leaned on machinery in a textile mill. Presumably it was a surviving Fruit of the Loom plant, although the photo was only a public-relations ploy. The real picture was more bleak. Every year since the sale of Velsicol, the report pointed out that the firm retained responsibility for certain environmental liabilities. It never discussed NWI Land Management, the odd first subsidiary listed in the SEC filings, which only owned superfund sites and a nuclear waste dump. Likewise, the report failed to elaborate on why the new owners of Velsicol had refused to accept the environmental liabilities when they left Fruit of the Loom. These details—as well as the various compensation packages for the officers, and the overwhelming indebtedness—were hidden behind ingenious presentation and positive projections.[37]

Of course, the document did not report the responses of former workers as they left the shuttered factories. Peggy Johnson from the Franklin, Kentucky, plant south of Bowling Green "vowed never to apply for another manufacturing job again. It wasn't that the 54-year old Johnson disliked her work as a quality-control inspector. She had simply grown frustrated watching the last three factories where she worked in her home town shut their doors."[38] Up the road in Bowling Green, the 1940 plant cut 30 percent of its work force.[39] When another five hundred jobs at the old facility were terminated at the start of 1996, and the first twenty at the corporate offices, John Holland announced his resignation. With that step, the last major link to the past disappeared, and in his place came a Farley ally, Richard Lappin. He had served Farley as president or CEO of Farley Industries, Acme Boot, and Doehler-Jarvis. Obviously he knew little about underwear manufacturing, but in Fruit's condition, he mostly needed to lay off people and steer a company that had once led the world in a basic product into insolvency.[40]

In 1997 the U.S. layoffs continued. In August the company terminated 5,100 jobs, including 2,000 in Kentucky. The giant Campbellsville plant became the focus of attention. At its peak, it hired over 3,000 workers. On August 7, a total of 1,480 workers were dismissed, leaving 1,700. Every community in which Fruit owned a factory was filled with stories such as: "David and NaDena Agee have a mortgage on a house they bought two years ago when they were both making good salaries at the Fruit of the Loom Plant in Campbellsville. They also have a 19-month-old son who is growing up fast. . . . They are worried about how they will live and how they will provide for their son."[41] More generally, Mayor Bobby Miller said, "We are going to decrease the quality of life here." His only consolation was to think, "I'm just glad the whole plant didn't leave."[42] He would not have been so optimistic if he had known what was going to happen in 1998.

Ignoring Farley's massive indebtedness, his huge interest payments, and other signs of mismanagement, company officials excused the layoffs as a consequence of globalization. Mark Steinkrauss, a Fruit of the Loom vice president in Chicago, asserted, "We're not insensitive. But if we didn't do this, we wouldn't be competitive." Larry Martin, director of the American Apparel Manufacturers Association, which had lobbied for liberalized trade, defended the repercussions, saying, "From our viewpoint, it's better for jobs to go to Mexico and Central America—instead of the Far East—because those are American companies operating in those places."[43] The same line was fed to the Louisiana press when 4,200 layoffs hit that state. "The company said the North American Free Trade Agreement and other global trade accords have forced it to search for cheaper labor."[44] Steinkrauss cushioned the blow of the 1997 cuts by lying about the future: "This gets us close to 100 percent offshore, so we don't anticipate any more reductions."[45]

Typically, the stock market and investment managers, divorced from productive communities, used euphemisms to explain the closures. One industry analyst in New York captured the deception in one sentence: "They can no longer afford the luxury of sewing in this country." He failed to note that Farley accepted salary and compensation from Fruit equal to the annual wages of 1,175 Kentucky workers. The business press ignored the stark injustice of providing more than $20 million in compensation to a CEO leading a world-class manufacturer into bankruptcy. The New York analyst added that the termination of 4,800 "is somewhat of a nonevent for investors."[46]

In 1998, the news from Fruit grew both worse and absurd. In January, Steinkrauss announced the layoff of twenty information-systems workers from headquarters. The firm decided to outsource computer work. He promised that the cuts were not a "harbinger of further terminations. . . . It doesn't portend anything more dramatic that I'm aware of. . . . [I] can't guarantee that it won't happen, but I'm not sitting on anything."[47] Less than three weeks later, he reported four hundred layoffs at Campbellsville, and suddenly Richard Lappin resigned as president "to spend more time with my family." Steinkrauss explained, "The company is so well through its restructuring now. We're upbeat about 1998 and our prospects for turning the company's fortunes around." When asked specifically about Bowling Green, Steinkrauss said, "I don't know of any other significant changes that are imminent. It's back to business as usual for us."[48] Of course, that comment should have worried any remaining employees.

The surprising result of a review of the company's behavior at this time was not the lies about general financial status and future employment, but that each deception was reported as unique even after regular repetition. Throughout 1998, every round of cuts was said to be the last. In March, five senior management jobs were discontinued.[49] In April, the company announced a complete shutdown in Campbellsville. In May, Fruit dismissed thirty-five in customer service, production planning, and product development. While clearly downsizing its operations, the press took a Fruit spokesperson at her word when she stated she was "not aware of any additional layoffs."[50] She had said the same thing three weeks earlier with the announcement of the Campbellsville closure.[51] Finally, those close to the terminations began to challenge the news. Mayor Miller in Campbellsville complained that the city had just undertaken a five million dollar water-pumping upgrade for Fruit of the Loom. Miller did not dwell on the lost money; he focused instead on the people. "It's broke [*sic*] the hearts of a lot of families."[52] Labor and its supporters also attacked Farley's compensation package, singling him out for a "U.S. Workers Betrayal Award."[53]

Bowling Green officials had multiple reasons to worry about Fruit's intentions in 1998. The original factory in town had closed. Every month seemed to bring more cuts at headquarters. In the summer, three other bits of bad news emerged. "A source within the company's hierarchy" let out word that another two hundred headquarters staff would have to go, cutting in half what once had been over one thousand employees. More dismaying was news that Farley had asked stockholders to approve transferring assets to a Cayman Islands holding company. Silence now replaced the earlier openness about reductions. While the previous lies were disconcerting, laid-off employees now had to sign nondisclosure statements essentially selling their First Amendment rights for severance pay. Finally, word surfaced that Farley wanted to sell and take his managerial talents to "fast growing areas like technology." An official at a Chicago "options-strategy firm" spread the usual untruths to cover Farley's exit strategy, saying, "Now the company is showing signs of turning around, it might be a good time to sell."[54]

TRADE WARS

While Farley deceived many workers and communities as Fruit reduced American operations, in his defense, he did not begin the decade with a plan to abandon all U.S. factories. In fact, the company fought a rear-guard battle until national policy circles ridiculed it. Whether because of his political ambitions, lingering management loyalties to Kentucky, or his lack of knowledge of industrial trends, Farley and Fruit did not lead the campaign to end underwear production in the United States. The firm became trapped in a no-win situation with a desperate need for cash that restricted maneuvering room. However, the company's conduct in communities with its factories, as it incrementally abandoned them, left it few allies. Yet, Farley and Fruit realized no benefits from its global realignment.

Clearly by the middle of the 1990s, the company fell into a trap because of both debt and changing national trade policy. As early as 1985, President Joe Medalie criticized national trade policy and supported the Textile and Apparel Trade Enforcement Bill that President Reagan vetoed. Still, Medalie noted positively that "By virtue of what we produce, we have been spared."[55] He wrongly assumed that since underwear was less labor-intensive than other apparel, no temptation existed to move offshore. Nonetheless, he sympathized with those in other parts of the apparel industry. John Holland, the CEO, helped organize a massive letter-writing campaign to Reagan that produced 2.7 million requests asking the president to sign the bill.[56] However, Reagan, a supporter of the neoliberal ideology of free trade unburdened and liberated from labor standards, defied the industry and its workers. Neither his Republican nor Democratic successors abandoned that policy.[57]

As the national ideology shifted from production-oriented to consumption-driven, Fruit of the Loom repeatedly found itself battling the retail industry as well as laissez-faire politicians. Farley once said he was not an industrial scavenger, but rather someone who perceived hidden value in industry. Unfortunately, the acquisition debt did not allow him to fully test whether, with dedication to sound management and product quality, he could have contained the relentless cost pressures inherent in the new consumerism. To his credit, Farley did try to transform the policy process in one last, glorious fight.

Having toyed with the idea of running for the presidency as a Democrat in 1988, Farley must have felt vindicated in his party affiliation when George Bush promoted progress on NAFTA. However, after the election of Bill Clinton, Farley leaned more Republican for a time. On the eve of the inauguration of NAFTA in December 1993, the Democratic president gave wholehearted support to the new "world trade agreement" under the General Agreement on Tariffs and Trade (GATT).[58] Farley immediately announced formation of an anti-GATT coalition. Thomas Friedman, the *New York Times* neoliberal international affairs reporter, referred to Farley as "one of the most vocal corporate opponents of the GATT."[59] Farley's coalition included labor, and Democratic and Republican political leaders from textile and apparel districts. For example, Democratic senator Fritz Hollings from South Carolina joined with his Republican counterpart, Strom Thurmond. Unfortunately, retailers who excelled at marketing led the opposition. They claimed that import restrictions on apparel cost each family $200 per year. The retailers and their allies did not discuss the working conditions that allowed for the $200 savings.[60]

By 1995, as Fruit began to lease plants in Central America, Farley modified his strategy. Realizing he needed to cut his costs, he became a major supporter of expansion of the Caribbean Basin Initiative (CBI)—a deal that essentially extended the *maquiladora* or assembly plant concept from Mexico to Central America. Under the CBI, firms received a break on import quotas if they manufactured products with American fiber. He realized he could ship precut textiles to the region for assembly and continue most U.S. operations other than labor-intensive sewing. Because the Clinton Democrats shifted to support free trade, Fruit of the Loom's political action committee (PAC) became the leading first-time donor to the Republican National Committee.[61] Fruit faced a major challenge in trying to catch up to rival Sara Lee's mass production of Hanes underwear in Central America. Farley believed it more productive to invest scarce Fruit funds in lobbying than in new plants and equipment in the United States.

His victory on the CBI expansion brought Farley to the attention of "good government" reformers. Common Cause singled him out for criticism. Much like earlier "Mugwump" reformers, the group possessed more interest in lobbying purity than in retaining jobs for garment workers, or in global labor rights.[62] Farley did receive some praise, though, from other progressives. Respected reporters for the *Philadelphia Inquirer*, Donald Bartlett and James Steele, defended Farley for protecting jobs. They chronicled how he had won the "Fruit of the Loom Amendment" to strictly enforce import quotas on country of origin for apparel. In 1995 Fruit of the Loom led the battle to get the U.S. Committee for the Implementation of Textile Agreements to impose quotas on Central American underwear imports, because many businesses were not following the rule to use U.S.-made fabric. However, this effort was costly, lonely, and futile.[63] The first ruling from the new Textile Monitoring Board (TMB) of the World Trade Organization rejected the quotas. Fruit's only glimmer of hope was that the TMB ordered the U.S. and Central American exporters to negotiate an agreement to replace quotas.[64]

Reflecting the dominance of laissez-faire consumer ideology, the retailers hired Doral Cooper, a former Reagan administration trade official, to oppose Farley.[65] Her victory essentially required that Farley speed up the transfer of his production offshore, since Sara Lee could now flood the market with underwear made in Costa Rica. Bartlett and Steele poignantly described how the triumph of Cooper and her affluent allies cost the workers at Farley's Batesville, Mississippi, plant their jobs.[66] But most consumers, along with the "good government" backers of Common Cause, were not concerned about helping workers in Mississippi.

Having received a fair deal in the Senate Finance Committee as well as aid from Senator Dole, Farley became a leading proponent of Dole for President, serving as a national co-chair of his campaign.[67] Sara Lee supported Clinton in 1996. As soon as Dole lost, Farley engaged in one final battle. He hired Dole as his lobbyist in an effort to preserve minimal restrictions on Caribbean imports. Sara Lee recruited a former close aide to Reagan, Ken Duberstein, to push the open import bill. Sara Lee possessed an additional ace in the hole. Its CEO, John Bryan, had been one of Clinton's top fundraisers and received the president's backing of "Fast Track" on free trade. He persuaded Clinton to make a personal call to Trent Lott, the Senate majority leader, to affirm bipartisan support for Hanes. Although Farley had won several delays in shaping CBI rules similar to those of NAFTA, his victories were ending.[68] Fruit needed a different strategy.[69] Shortly after Sara Lee succeeded on CBI, Fruit announced more layoffs.[70]

The layoffs served as a cruel way to effect one small achievement. Just as Fruit publicized the transfer of the last 2,900 sewing jobs out of the United States, the House of Representatives rose in rebellion against Clinton and defeated his request for "Fast Track" trade agreement ratification.

Washington columnist E. J. Dionne captured the vote's underlying message: "[The House was] willing to hold globalization hostage to equity. . . . By embracing open trade linked with real social protections, Clinton could give himself a cause and the legacy he keeps ruminating about. He would give Americans something to chew on: not protectionism vs. free trade but free trade with equity vs. free trade without rules."[71] Fruit's last major job cuts had a brief impact for good.

By 1998, with the company in deep financial trouble, Farley made his peace with Clinton's commitment to free trade without equity. In a remarkable political turnaround, the Dole co-chair of 1996 hosted a Democratic fundraiser at his Chicago condominium during which Clinton played the saxophone as Farley's wife sang along.[72] Farley now unleashed his lobbying efforts to attain complete openness in Caribbean and African trade. Again the "good government" types criticized Fruit. Charles Lewis of the Center for Public Integrity said, "It's a company in bad shape giving money fairly lavishly to the [political] process, with incredible things to gain."[73]

By this time, Fruit also reconciled itself with Sara Lee and sought an arrangement that guaranteed each a share of the import market. The combined African-Caribbean Basin Trade Bill passed in 2000 eliminated duties on apparel as long as it was made with U.S. fabric. In contrast to 1985, when Fruit expressed opposition to open global apparel trade, now only small manufacturers in places like the New York garment district opposed trade liberalization; the big firms embraced it. These divisions resulted in odd splits in Congress. For example, the two Republican senators from Kentucky came down on opposite sides. Both received contributions from Fruit managers and the Fruit of the Loom Good Government PAC.[74] However, Jim Bunning firmly rejected the bill, saying, "How many more jobs do we have to lose until we wake up and smell the Caribbean coffee?"[75] Senator Mitch McConnell supported the new Fruit position.

THE CAMPBELLSVILLE COMEBACK

Neoliberals always faced the potential rejection of their agenda when mass layoffs occurred. The elimination of Fruit of the Loom jobs in Kentucky forced them to find a public-relations slogan to distract voters. Having switched from a defender of domestic manufacturing to an advocate of off-shore production, Fruit of the Loom acted quickly to boost its productivity in the "developing" world. Between 1995 and 1998, the company increased foreign sewing from 12 percent of all garments to 95 percent. Because of its slow start overseas, half of the 1998 labor came from "contract facilities," not from factories owned by Fruit.[76] It claimed it hoped to reduce that amount as soon as possible.[77] By 1999 the company operated five subsidiaries in Honduras, three in Mexico, two in El Salvador, and one each in Costa Rica and Jamaica.[78] The firm owned about 10 percent of these plants; the others were leased.

Senator Bunning's brief rebellion against so-called liberalized trade policy revealed that too much suffering existed in Kentucky and the bayou country of Louisiana, as well as in other American towns, to celebrate this expansion of production. The supporters of "liberalized" trade policy claimed that plant closings were inevitable, while job losses provided an opportunity for growth, not decline. They developed a strategy of celebrating the opportunities that replaced the lost jobs. In the case of Fruit of the Loom, they coined the term "the Campbellsville Comeback."

The potential marketing difficulty for proponents of free trade grew worse in 2001 when Jonathan Moore, a basketball hero from Campbellsville, filmed *The Factory*, a documentary about the crisis in town. The son of the high school basketball coach, Moore had gone to college on an athletic scholarship, concluding his playing career at Kentucky Wesleyan. When he returned to town unable to find work that matched his training, he took a job for two years at Fruit of the Loom's plant in Greenville, Kentucky, northwest of Bowling Green. After that plant closed, he left Kentucky to attend graduate school at UCLA, majoring in film and television. His memories of Fruit of the Loom and its impact on Kentucky towns came to mind when he learned of the Campbellsville closure. He recalled, "Like most of Kentucky's creative people, I feel a spiritual connection with this place . . . I only worked for the company for two years—not 35 like some of the other people. But when I left, I had a soft spot for those people in the factory because I had learned a lot about life there."[79]

Kentucky Educational Television broadcast *The Factory* statewide in early 2002. In it, Tony Young described the "anger, shock, and frustration" at its closing. Terrey Davis said of the last day, "And you want to talk about a tomb. Now it looked like a tomb." Using a similar analogy, Laurel Blevins observed, "It was like someone had told me maybe someone had died." As at West Point, Georgia, and Clarksville, Tennessee, a number of the former workers focused on Farley and the general lessons his misconduct supplied about the U.S. policy process. Clarence Davidson said, "This fellow Farley has a reputation of going in and buying a business and dismantling putting money in his pocket." Even the head of the pro-business Campbellsville industrial development agency, created in response to the closure, joined in the reproach: "Farley had no reservations whatsoever of looking a U.S. Congressman or the Governor of Kentucky or anyone else in the eye and saying, 'This is it. I'm going to lay off 100. I'm going to lay off 500.' I believe in the free-enterprise system, but I also believe in social and corporate responsibility. And Bill Farley has none."[80]

However, the inherent commitment to individualism and neoliberalism in the modern South blunted much of Fruit's condemnation. *The Factory* did not end on a depressing note. The movie's finale announced that the book distributor Amazon had bought part of the old Fruit of the Loom facility for use as its Midwestern distribution center. Moore stated at the end, "Corporate America can devastate a town like this, but not for long. And while all the jobs have not been replaced, there is a sense of hope and encouragement."[81] Truly, the community modeled a process to address the crisis of job loss. The town and Taylor County, in which it was located, instituted an economic development program when Fruit closed. Following the cutbacks, small business leaders and officials combined to form Team Taylor in order to promote growth. Establishing a small "occupation tax," the county raised funds for redevelopment. The small Baptist college in town, Campbellsville University, stepped forward to offer occupationally related courses, and workers were encouraged to use NAFTA "Transitional Adjustment Assistance" to support themselves and pay for retraining tuition.

Soon, the economic-development initiatives brought some results. In addition to Amazon, some former managers adopted Fruit's function of supplying the military with underwear. Because the military continued to "Buy American," a new small underwear factory, Campbellsville Apparel, received Defense Department contracts.[82] Team Taylor director Kevin Sheilley worked with smaller manufacturers and service employers to add as many jobs as had existed at Fruit. The unemployment rate fell from 27 percent immediately after closure to 5.2 percent in early 2004. Mayor Paul Osborne, who had just been elected when the closure occurred, said the community succeeded because local leaders "had to unite. The community had to be 'brutally honest' in assessing strengths and weaknesses. It had to come up with a plan to fix problems and develop the economy and go after whatever help was available."[83]

The turnaround in Campbellsville, like the recovery from Velsicol in St. Louis, Michigan, provides lessons about how local civil society can confront and overcome large problems. Kevin Sheilley, of the local development office, pointed out these insights for members of the U.S. House of Representatives Ways and Means Committee during a May 2004 visit to

Campbellsville. "[The] government programs that work best are ones with local control. . . . He also felt it critical for federal lawmakers to keep some discretionary money, which could bypass state and regional control."[84] At that forum, many emphasized taking advantage of local institutional assets available from a place like Campbellsville University. As Alma College faculty worked with St. Louis, Michigan, Campbellsville University gave the town more resources and expertise than a community of that size "should expect." When providing such help, the institutions demonstrate the wisdom of a decentralized higher education system in the United States.

The Ways and Means Committee members came to Campbellsville primarily because the region's congressman, Ron Lewis, a strong supporter of free trade, wanted to counter *The Factory* with a new story about what he called the "Campbellsville Comeback." Rather than focus on negatives, Lewis repeatedly told the tale of the community's recovery. His ally on the committee, Wally Herger from California, said, "You gave the example of this community—you gave the example of insourcing. . . . It's an incredible story, a secretive story. . . . We rarely hear about insourcing. . . . The results have been nothing short of phenomenal." Congressman Lewis said the events in Campbellsville revealed the problem of big employers: "The worst thing a community can do is have all its eggs in one basket."[85]

While the community's success in attracting new investment may have been a model, some critics emphasized the limits of the accomplishment. First, many of the new jobs paid less than the old jobs at Fruit. In constant dollars, the incomes in Campbellsville declined through the first five years of the new millennium.[86] As typical in the country, the community witnessed the construction of a number of luxury homes while the plant shutdown drove down the incomes and options for many others.[87] When Congressman Lewis arrived in Bowling Green in April 2004 for a "Forum on International Trade Policy" and discussed positively his favorite term, "insourcing," he received skeptical questions from the son of a factory worker. Kyle Gott asked how his father could replace his lost income when his job went overseas. Falling back on the neoliberal ideology, Lewis reminded Gott, "Some win and some lose; our economy is dynamic." The nation needed to respond with reduced business taxes, lower health-care costs, and tort reform.[88]

Generally, discussion of the "Campbellsville comeback" ignored several objections. First, behind the rhetoric of community coming together, the reality was a rather unequal struggle, pitting small towns against global structures favoring powerful economic institutions. Second, the discussion neglected to explore who won and lost in places like Campbellsville. In West

Point, Georgia, newspaper staff perceived that some managers with stock options gained windfall benefits when Farley bought their shares at inflated prices. However, most people lost their jobs or faced income declines. The Bowling Green paper gloated that "3,475 jobs relocated, but [the] U.S. export market has expanded."[89] Despite Congressman Lewis's claim that the pro-trade message was a secret, even the local press and elite endorsed without question the assumption that the transfer of jobs to the lowest-cost labor market was inevitable and good. As Western Kentucky University economics professor Bill Davis rejoiced, "Many of the goods that we sort of take for granted that are produced less expensively abroad means that we enjoy a higher standard of living. . . . The average American standard of living is secured in allowing our economy to flourish."[90] Kentucky senator Mitch McConnell's office distributed a "fact sheet" to those who asked about trade, which showed that Kentucky gained 15,000 jobs from NAFTA between 1993 and 1998. The U.S.-Mexico Chamber of Commerce claimed that Kentucky's exports to Mexico under NAFTA soared 122.4 percent in the period.

Unfortunately, the qualifications or questions about these positive assumptions and the numbers on which they were based were ignored. First, the figures of jobs lost under NAFTA published by the U.S. Department of Labor did not include all the lost positions, since it only counted the numbers of those who applied and were certified for retraining under the Transitional Adjustment Assistance. Janet Hoover at the Kentucky Workforce Development Department asserted more honestly, "There's no way for us to know the total number of people who have lost their jobs due to NAFTA. We don't track that number."[91] Neither did trade advocates have data showing net economic gains in those countries receiving Kentucky's old jobs. There were a number of studies showing millions of rural poor being displaced under NAFTA.[92] Ignoring neutrally generated data, the press gave advocates of neoliberal trade policy the last word.

Considering the ambivalent conclusions in the 2001 series on "NAFTA and Kentucky," the Bowling Green newspaper, nevertheless, did not shrink from editorializing:

> It is apparent from reporter Deborah Highland's recent three-part series that despite the noticeable losses in the apparel manufacturing industry, NAFTA has proved to be a big winner for Kentucky. . . . As individuals increase their earning power in our fellow NAFTA countries, there will be an increased demand for better technology and more convenience, meaning they presumably will buy products—primarily high-tech—that Americans have excelled in making. . . . With or without NAFTA, these changes in the economy were bound to happen.[93]

As with much of the discussion of trade in Kentucky, the nation, and the world at this time, the editorial made several key assumptions. First was the word "apparent." Little in the trade data and trade stories was either clear or authenticated. Second, the story assumed increased earnings in the "fellow NAFTA countries," which presumably meant Mexico as a surrogate for the developing world. Third was the inevitability. Trade-policy discussions contained seemingly contradictory combinations of fate and choice. Inevitable changes did come to places such as Campbellsville, but its "comeback" proved that little towns had choice. Finally, trade deals could spawn revolutions. The elevation of materialistic economic change to the level of "greatest revolution," regardless of whether it benefited the common good, confirmed, as did the references to economic inevitability, what one historian called the emergence in America of the "Wall Street Journal school of deterministic materialism."[94]

RATIONALIZING GLOBAL EXPLOITATION

Like Kentucky's uncorroborated trade data, the numbers claiming increased earnings in "fellow NAFTA countries" remained unconfirmed. In fact, the Bowling Green newspaper also erred in assuming that all city jobs migrating south went to "fellow NAFTA countries." By the late 1990s, Fruit of the Loom had cut back on Mexican employment, and certainly never considered moving employment to Canada to escape high wages.[95] The countries that acquired the Fruit jobs were the old banana republics in Central America. Inconsistent data from that region certainly raised questions about how many high-technology products from the United States would be purchased south of the border. Even if new Fruit of the Loom workers desired something "high-tech," it probably would not be produced in the United States.

Reports from Central America indicated fundamental problems for workers at Fruit facilities. Conditions appeared especially bad at many subcontracted plants used in the mid-1990s when the firm rushed all sewing offshore. The media ignored these wage issues, according to Charles Kernaghan. He organized the National Labor Committee to fight for workers' rights in the region, and brought employees from Central American garment factories to the United States to expose the situation. One of them, Claudia Molina, worked for Orion Apparel, a Korean-owned subcontractor, in Honduras. Orion produced shirts for Fruit of the Loom's Gitano subsidiary. Ms. Molina described receiving 38 cents an hour while working 15-hour shifts Monday through Friday, with 22.5 hours from Saturday morning until dawn, Sunday. Bob Herbert of the *New York Times*, in a piece called "In America: Children of the Dark Ages," quoted Kernaghan: "The companies make no secret of their preference for young females. A common explanation is that girls at about the age of 16 are at their peak of hand and eye coordination, perfect for the factories. A more persuasive explanation is that young girls are the most docile of all workers, less likely to object to abuse or to fight for any rights."[96]

Kernaghan arranged a national speaking tour for Molina and Judith Viera, another sweatshop worker, which shocked audiences at liberal churches in upscale communities, such as Columbia, Maryland. One executive who heard them responded, "It makes me embarrassed to do what I do." Viera, who worked for a contractor who produced goods for Eddie Bauer and other brands, explained that her family in Columbia had been threatened because of her trip. A State Department report to Congress at the time confirmed "corruption in the Honduran judiciary and Ministry of Labor, harassment of union workers and 'credible evidence' of blacklisting of union employees in Honduras. The report said the allegations of forced overtime in Honduras are credible and that the government does not enforce child labor laws."[97]

After meeting Kernaghan, Molina, and Viera in Canada, a thirteen-year-old suburban Toronto crusader, Craig Kielburger, came to Washington to speak to the Democratic Party's policy committee about trade and child labor. He also tried to form an affiliate of his anti-child-labor group Free the Children. He worked fruitlessly to overcome the unvarying U.S. references to economic laws and other concepts based on "economic determinism" that excused sweatshop labor of children. Kielburger reminded the assembled "liberals" that "The United States is one of the superpowers of the world; the United States also can put pressure on countries to take action."[98] Charles Kernaghan joined Kielburger at the hearing and singled out Fruit as one of the violators of basic labor standards.

This mid-1990s campaign resulted in pressuring firms, not the U.S. government, to adopt "codes of conduct."[99] The challenge, however, remained because many producers were "independent contractors who ignore such codes of conduct."[100] Highlighting anew the contradictory combinations of fate and choice, although The Gap adopted a code of conduct, it did not influence the actions of its subcontractor. One of the most notorious firms, Mandarin, did not produce primarily for The Gap but for Fruit of the Loom. It was easy for companies such as The Gap that had few or no employees in a country to pledge to observe its labor laws. It was necessary to force companies to reform practices throughout their apparel supply chain.

To avoid continued criticisms, Secretary of Labor Robert Reich persuaded

President Clinton to form a task force called the Apparel Industry Partnership (AIP).[101] In 1997, the AIP established a Workplace Code of Conduct and Principles of Monitoring. After the withdrawal of nonbusiness organizations, the AIP developed the Fair Labor Association (FLA) to monitor the standards created. However, this process had several inherent weaknesses, especially an impotent enforcement mechanism. Once the codes were established, concerned consumers may have assumed wishfully that they had fixed everything.[102] The industry certainly supported that view.

In the summer of 1997, the Asociación Hondureña de Maquiladores (AHM), which referred to itself as "a nonpolitical trade association from the Honduran private sector," sponsored a Congress of the Apparel Manufacturing Industry. The event received much coverage in the business press, which excused labor problems by noting, "Because the *maquila* industry is young yet massive enough to employ almost 87,000 workers, there are inevitably growing pains in the industry's labor relations. . . . Acting on its own initiative, the AHM chose to organize the congress so that the code of conduct for the Honduran apparel industry would not be mandated from above, but rather achieved through a consensus among the interested parties."[103]

Those who developed the code included leaders of the apparel companies, industrial parks, free-trade zones, the media, the Honduran government, but not organized labor. After explaining the code efforts to the media, the AHM took the reporters on a tour of factories. "Contrary to all the reports in the U.S. media, these were not Dickensian sweatshops, but well-equipped, modern facilities operating under all of the relevant laws governing employment practices."[104] The plants the reporters toured were clean and bright, and included support facilities such as cafeterias. By contrast, when a group of Duke University anti-sweatshop students visited Honduras, they reported that they could learn little about conditions and whether they should try to strengthen a code of conduct imposed by their school on the apparel factories. "A potent combination of secrecy, deferred responsibility and militarism provides a protective barricade around much of the global garment industry."[105] More than a year later, a sociologist from the Women's Studies Research Center at Brandeis University reported that Fruit's subcontractor had "women working in very inhumane conditions . . . experiencing severe violations of dignity, of self-esteem, and of human rights."[106]

In fact, the AIP Workplace Code and Monitoring Principles had not succeeded. The reasons for disappointment entailed both the principles underlying the code and the procedural or enforcement weaknesses implicit in its voluntarism. First, the code neglected to limit hours of work, enforce a living wage, or provide an alternative to work for child laborers and their families. Second, the process of monitoring compliance lacked accountability and transparency. As one critic noted, "Even skeptics of the FLA [Fair Labor Association] might be willing to overlook the loopholes contained in the Charter's language and trust in the good faith pledge undertaken by the participating companies and their monitors were it not for the secrecy and confidentiality requirements appearing on virtually every page of the Charter [of the FLA]."[107] In addition to the secrecy, exceptional barriers prevented bringing a complaint of Workplace Code violations to the FLA. Workers needed to support their complaints with reliable, specific, and verifiable evidence. Complicating the collection of information, firms did not have to reveal the locations of their factories, and whistle blowers could be fired before receiving any redress.[108]

Although it was possible to modify the code and principles, it was only likely if they had been developed in good faith. Given the reasons the companies had transferred production to Central America—to reduce wages, avoid unions, and generally cut costs through labor exploitation—it was folly to assume that self-imposed standards would help workers. The code was approved primarily to counter the bad publicity about child labor and other extreme forms of abuse. The reforms enacted resulted from the inspiration of articulate children such as Craig Kielburger or college and university students. Only these rare young people recognized the exploitation implicit in their wearing "sweated sweatshirts" produced by youth—often younger than themselves, with none of their opportunities.

The possibility of moving beyond child labor to the equally inhumane denial of basic rights to women and men in assembly plants required a prophet more powerful than Charles Kernaghan. Few adults in the United States bothered to probe beyond their consumer rights to think of labor abuses. Some students, not yet enculturated to the compromises of adult consumerism, investigated—such as one Duke University student, Tico Almeida, who observed, "Students have a unique connection to this issue because many sweatshop workers are young. You have two groups of people, roughly the same age, who are getting such different experiences out of the same institutions."[109] Fruit of the Loom, U.S. trade negotiators, and their partners gambled that the vast majority of people would not notice the human injustice suffered to produce the material goods they consumed. Worse, when they did notice, adults rationalized the exploitation as part of globalization. The students were not so gullible. In 2009, after one of Fruit's subsidiaries closed after a successful unionization drive, United Students Against Sweatshops launched a massive nationwide campaign against Fruit and forced it to reopen the plant.[110]

SO WHAT?

One familiar rationalization predicted that everything would get better in a few generations. Central America was simply in the position of Britain in 1750, New England in 1820, or Japan at the start of the twentieth century. In those countries, which eventually developed into First World nations, years of low-wage textile production, with few or no worker rights, built the wealth that shaped an advanced economy. Under this logic, the United States should choose to remove all barriers to the manufacture of apparel.[111] This was Central America's opportunity. One study found that workers in the maquiladoras of Honduras received somewhat higher incomes than a comparison group not hired into the factories. In addition, the report noted that *maquila* workers had better household relations and an enhanced sense of political power. However, the same workers appeared worse off than the comparison group in health, membership in labor unions, and social life, which was dubbed a necessary tradeoff.[112] Those with some history of the textile industry in the American South also understood that firms moved there in flight from high wages and labor-standard protections implemented in the North.[113] Common to all these interpretations was the inevitability of suffering in Central American communities in the 1990s as entrepreneurs arrived bringing creative destruction.

A slightly different emphasis lay at the heart of various critiques of conditions in the region's garment factories. A report on El Salvador's working conditions in 2003 titled *Deliberate Indifference* captured this attitude. In that investigation, an attorney with Human Rights Watch, Carol Pier, discovered that nothing had changed since the U.S. Agency for International Development launched a four-year program in 1997 to protect core labor standards.

> Labor laws are weak and government enforcement is often begrudging or nonexistent . . . employers who flout the law have little worry that they will suffer significant consequences. Aggrieved workers, confronting intransigent employers, an unresponsive Ministry of Labor . . . , and slow and cumbersome labor court procedures, often settle for minimal, one-time payments. Out of necessity, they exchange their human rights for a meager sum to help temporarily support themselves and their families.[114]

A report from the Congressional Research Service identified a different reason at the heart of the textile shift to Central America.

> In the case of Central America, lower wages are part of the strategic alliance that U.S. businesses have developed in the region, specifically to meet competition

based in part on similar low-wage production in other parts of the world. If the tradeoff is between losing jobs to Central America that support other employment in the United States because the firms use U.S. materials, and losing jobs to Asia that use only Asian inputs, then the labor argument [against moving to Central America] becomes more complicated, but the depressing effects on employment and wages of lower-skilled workers presents something of an adjustment issue for the United States.[115]

While some of those who offered these explanations hoped and expected to reverse recent trends, an American textile executive could endorse their words with no fear that corporate behavior would change. Without thought, industrial leaders and their allies in government distorted the history of previous textile industrialization, which was why Carol Pier labeled the process "deliberate indifference." The progress gained by factory workers of the developed world included much conscious conflict and policy advancement that allowed laborers to extract a greater share of the wealth from their employers and shift it into other enterprises or the acquisition of education and skills. The fight over the National Recovery Administration codes and the Fair Labor Standards Act during the Great Depression were guided by a successful effort to force the South to comply with many of the labor standards of the North.[116] While massive resistance to these labor standards continued in the South through the remainder of the century, its long-term weakening was reflected by the fact that organized labor eventually succeeded at most of the companies that were part of Fruit of the Loom, such as Acme Boot in Tennessee and Universal Manufacturing in Mississippi. Even West Point Stevens had some unionized plants.[117]

A core problem in the modern exploitation in Central America was that it resembled economic colonialism more than economic development. By contrast, in earlier U.S. or Japanese industrialization, the exploiters and exploited tended to share the same nationality. Eventually, the exploited used the common political and social system to reduce their manipulation and participate more fully in the common economy. Jacob Goldfarb was a patriotic World War I American veteran. Even if he preferred life in New York to Bowling Green, Kentucky, he visited regularly and approved of the gains enjoyed by the employees of his firm. Jessel Cohn actually moved to the town of his adopted headquarters. Bill Farley had no interest in relocating to San Pedro Sula, Honduras, where Orion Apparel produced his Gitano shirts. How could he care about its inhabitants as Cohn had worried about Clarksville and its citizens?[118]

Investment in Central American maquiladoras at the expense of mills in the South completely differed from earlier industrialization. Not only was there no interest in Honduran prosperity on the part of American investors and their Korean subcontractors, such as Orion, but also there was no meaningful concern with the fate of workers and communities back home. As Robert Reich so succinctly phrased the contemporary problem, "As borders become ever more meaningless in economic terms, those citizens best positioned to thrive in the world market are tempted to slip the bonds of national allegiance, and by so doing disengage themselves from their less favored fellows."[119] The new globalization of the late twentieth century, such as Farley's work on behalf of the African and Caribbean Basin Initiatives, usually was portrayed as a development program. Of course it was nothing of the sort. It represented a policy to reduce production under conditions of fair employment practices and locate in places where labor could be mercilessly exploited.[120] The elite supporters of free trade possessed the same detachment from their fellow human beings as the business owners. For example, after Fruit of the Loom's cutbacks in Louisiana, the director of business research at Louisiana Tech University considered the three thousand lost apparel jobs "not a significant hit to the state economy."[121]

Those who led the shift of production south were not concerned if this growth corresponded to what the UN's Michael Mortimore labeled the "Stylized History of the Economic Growth of Nations." They asserted that the host countries would profit from industrialization by building schools and other infrastructure, opening the door for growth into more technologically advanced production tied into the local economy. Yet, the export-processing zones neither produced sufficient tax revenue to build the schools and other infrastructure nor developed any supplier linkages to the local economy. The free traders perpetrated the cruelest fraud in telling Central American labor that the solution to its problems was "Education: The People's Asset."[122] However, the municipalities in which the workers lived were unable to build the schools and hire the teachers needed to overcome their problems. Secondly, the suppression of labor rights in the maquiladoras made it unlikely that the workers could improve their own lot. There also seems to be a disturbing linkage of employment in maquiladoras and participation in fundamentalist Christian religious activities. In an overwhelmingly Catholic country, such as Honduras, maquiladora employees were more likely to be involved in evangelical religious meetings that, as in the U.S. South, tended to preach a Christian heresy hostile to labor rights.[123]

Generally, rather than further economic opportunity, the growth of apparel production in maquiladoras may imperil economic development in Central America. As Mortimore of the UN Economic Commission for Latin America and the Caribbean framed the problem:

> The Caribbean Basin assembly model of exporting apparel to the U.S. market simply does not meet the requirements of the stylized view of the growth of countries. . . . It is evident that apparel assembly in the Caribbean Basin resulted in an impressive explosion of exports. However, given the characteristics of this particular manner of exporting, this phenomena did not represent an intensification of the national industrialization process. Instead, it truncates it. The exports are not the extension of the national apparel industry into the international market; they simply represent the localization of the assembly function itself.[124]

Of course this should not have been surprising, as earlier industrial revolutions driven by exploitation from afar returned few benefits to the subject peoples. For example, the Irish and Indians at different times under British rule did not experience the rewards of industrialization that English factory workers received.[125] Professional detachment permitted a few to possess enormous power over global production while the least skilled were left in a state of utter helplessness. The managers of firms such as Fruit epitomized what Burton Bledstein described as the core element of modern professional culture. "The professional excavated nature for its principles . . . allowing him both to perceive and predict those inconspicuous or unseen variables which determined an entire system of developments."[126] The leaders of companies such as Fruit of the Loom had influence over the national policy process, participated in international trade forums, and knew the leaders of global financial institutions and retailers who shaped investor and consumer wishes. Yet, when they moved a business from Kentucky to Honduras or from Tennessee to Mexico, they suddenly became victims to economic laws over which they had no control.

ONE MORE RELOCATION

As the transfer of Fruit of the Loom production from Kentucky and Louisiana to Honduras and El Salvador demonstrated what Reich called slipping "the bonds of national allegiance," Farley's next action further severed his bonds to the United States. In the spring of 1998, he asked shareholders to approve the move of Fruit of the Loom's headquarters from the United

States to the Cayman Islands. With the company running a large deficit, he proposed to save revenue by reducing taxes paid to his homeland.[127] The relocation also enabled Farley to sail to his Cayman post-office-box head-quarters in his yacht *L'acquisition.*[128]

Of course this was merely a tax ploy, as Farley's office remained in the Sears Tower in Chicago while the headquarters of Union Underwear did not leave Bowling Green. While the move reduced the taxes paid by Fruit to the U.S. government, it saved even more for Farley as he switched his old shares for the new. As part of the transfer process, he altered his "class B" shares in the old company into new preferred shares, which paid a dividend, only to him, of about $5 million. In order to get the deal approved in his favor, he first needed to buy out Leon Black, his old financier from Drexel. Black possessed about $25 million in stock in Farley Inc. Where did he get the money? He obtained a $26 million loan guaranteed by Fruit. Of course, Farley's hand-picked board approved this arrangement. One director explained, "If the board didn't consider it appropriate, we wouldn't have done it." Writing in the detached, powerless persona that protected a leader from criticism, Farley said, "The board approached me and suggested that it might be appropriate to buy out Leon."[129]

While the conduct of the board was predictable, the actions both of the stock market and of economists were baffling. On news of the Cayman plan and Farley's recent huge bonuses, the stock rose from $24 per share to $36⅝ in May. What were investors thinking, investing in a firm that lost $287 million in 1997 with a 12.6 percent decline in sales? Economists at Western Kentucky University, who should have been most attuned to Farley's manipulations since his company victimized their region, provided a free-market critique of the United States for bringing on this development. One said, "The U.S. tax structure is too high and, if other companies do as Fruit of the Loom plans to, there may be more of an impetus to re-evaluate that structure." A colleague added uncritically, "There are obvious tax advantages to the corporation or to Mr. Farley." A local stockbroker emphasized the near record value of the shares and offered, "We have an emotional attachment to the company because they have been a good employer."[130] He did not connect his comments to the fact that the company had laid off thirty-five more people at the Bowling Green offices the day before, and several weeks earlier had shut down the Campbellsville plant.[131]

Business Week questioned somewhat the financial community's response to Farley's initiative: "That shareholders will . . . follow Farley abroad surprises some. 'If you saw *Men in Black*, they had that little device that flashed a light and the person's memory was erased,' comments a consultant for a rival apparel maker. 'It's like Farley has the same effect on Wall Street.'"[132] One of Farley's compatriots from a New York firm transferring its head-quarters to Bermuda spoke for all the executives: "The benefit of avoiding U.S. corporate tax . . .'is at the heart of what we are paid to do—increase shareholder value.'"[133] He did not add that the advantages of avoided taxes went primarily into executive compensation and not to the shareholders.

THE DÉNOUEMENT

The strategy of shifting jobs to Central America or the headquarters to the Caymans did not rescue Fruit of the Loom for long. By mid-1999, the com-pany was in free fall. The foolish expansions became burdens, not assets. The movement of manufacturing to developing countries undermined product quality and created distribution nightmares for modern "just in time" retail warehousing. Despite the payment of millions to rename the Miami base-ball stadium for "Pro-Player," the division lost $31 million. Farley sold the naming rights at a loss in 2000.

News of Farley's remaining empire continued to worsen. In the sum-mer of 1998, the New England Health Care Employees Pension Fund filed a federal lawsuit accusing Fruit of the Loom of deceiving investors. They alleged that Farley had expressed confidence in the company in 1996 and 1997 to inflate the stock price while selling his holdings.[134] But, even with this report and that of the self-serving consequences of the Caymans move, prices for Fruit stock soared in late 1998 on rumors that Farley would sell.[135] The following spring, he still owned Fruit and promoted the dying brand. At the end of the annual meeting, the president of the retail divi-sion said T-shirt sales were "extremely hot. It's been difficult to supply the demand." He neglected to inject that the same situation had marked Fruit's ex-subsidiary Doehler-Jarvis the year before. Always the promoter, Farley declared the "active wear area" was one where Fruit could "move from a very powerful player to the dominant company."[136]

Given this apparently good account in the midst of calamity, the press began to use a *Titanic* metaphor when discussing the company. In Sep-tember, the board of directors announced losses for the first three-quarters of the year and then removed Farley as chairman of the board and CEO. Since Farley controlled the company, his dismissal indicated a crisis rather than a careful turnaround. The shares tumbled to seven dollars. One retail consultant described Farley's fate: "If you're the captain of the ship, and you run into icebergs, that's what happens."[137] A shareholder skipped the *Titanic* metaphor, saying, "He ran a great business into the ground."[138]

By October, matters were growing out of control. The new CEO, Dennis Bookshester, a retail executive and director of Playboy Enterprises, had no apparel manufacturing experience.[139] He begged John Holland, who had been president and CEO of Union Underwear for years, to return as a consultant.[140] As deliveries continued to falter and losses increased, the stock fell to $3 per share in early November. By Christmas, the board was so desperate that it hired Holland as executive vice president.[141] Nevertheless, on December 20 the stock plunged to 1⅜. Holland vainly hoped, "We want to try to restore this thing to profitability."[142] After destroying the company, the board conceded that they had to stop mining its assets for its survival. Only then did it pass leadership back to people with technical knowledge, who were "long-time Warren County residents and are active in civic and church organizations."[143] Here was powerful proof that "community-based capitalism," as one author has called it, was the sustainable form.[144]

Holland faced an impossible situation. Like the other leading components of Northwest Industries, Fruit had been stripped of every possible resource. The coup de grâce came on December 28 when the California cotton-growers cooperative filed suit for $4.3 million in unpaid bills. The next day, the company filed for bankruptcy protection.[145] During this upheaval, Farley sat quietly on massive severance pay and a $65 million loan guaranteed by the board of directors. Meanwhile, in Osceola, Arkansas, workers at one of the last Fruit factories learned the local banks would not cash their paychecks on New Year's Eve.[146]

After these events, Susan Harte of the *Atlanta Journal and Constitution*, who had trailed Farley since he destroyed West Point-Pepperell, finally succeeded in getting her exposé into print. On the day of the bankruptcy, she phoned Fruit's Cayman Island headquarters, but no one at the bogus facility answered the call. She only received a recorded message to call El Salvador. She described Farley's ostentatious and self-serving behaviors, such as his boast that from his Kennebunkport mansion, "I can see right into George Bush's beach house."[147] More importantly, she exposed and explained how he tricked unsuspecting investors. Unfortunately, her story was too early to depict the silliness of professional baseball's efforts in 2000 to remove Pro-Player from the Miami Marlins' stadium. Baseball leaders wanted no one to remember business foolishness on the way to a ballgame.[148] Meanwhile the Fruit Board appointed Bryan Wolfson as CEO, another ex-director of Playboy and an English friend of Farley's who liked to be called "Sir." Farley's old friends at Credit Suisse also persisted in working out deals in early 2000 to extract themselves from his final $65 million loan.[149] Of course all these former deal makers wanted to distance themselves as far from the company and Farley as possible. By April, nine federal suits had been submitted against the former officers.[150] In the summer, the Pine River Task Force filed its $100 million claim in bankruptcy court.[151]

Eventually, with John Holland in charge, and purchased by Warren Buffett, the underwear portion of Fruit emerged from bankruptcy in 2002. The Bowling Green newspaper headline said it all: "Fruit of the Loom Revival is Nothing Short of Amazing."[152] However, that good news was not shared in many of the former company communities. In order to get out of debt, the firm had continued to shut the last mills in Arkansas, Alabama, Texas, and Kentucky.[153] In each case there were poignant stories, often of generations from the same families working for the company. As Smiley Sams, a forty-three-year-old employee at the Fayette Mill, said, "My mama worked here. I got nine brothers and sisters, and all but one works here. I got on Sept. 20, 1974. I don't know nothing else. I've never even filled out an application. This is all I've ever done."[154] Still unable to detach the town from low-wage and low-skill work bought by public subsidies, the mayor of St. Martinsville, Louisiana, went to China in the spring of 2004 in hopes of recruiting a company to replace Fruit of the Loom. He recalled that in 1969 his predecessor had succeeded in recruiting Fruit with a visit to Bowling Green. Robert Karp of the U.S.-China Business Council said, while he did not know of any companies planning to expand in the United States, "it was only a matter of time before Chinese investors would be spreading out across the country."[155] Such were the fruits of globalization.

The southern towns where Fruit of the Loom had operated failed to respond effectively to their loss. Their inability teaches lessons about the proper civic role in confronting excessive private pursuit of self-interest. The St. Martinsville hope in China probably confirmed what Howard L. Preston said about southern efforts to recruit businesses and relieve regional poverty:

Despite the South's new-found national dominance in manufacturing, the economic well-being that contemporary southern recruiters hoped would accompany industrialization, not unlike the experiences of their late-nineteenth and early-twentieth-century predecessors has been thus far only a pipe-dream. Recent industrialization has not re-made the South into the vision of independence and progressiveness that prophets of the New South, past and present, said it would. If trends continue, this new industrial livelihood promises to make the region even more dependent, not only on the North for economic development, but also on foreign countries. Perhaps too much has been given away in terms of low wages, guarantees against unionization, and tax exemptions to achieve . . . the prosperity found elsewhere in the country.[156]

It is the conclusion of this book that policies to achieve the long-term public interest require commitments to community well-being and respect for their civic capacity that few outside interests bring. This study maintains that contemporary liberal and neoliberal development ideologies are fatally flawed from a community perspective. Their emphasis on welcoming economic actors demanding unregulated tolerance of the pursuit of self-interest inherently demeans the civic and individual citizen capacity essential to protect the common good. One study of textile workers in the South by Jeffrey Leiter found that the Albert Hirschman concept of "exit" applied to many of the negative encounters of citizens and workers with business. Except for the rare organizational efforts in Campbellsville, most people decided to "exit" a problematic situation rather than "voice" their grievances and seek change.[157] This approach to economic exploitation found endorsement in the laissez-faire ideology. Southerners did not easily adopt behaviors from the alternative ideology that celebrated democratic civic collaboration and citizen empowerment to force development in the long-term public interest. We have to turn to the struggles of the people in Michigan with Velsicol for an example of effective collaboration.

CHAPTER 7

Experts and Local Knowledge

Fruit of the Loom and its subsidiaries, by transferring work overseas, instigated the theft of jobs at home and the loss of technological expertise to benefit a few incompetent managers. In contrast, Velsicol symbolized the looting of natural resources in the name of technological innovation. Consequently, while the former firms received a pass for many years from the business press and the finance community, Velsicol was overlooked for criticism by fellow technical experts, including those in regulatory agencies. If its exploitation of the community had ceased in 1986, the common link would have been ownership by Bill Farley and his predecessors. However, since pillage persisted after 1986, the year Farley arranged to sell the core of Velsicol to its management, other explanations for its abuse must be found. Like Doehler, Velsicol manufactured products that were demanded at home and overseas. It would be the victim of neither adverse trade policy nor loss of markets.

Velsicol's story in the years after 1987 exhibited three differences from Farley's other firms. First, its new managers possessed some characteristics that resembled Herman Doehler or Jessel Cohn more than Bill Farley or Vince Naimoli. Those who purchased the company in 1986 included technically trained people interested in operating a successful business. They did not underfund the pensions or fully ignore responsibility to host communities. Second, despite this improvement, their problems with communities arose from an inability or unwillingness as beneficiaries of the post–World War II chemical revolution to address the concerns of citizens. Unlike the cockiness of leaders like Farley, who felt the manipulators of finance earned a right to exploit the public, the owners of Velsicol had an arrogance and disdain rooted in real expertise. They knew more about the substances they produced than anyone else and could not acknowledge that the lay public had an equal right to interfere.

However, the third and most significant part of the modern Velsicol story is its conclusion. Despite management's efforts to prosper, its arrogance caused community unrest and opposition so intense that it threatened the dominance usually assumed by global experts. Surprisingly, the ferment originated primarily from a small corner of Velsicol's world, one that it thought it had put behind it—St. Louis, Michigan. A community that the company had abandoned rose up at the turn of the century to demand some justice from those who had exploited it. St. Louis created an important legacy for places that insist that those who harm a community assume responsibility to past and future generations. St. Louis also engaged with other communities to stimulate companion demands on Velsicol.

Despite some impressive efforts, the failure of Velsicol's management to shift the company from the practices of the past teaches much about modern leadership. Technical arrogance can be as destructive as the financial swagger exercised by Farley. In fact, technical arrogance can be more offensive to the general public. Farley and Naimoli tended to hide their real intent from people and host communities. The technical experts at Velsicol were unable to hide their superiority. Extensive literature exists on the interaction of technical experts and citizens.[1] The importance of the St. Louis story that unfolded after 1990 is not that local residents felt threatened by corporate and government experts. That is a given, especially in small rural communities. St. Louis is significant in that instead of adhering to what James Scott has called a hidden transcript, the citizens of that town decided to take control of the public transcript.[2] No doubt they learned from the total disregard for the community in the settlements with Velsicol in the 1970s and 1980s, physically symbolized by the tombstone on the plant site. This insensitivity to the people implanted in them what community-organizing literature labels "cold anger."[3]

There are innumerable examples of community opposition in St. Louis. However, this study will cite three: the community's aggressive legal response to its problems, its growing resistance to the public-health establishment's indifference to their requests for help, and its willingness to confront international lobbyists for the pesticide industry. By the end of the first decade of the new millennium, people in the community recognized that the environmental phrase "Think globally, act locally" needed to be reversed.[4] They acted to prevent global and national experts, in league with special interests, from making local success impossible.

THE NEW VELSICOL

At the end of the 1980s, no one in St. Louis anticipated that their interaction with Velsicol would resume. The company had abandoned the community, leaving only its contaminated sites. Residents did not know that Velsicol had transferred ownership of the Superfund properties to its former parent, Fruit of the Loom. In fact, no one realized there had been a link to Fruit of the Loom. Former subsidiary managers charted the political and environmental engagement of the company with the wider world while working to achieve profitability for a much-scaled-down corporation. After stabilizing operations and sales in the first three years, the new management enjoyed a 12 percent growth in profits during the first half of the 1990s. The company regularly acquired new lines of production.[5]

In many ways, after separating from Fruit the firm seemed to become a model corporate citizen. The "new Velsicol" boasted of both its respect for employees and its investment in research and development, two of the strengths often found in firms managed by production experts rather than financiers.[6] Velsicol CEO Arthur Sigel even embraced speaking about cleaner production, pollution prevention, and management improvement. Under Sigel, Velsicol targeted a 35 percent cut in hazardous wastes, 30 percent reduction in all wastes, and 30 percent less water use.[7] Increasingly during the 1990s, Sigel assumed a highly visible role with the Chemical Manufacturers Association's Responsible Care Initiative.[8] Recognizing that local communities had varied expectations of Responsible Care, Sigel advocated forming community advisory panels as a way to improve the industry's image.[9]

Balancing this positive account was Velsicol's fierce defense of two of its most toxic and profitable products: chlordane and heptachlor. In the winter of 1986–87, Velsicol stonewalled all EPA efforts to police the two chemicals, and through lobbying forced the agency to identify them as only of "regulatory concern."[10] The company utilized its lobbyist, Charles Frommer, a former New York environmental official, to orchestrate critiques of studies from hostile parties.[11] The company finally agreed to stop sales on November 30, but only if use on the perimeters of homes could continue until April 30, 1988.[12] Donna Jennings, the firm's public relations head, absurdly dismissed the large body of health research: "There is no scientific evidence that there is human health hazard [from chlordane]."[13] Even after a U.S. ban, the company won support from neoliberals in the Republican Party, including former exterminator Tom DeLay, to allow it to continue overseas sales. One of the company's major critics in the era was Samuel Epstein, professor of environmental medicine at the University of Illinois Medical Center.[14] Jennings lashed out: "[Epstein has] never done a piece of original research [on the chemicals]."[15]

The battles over chlordane were not the end of the "new Velsicol's" defiance of critics. One of the firm's major products after 2000 was hexachlorocyclopentadiene (HCCPD), a chemical used in the production of endosulfan, a potent insecticide shown to have serious environmental and human-health effects. As the company made transitions in ownership after 2000, it restructured its staff, with one partner, Chuck Hanson, moving to become executive director of a seemingly benign nonprofit, the International Stewardship Centre (ISC). According to its website, "[The ISC] was created for the purpose of educating, promoting and encouraging safety in the manufacture, transport, distribution, storage, use and disposal of chemical substances."[16] Despite such positive claims, and the ISC's seeming independence from Velsicol, its primary function seemed to be opposing global regulation of endosulfan and Velsicol's loss of a market for HCCPD. Hanson and Dennis Leu, Velsicol's new president, were joined by other executives of Velsicol and the endosulfan industry in resisting bans on the insecticide. They made repeated lobbying trips to oppose addition of endosulfan to the UN's Stockholm or Rotterdam pesticide conventions.[17] While more than sixty countries banned the insecticide, the ISC and Velsicol executives worked with the leading Indian manufacturer, Excel Corporation, to prevent either a universal global ban or restrictions in the subcontinent. As with the firm's record a half century earlier, Hanson and the Velsicol leaders dismissed the evidence of health effects. Yet, even more so than DDT or PBB, endosulfan seemed clearly related to endocrine disruption, especially high rates of male reproductive-system birth defects in places of heavy use, such as parts of Kerala.[18] As with chlordane and heptachlor, appearance of endosulfan in the Arctic and Antarctic showed its persistence.[19] Despite Velsicol's repeated failure to concede the harm of its products, the firm's lobbying of experts, such as at the UN Environmental Program or the US EPA allowed for repeated delays in banning the products. Not surprisingly, communities with direct contact with the company often distrusted government experts as much as Velsicol. The regulatory agencies, when not siding with the company, repeatedly approved cheap remediations that failed.[20] Often the only recourse was suing, rather than expecting help from the compromised regulatory process.[21]

RENEWAL OF STRUGGLE IN ST. LOUIS

The expert-to-expert negotiations between Velsicol and the EPA led to delisting several of the firm's sites in the United States: North Hollywood in Memphis, and the Gratiot County Golf Course in St. Louis, Michigan. In St. Louis, the EPA and state environmental professionals worked closely with former colleagues at Velsicol to quickly transfer the firm's contaminated properties in town to Superfund, and then celebrated their remediation. The two biggest sites in town were among the earliest at which supposedly "construction was completed," the definitive measure of program success short of delisting. It would take years of local opposition to admit these cleanups had failed—in the case of the Gratiot Golf Course, twenty-seven years.[22]

By the late 1990s, the county landfill's contaminants were seeping into the aquifer used for residential wells. Likewise, the Pine River next to the old plant site required massive remedial action, while residents correctly predicted that the plant site itself was leaking contaminants into the city's aquifer. In 2010, the government again added the golf course to the National Priorities List, admitting, "Cleanup completed in 1984 has been determined to have not been implemented correctly."[23] Of course, focusing on the EPA's mistakes in the 1980s masked the real culprit. Velsicol was the cause of the problems that led to the flawed 1982 legal settlement and subsequent "cleanups."

The three Superfund sites in town were in different stages of the remediation process throughout the 1980s and early 1990s. In addition to the "delisted" golf-course site, the old plant site was surrounded by a fence, contained under a cap and within a slurry wall, and monitored by wells to measure the level of contaminated water under the cap. When water rose significantly above river levels, it had to be removed. The water removal fell under the routine "operation and maintenance" (O&M) necessary at many Superfund sites. Even though the Fruit of the Loom subsidiary, NWI Land Management, owned the site, Memphis Environmental, Velsicol's subsidiary, performed the work under the A&I Agreement with Fruit. Finally, a mile south of town at the former Gratiot County Landfill, owned by the county, the state performed O&M under the 1982 consent decree.

While problems developed at each of the sites and at the old radioactive waste dump near Breckenridge, Michigan, the trigger for reopening intense intervention in the community was the failure at the plant site. In 1993, Velsicol pumped 1.25 million gallons from the old plant site's containment system. The next year, the company had to remove another 1.28 million. When the EPA and Michigan Department of Environmental Quality (DEQ) learned of these withdrawals, they asked Velsicol to study what was going on.[24] The firm verified that about 10 million gallons of water that had entered the containment system was missing.[25] The Agency for Toxic Substances and Disease Registry (ATSDR) of the U.S. Public Health Service, a specialized federal agency established under Superfund to provide health assessments for the EPA, raised fundamental questions about the site: "The potential may still exist for humans to be exposed to contaminants via the ingestion of contaminated fish and wildlife. The potential also exists for human exposure to occur by direct contact with river sediments."[26] The assessment specifically noted that some people had previously ingested the area's contaminants by consuming wildlife, and that contaminants had been detected in soils "throughout the entire St. Louis area."[27] Two other studies presented proof of active and effective monitoring. In 1991, the EPA requested an assessment by the U.S. Corps of Engineers that reminded the agency that earlier studies had documented DDT residues in sediment near the old plant site of 3,570,109 parts per billion.[28] Two years later, the ATSDR found potential deep aquifer contamination to be "somewhat troubling." The ATSDR also observed, "Various fish species taken from the Pine River . . . in 1989 have shown that unacceptable levels of DDT (and its metabolites) and PBB are still present in the fish."[29]

Working in conjunction with the EPA, the Michigan Department of Environmental Quality and its predecessor agency collected fish samples in the river in 1983, 1985, 1989, 1994, and 1995. Review of that data disclosed a disturbing trend.[30] In common carp caught downriver from St. Louis, the mean DDT concentrations increased from 9.7 parts per million in 1985 to 22.2 in 1994. In samples collected by the old plant site in 1989 and 1995, concentrations rose from 5.7 parts per million to 9.0 in black crappie, and from 9.6 to 15.6 in carp. The DEQ concluded, "The results from this survey . . . [suggest] either the erosion of sediments and exposure of older contaminated sediments, or recent releases of DDT."[31] These conclusions, added to the earlier reports from the ATSDR and the Corps of Engineers, prompted the resumption of work at the St. Louis sites. While the reopening demonstrated an achievement of the current policy process, any celebration of success was tempered by observing the effects of the earlier failure. Since responses to the new data took more than twelve years, the failure in the 1980s to properly remediate St. Louis both exposed residents to contamination longer than necessary and allowed Velsicol to restructure itself to avoid linking liability with most of its former assets.

Of course, if the regulators had sought local knowledge from the community in the late 1970s and early 1980s, they would have learned that the river was the most contaminated area in town, and that dumping golf-course pollutants amid the rubble of the old plant was unwise. It was also foolish to pretend a cap could contain the mess. Two studies that grew out of Alma College student interviews with residents in 1992 verified that the public knew well before the regulatory agencies that the earlier cleanup was flawed. Four years before the EPA and DEQ admitted that the policies of the early 1980s were inadequate, the students summarized citizen opinions, saying, "[The policy response] demonstrated the inequity of policies established by outside experts. In the opinion of outside officials, the system worked. . . . Yet the Pine River in St. Louis received no attention. It remained heavily contaminated and area residents, especially children, had regular contact with it."[32] The accuracy of this citizen assessment demonstrated the value of local wisdom in the policy process.[33] It was common knowledge in St. Louis that the river represented the most likely path to impacting human health, yet the 1982 settlement with Velsicol exempted it from attention. To the credit of the policy process, especially at the EPA, the agency and Congress learned from such mistakes. In 1986 they modified the law in the Superfund Amendments and Reauthorization Act to fund the technical assistance grant (TAG) process that provided communities with access to modest amounts of funding to pay for second opinions on technical questions.[34] By executive order in the early 1990s, the EPA established a procedure for forming and recognizing independent community advisory groups (CAGs) at any EPA site. Developed to address the problems of environmental injustice, the CAG process became a method of assuring both justice and wise decision making.[35] St. Louis provided an excellent test of both reforms, since the community formed a CAG, and the CAG received a TAG.

As a result of the negative reports, the EPA formally returned to the community at a public meeting on October 29, 1997. The EPA did not anticipate the huge citizen response.[36] On the appointed evening, the hall had standing room only as many people came to express their anger at the failure of earlier agency actions. The St. Louis citizens focused their hostility on Velsicol and only criticized EPA and state regulators when they tried to be neutral in the conflict between the town and the company. After EPA project manager Beth Reiner explained that regulations require the agency to consider one option as "no action," people were outraged. Community members claimed that they did not need to await a new report on the contamination, saying, "We told you it was polluted in 1982, but you wouldn't listen." Although the majority of EPA officials were too young to have participated in earlier policy mistakes, they had to take the criticism meant for their predecessors.[37]

As the October meeting approached four hours in length, the EPA community involvement coordinator, Stuart Hill, offered to facilitate organization of a CAG.[38] Of course, local opinions varied. City councilman Bill Shrum said, "This is a problem for the city, and now we have a chance to do something about it." Fellow council member Nancy Rusch dissented, expressing widely shared concerns: "Is [it] worth . . . putting St. Louis on the map for being the most polluted place in the U.S.? . . . Workers in white suits and masks [would reflect negatively on the city]."[39]

CITIZEN EMPOWERMENT

By January 1998, the CAG formulated a constitution and written bylaws and elected officers.[40] Neither the EPA nor Velsicol, and certainly not Fruit of the Loom, suspected that a functioning CAG in a community that had been completely abandoned would present many challenges to their decisions and plans. By that time, with DDT levels in fish and sediments higher than any recorded in the United States, the EPA acted quickly to launch an emergency removal of 21,000 cubic yards of river sediments with over 3,000 parts per million (ppm) of DDT. The agency also announced they would conduct a feasibility study for disposal of the remainder of contaminated sediments. However, instead of quiet gratitude for these commitments, the CAG members asked questions. When the EPA proposed opening the downriver dam and returning the river to its old course while they removed the contaminants, community members challenged the wisdom of such a drawdown since it first would scour out many of the contaminants and wash them downriver.[41]

Unfortunately for the EPA, it was trapped between the community and Velsicol. As the agency searched for a staging area for its work, the only vacant space by the river was the abandoned Velsicol plant site. An EPA attorney and Beth Reiner met with Velsicol's vice president for the environment, Chuck Hanson, on April 7, 1998, to review possible use of that location. Hanson objected, but favored paying the company to use the grounds as a disposal site for the contaminated sediments.[42] At an April 8 community advisory group meeting, Beth Reiner briefed the Pine River Task Force on Velsicol's position. The task force unanimously rejected on-site disposal, believing Velsicol's proposal would result in the EPA paying what was a company responsibility—reopening and then recapping the old site.[43]

Throughout the spring, the community sought more information about the status of the plant site. The EPA advised residents that Velsicol still maintained it. One task force member initiated a big commotion over the likelihood of leakage from the site. When the agency revealed that it had met privately with Velsicol, the CAG demanded to be included in future meetings.[44] When the EPA finally offered to bring Velsicol officials to meet with the CAG, the members insisted the discussions be held with the executive committee.

The Velsicol meeting on July 16, 1998, represented a confrontation between the community and Chuck Hanson. Beginning the conversation with his folksy southern accent and acting like only a mid-level corporate official, not one of the five "principles," Hanson asked, "What do we have to do to make it right with the community?" He apparently thought the terrain had been prepared for this opening, since the company had made known to the city that it might build a small recreation center or provide another "gift" to the city. Expecting a grateful response to his question, he was shocked when the future CAG chair, Jane Keon, said, "Look. We're all busy people. If you want to make it right, take out your check book and write a check for $30 million." Communication between Hanson and the executive committee deteriorated from there. At one point, when Hanson asked, "Tell me when we have ever lied to the community," Jim Hall responded, "We don't have all day." Hanson did not appear in the community again.[45]

Other problems arose when the remedial project manager learned that an Indian tribe had a reservation in the region. At the October 1997 meeting, she had made an offhand remark that "It is too bad you don't have a tribe nearby. Then you could expect a thorough clean-up." While that assessment of tribal power might surprise some indigenous environmentalists, the CAG made a concerted effort to include the nearby Saginaw Chippewa Tribe in its membership. Beth Reiner seemed intrigued by inclusion of the Indians, at least to a point. The tribal environmental engineer, Bill Snowden, soon joined the CAG and attended several meetings. The CAG became especially interested in tribal involvement when it realized that the tribe possessed treaty rights to fish the river, which had been lost because of the company's pollutants.[46] In the summer of 1998, the different perspectives of Reiner and the CAG on Indian involvement surfaced. The tribe sent a letter to Velsicol on July 15, 1998, asserting its treaty right to fish in the Pine River, and protesting its exclusion from meetings between the company and the EPA.[47]

Reiner responded with ceremony rather than substance. She invited the tribe to hold a pipe ceremony on the riverbank to bless the river cleanup.[48]

Unfortunately, the tragic workings of tribal politics began to undermine Snowden's role. The Saginaw Chippewa tribal government, with revenues flowing into its coffers from its massive casino operation, faced intratribal disputes set on purging the ranks of factions that disagreed with the leadership.[49] After Snowden tried to use EPA regulations to force improved operation of a tribal water-treatment plant outside Mt. Pleasant, the tribe removed him from membership.[50] The EPA did not come to his defense, but supported those in tribal leadership. Eventually the tribe dismissed its Seattle law firm, abandoning its stewardship of ancestral resources.

In another effort to divert the community from effective opposition to agency plans, the EPA proposed funding a facilitation process to resolve what they saw as local divisions.[51] When the CAG decided this was an unnecessary waste of resources, the EPA remedial project manager responded that the facilitation would focus on the community's "movers and shakers." She was right in her assessment of the need to include local officials in organized dialogue. Local elected officials were so enamored of powerful federal and state officials and corporate representatives offering "deals" that they were unable to share the confrontational approach of the CAG. Only four of the fourteen people who became the core team for the facilitation meetings were not public officials, and none were CAG officers. When the last facilitation sessions were held on April 8–10, 1999, the CAG countered with its own event, a fund-raiser showing of the movie *Civil Action*, which attracted almost two hundred residents.[52]

The St. Louis CAG brought a distinctive group of participants into the policy process. First, several of the members were active labor-union members from the nearby Total Petroleum refinery. They added a no-nonsense attitude about corporate America to any discussion. Second, a number of CAG leaders were active church members with a strong conviction that there was a right and wrong way to steward the world. In negotiations, they were not prepared to alter their principles just to make a deal. Third, the CAG had the good fortune to be near a college that had long struggled to balance practical education with the liberal arts. It was committed to service learning and found working on local environmental issues an excellent way to apply specialized learning to a real-world interdisciplinary problem. Unlike university faculty, co-opted by grants from major corporations or government, the faculty from Alma College were interested in sharing with the community their knowledge of science and politics. Finally, a number of community retirees with remarkably relevant expertise gave of their time and talent. For four years, the CAG secretary was a retired science teacher from St. Louis. His replacements were, first, a college student, and then co-secretaries, a chemistry professor from the college, and a leader of local businesswomen. The CAG also attracted two exceptionally well-trained retired scientists from Dow Chemical: Fred Brown and Gene Kenaga. They became core members of the CAG's technical committee, chaired by college geologist Murray Borrello.

These two retirees supplied the CAG with expertise to match the best of the EPA and Velsicol, and succeeded in winning the technical arguments. Fred Brown had long been a leader in state environmental affairs, directing for a time the Michigan United Conservation Clubs. In retirement, he regularly reviewed and challenged the details of pollution discharge permits. Fortunately for the CAG, Brown had the willingness and time to read detailed EPA feasibility studies and related remediation documents.[53]

Gene Kenaga provided the CAG with a world-class proficiency in toxicology. A founder of the Society of Environmental Toxicology and Chemistry, he had specialized expertise in DDT. Having worked with this chemical since World War II, he recognized its beneficial uses, but also realized its usefulness was dependent upon limited application. He was appalled with the careless overuse encouraged by firms such as Michigan Chemical that spawned DDT-resistant pests and threatened the utility of DDT as a public-health tool. The scholarly yet humble Kenaga embarrassed imprudent young EPA and DEQ officials who, not knowing to whom they were speaking, made hasty and cursory arguments about what was known of DDT and its remediation.

The CAG, free of EPA control, demonstrated the value of a large and diverse group to second-guess experts. Its three main contributions included its role in seeking river remediation from an upstream refinery, its challenge to the Fruit of the Loom bankruptcy settlement, and its insistence on investigating the effectiveness of the plant site containment system. These second guesses prepared it for its biggest challenge, confronting global pesticide lobbyists. Merging independent judgment (unburdened with budget concerns), specialized skill, and invaluable local knowledge, the CAG repeatedly proved that a second opinion helps the decision-making process. The CAG's independence from regulators may have been its most valuable feature. The EPA's community involvement coordinator, Stuart Hill, wisely advised the Pine River group to become a private nonprofit agency capable of setting its own agenda.[54] One of the first decisions of the Pine River Task Force was to reject EPA advice to name the CAG for the Superfund site for which it was created. Rather, the group decided to address broader watershed remediation and place that goal in its name.[55]

The fight over the plant site containment system began when Fred Brown and Murray Borrello challenged EPA assumptions about the integrity of the site wall. At the February, 1999, CAG meeting, Brown initiated the dispute by questioning the regulators about the problem of increasing amounts of DDT in Pine River fish. He said:

> I don't know why and neither does the DEQ or the EPA know why there has been a sudden increase in the DDT levels of fish in the past few years. . . .
>
> They have said they don't know, yet are moving forward with cleaning up the sediment at a cost of millions of dollars without knowing for sure if that will solve the problem. . . .
>
> If I was spending millions of dollars I would want to know whether it is going to work, wouldn't you?[56]

Borrello, the Alma College geologist, reminded the agency that the river bottom was covered by decades of deposits from other polluters, including the nearby refinery: "It may need to be $60 million or $80 million, but if we are going to do it, why not do it right?" When criticized for asking questions at a critical stage, Borrello responded, "If this is not the right time to raise questions, then when is a good time?"[57]

THE LITIGIOUS TOWN

In the beginning of the site's remediation, the EPA and Velsicol acted as if the old chemical company continued to be the responsible party at the former plant. In preparation for one of the CAG's earliest meetings, a city official and CAG member checked tax records and found the name NWI Land Management attached to the property. Further investigation discovered the e-mail address of a company representative, John Hock, to be "fruit.com." Recognizing the former link between Velsicol and Fruit of the Loom through Northwest Industries, the CAG learned that Fruit of the Loom had retained ownership of the site. Just as that relationship was identified, the news of Fruit of the Loom's bankruptcy surfaced.[58]

Fruit of the Loom's annual reports disclosed that the company retained $33 million in cash and $67 million in insurance reserved for environmental liabilities.[59] Later investigation showed that both Fruit and Velsicol bought companion $100 million environmental policies from AIG insurance, apparently in an effort to reassure investors in Fruit stock. Aware that the cost of the river cleanup was estimated at $60 million, and that the plant site and other contaminated spots in town could add to these totals, the CAG executive committee filed a claim against Fruit in bankruptcy court for $100 million. By this time, the CAG had received a technical assistance grant that supported some of its research. However, the EPA regulations explicitly forbade using TAG money on legal matters. The CAG appealed to the EPA Office of General Council for permission to spend the TAG grant on legal research.[60] When the request was granted, the CAG secured low-cost but expert legal assistance from the University of Michigan Law School.

Over the winter of 2002, University of Michigan law students working for the CAG drafted a long plea for intervention from the Justice Department, the National Oceanic and Atmospheric Administration, the U.S. Fish and Wildlife Service, and the Nuclear Regulatory Commission to secure Fruit of the Loom assets to finance natural-resource damages at the Michigan sites.[61] A few months later, the CAG learned that the settlement secured as much of the Fruit of the Loom assets as possible and mandated the sale of Velsicol, with half the proceeds going to remediation.[62] Continuing to seek transparency as in the refinery agreement, the CAG next insisted on a public hearing on the settlement.[63] It also contacted community leaders from Memphis, which was mentioned in the settlement, along with people near other sites in Tennessee, Illinois, and New Jersey. While efforts to reach three of the communities were fruitless, members of the sister community group in Memphis, led by Rita Harris, agreed to come to the hearing in Michigan.[64]

The CAG and Memphis leaders asked and received two important changes in the consent judgment. One explicitly recognized the CAG's inclusion in the planning process related to the site. The CAG also sought an explanation that stated any remaining funding after remediation would be used to pay for natural-resource restoration.[65] Although Farley had depleted all his companies of most of their resources, the settlement nonetheless made explicit that nothing would survive if environmental responsibilities went unmet. Furthermore, the requirement to sell Velsicol and distribute its assets satisfied some CAG members' sense of justice. The company would not endure if it left an unremediated contamination.

Despite the large potential settlement with Fruit of the Loom and Velsicol in 2002, the cleanup in St. Louis exhausted any significant cost recovery from the firms. According to estimates at the end of the 2003 construction season in St. Louis, the EPA and DEQ had spent $51 million between 1999 and 2003 in removing 350,000 cubic yards of contaminated sediment from the south side of the river. In 2004, they launched a three-year remediation on the north side, with an estimated price tag of $47 million to extract 300,000 more cubic yards. While river remediation totaled just under $100 million, it did not entail spending at the plant site itself, the county landfill, the golf course, or the radioactive-waste site. That sum could easily exceed $200 million. Potentially larger costs also could be incurred in Hardeman County and Woodridge, with smaller amounts anticipated for the other sites in Illinois and Tennessee.

Meanwhile, Bill Farley continued to delay the flow of resources to the liquidation trust established to distribute assets under the settlements. He defaulted on a letter of credit to secure insurance policies for environmental remediation at his former properties. As late as May 2004, two years after the environmental settlement with Fruit of the Loom, such matters remained in litigation.[66] Yet, there was some good news about the company's environmental insurance. In the fall of 2003, the trustee for the former Northwest Industries properties recovered $750,000 from Fruit of the Loom's policies to reimburse the Nuclear Regulatory Commission for the initial removal of contaminated wastes near Breckenridge, Michigan. However, once the disposal began, the Nuclear Regulatory Commission learned that the company had underreported wastes buried at the site, and more money was needed. If the CAG had not persisted in calling attention to Breckenridge, EPA Superfund staff might have overlooked it.

Probably more important than the CAG's persistence with Breckenridge was its insistence on a comprehensive approach to all Superfund sites in town, not just the river, which the EPA seemed to be fixated on. Their efforts became easier after two investigations proved the migration of contaminants off-site. First, state and federal agencies discovered a syrup-like substance, called non-aqueous phase liquid (NAPL), leaking into the dried-out river bottom during sediment remediation in 2001. Further inquiry detected large amounts of highly contaminated NAPL and [dense] DNAPL migrating off-site, as well as almost one hundred feet below the surface. No one understood its impact, but the NAPL confirmed that the CAG was correct to fear that the plant site was continuing to contaminate river fish. In the fall of 2006, the state also found NAPL flowing from the delisted golf course. The scope and potential cost of any cleanup increased again in late 2005 when EPA found a byproduct of DDT production—para-Chlorobenzenesulfonic acid (p-CBSA)—in city drinking water. This discovery contradicted categorical statements in the early 1980s that St. Louis water contained no contaminants.[67] Worse, the EPA had not promptly informed the city of the problem, which was first detected in September 2004, reinforcing community distrust of federal officials.[68] After two years of wrangling, on April 11, 2007, the City of St. Louis filed suit in Michigan against the remnant of Velsicol, NWI Land Management, Fruit of the Loom, and the bankruptcy trustee for recovery of the costs of the new drinking-water system for the city.[69]

An indication of how the process can be stacked against average citizens and their communities occurred in December 2007, when the City of St. Louis faced a countersuit filed by the U.S. Department of Justice (DOJ) on behalf of the U.S. Environmental Protection Agency. The EPA considered the St. Louis lawsuit against Velsicol to be undermining the 2002 bankruptcy settlement with Fruit of the Loom. The Justice Department claimed that if St. Louis won funding for a new water supply from the underfunded Fruit of the Loom trust, New Jersey and Tennessee would not receive the proportion of funds promised in 2002. Meanwhile, the DOJ concluded a concurrent settlement with AIG, accepting $42.5 million, not the full $100 million that AIG was committed to pay under the Fruit policy. The CAG criticized the Department of Justice at a March 2008 public hearing. Instead of spending resources on fighting AIG and Fruit of the Loom, the Department of Justice joined the polluter in attacking St. Louis.

On a cold January day, St. Louis had its day in federal district court in Bay City. In a classic David vs. Goliath scenario, the St. Louis attorneys faced a team including U.S. Justice Department lawyers from Washington, an EPA attorney from Chicago, and attorneys for Velsicol and the Fruit of the Loom Trust. The scenario clearly did not portray the government as being on the side of the public. Ultimately, the local determination paid off. On March 18, 2011, Velsicol and AIG and their allies reached a settlement with St. Louis providing $26.5 million in additional funding from the Velsicol insurance policy. Calling the outcome a "monumental and historic" result, the St. Louis City Manager Bob McConkie joined leaders of the CAG and the city government at a victory party.[70]

The community had multiple reasons to celebrate. The settlement amount would fund much of the cost of a new water treatment system. Also, because the community's legal efforts had brought a significant amount of new funding to the remediation of Velsicol's wastes, the city had more bargaining power as EPA and the state finalized the decision on full regional remediation. Most importantly, the lawsuit promoted corporate responsibility. It successfully challenged the legitimacy of the sale by AIG of insurance designed simply to enhance stock price. In 1998, as Fruit and Velsicol tried to avoid a collapse of their stock price, they had purchased the twin $100 million policies for cents on the dollar, boasting that their environmental liabilities would not hurt their bottom lines.[71] As early as 2000, when the CAG had learned of the existence of the policies, it had sought to make AIG and Fruit's leadership pay some costs for their deceptive practices. Without success, the CAG repeatedly urged the EPA, state, and the bankruptcy trustee to seek the insurance funds.[72] It took the St. Louis lawsuit to force corporate responsibility.

By 2005, the sale of Velsicol under the 2002 settlement seemed to assure a corporate execution, since the new buyers resembled Farley.[73] Several of the firm's plants had closed in the preceding years as fuel costs soared and market prices for plasticizers and other products stagnated.[74] During the Fruit bankruptcy process, Department of Justice accountants fixed the value of the company at less than $3 million. Since the owner-managers had only put up half the price of the company in 1987, and Fruit received the other 49.999 percent, the trust created under the 2002 Fruit of the Loom settlement would realize $1.5 million at most from the sale, with the managers splitting the rest. Given that the rump version of Velsicol still had environmental liabilities at 159 sites, other than the seven biggest that had been transferred to NWI Land Management, the managers might expect significantly less than $1.5 million from the sale. Whatever assets management extracted from the settlement, the community had the satisfaction that the company would cease to exist independently, and millions in insurance premiums and some cash might support remediation.[75]

THE BATTLES WITH HEALTH EXPERTS

As the CAG fought to recover funds from bankrupt polluters and their failed insurance carriers, it also engaged in an effort to secure support for a comprehensive health assessment from the public-health arms of Superfund—the National Institute of Environmental Health Sciences (NIEHS) and the Agency for Toxic Substances and Disease Registry (ATSDR). The two health agencies each received annually more than $75 million from the Superfund budget, totaling more than 10 percent of all Superfund allocations. The CAG members questioned why the community obtained little health-services assistance from the two agencies. At an EPA meeting for communities receiving TAGs in October, 2000, the CAG learned that many other groups shared their frustration. As with the DOJ response to the St. Louis lawsuit, health officials often resisted community requests for support. Older residents of St. Louis remembered being part of numerous studies of the human health impact of exposures to PBB and DDT. In fact, Velsicol's earliest public-relations efforts to promote DDT arose from concerns that the lay public feared it. Yet, the most residents received in response to those studies were form letters advising them of the results of their blood work.

Few local residents read the rather voluminous literature in peer-reviewed journals that was generated from the earlier sampling and the continued follow-up of the "PBB cohort." These studies revealed that many with higher exposures to PBB had higher rates of breast cancer, digestive-system cancers, endocrine impacts, and non-Hodgkin's lymphoma. Likewise, related to the non-Hodgkin's lymphoma, there was the Breckenridge cancer cluster in the downwind town with the Velsicol's radioactive-waste burial dump. At the October 1997 public meeting on sediment contamination, residents asked about the links between the Breckenridge lymphomas and Velsicol, but were rebuffed by staff from the ATSDR state office. A subsequent Freedom of Information Act request filed by the new CAG uncovered a file on the cancer cluster in the Velsicol records.

With this background, the CAG formally made requests for a community health study. These petitions began at the suggestion of federal health researchers. In August 1999, as part of their academic responsibilities at Alma College, the CAG chair and a future CAG secretary attended the American Chemical Society meetings in New Orleans. Part of those meetings was a Symposium of Child Health and the Environment, at which the CAG chair learned about new methods of testing fetal exposure to DDT, and indications of impacts of exposure upon human health. At a break in the proceedings, he spoke with the presenters, several of whom came from the Centers for Disease Control (CDC), and received encouragement to seek support from the NIEHS "Community-Based Prevention and Intervention Research" fund.[76] Upon return to Michigan, the CAG secured the help of Dr. Wilfried Karmaus, then at Michigan State University's College of Human Medicine. They prepared and submitted a grant proposal in December, 1999, requesting funds for a study of environmental health in the region, and for future community-designed responses and preventive measures. Four months later, NIEHS rejected the proposal but encouraged resubmission, noting, "A major strength of this proposal is that the investigator proposes to work with a strong community group that spans three communities with quite different, but significant exposures."[77] Encouraged by the positive comments, Karmaus and the community revised and resubmitted the grant in 2001. Consistent with the commitment shared by Dr. Karmaus and the CAG to community empowerment in public health, the new proposal explicitly committed to assuring the community the core role in guiding health-risk research.[78]

On August 20, 2001, the CAG learned that the NIEHS review panel had given a poorer score to the revised grant proposal than to the earlier one.

Among the criticisms, one reviewer claimed that "community support for the revised protocol is unclear."[79] Incensed, the CAG and Dr. Karmaus complained to NIEHS, with the support of Senator Carl Levin. The CAG chair wrote to the director of research at the National Institutes of Health, assailing the review process and referencing academic criticisms of NIH by Daniel Greenberg of Johns Hopkins.[80] The CAG reminded NIH of findings that "Even when evaluating grants ostensibly targeted on communities with lower socio-economic status, [their process] results in awards to the well-connected medical research centers. It furthers traditional rather than innovative approaches. Furthermore, the process avoids support for significant lay involvement in the prerogatives of scientific specialists."[81] A few days later, Senator Carl Levin's office raised questions about the fairness of the process in correspondence to the acting director of NIH.[82] Over the following months, NIH and other components of the Department of Health and Human Services defended their decision and their version of community-based participatory research. The CAG did not withdraw from the fight with NIEHS, but rather found new ways to seek support for its agenda. Early in 2002, the CAG chair was appointed to the Superfund Subcommittee of the EPA's National Advisory Council on Environmental Policy and Technology (NACEPT). Since NIEHS was repeatedly criticized for a variety of alleged abuses, he and other community representatives on the NACEPT attempted to pressure the environmental health agencies to be more responsive.[83]

After the 1999 American Chemical Society child-health symposium, the CAG sought cooperation from federal and state public-health agencies with a study of fetal exposure to toxins in the St. Louis region. At the 1999 symposium, Virlyn Burse and Harry Hannon of the CDC had discussed the use of "infant blood spots" collected on all newborns to test DDT exposures.[84] Immediately after returning from the conference, the CAG learned that Michigan planned to destroy blood-spot cards archived since 1960. The group believed that they presented an ideal resource for assessing the impact of historic maternal exposures and their effects on children. Opposing the CAG's efforts were those who feared the cards could be used by health insurers to learn of hereditary conditions and thereby deprive individuals of necessary insurance and care. While this was a legitimate concern, the CAG fought to preserve use of the cards, with restrictions on any use that would be detrimental to the individual.[85] This battle demonstrated the sophistication communities need in struggling for a voice.

The blood-spot campaign and criticism of the CDC's health-research process came together in 2003 and 2004 after the leaders of the ATSDR and NIEHS scorned the NACEPT Superfund Subcommittee's efforts to ascertain more about their use of funds. The former CAG chair, now head of its legal committee, convinced the CAG to demand that agencies respond to the concerns in St. Louis, even if they were to ignore other parts of the country. On April 30, 2004, the CAG utilized a section of the ATSDR process to file a "Petition for Public Health Assessment." Referencing the rejected health grants, the petition requested a clinical response. Recalling the innovative plans in the health grant for community inclusion in research, the CAG stated explicitly in the petition, "We would like not merely to request this assessment but to propose working intimately with you through all steps of the assessment process."[86]

The CAG succeeded in several ways from this aggressive procedure. First, the ATSDR agreed to work with the CAG to host a community health conference as the first step in developing renewed human-health work in the region. In January 2005, a half-dozen ATSDR staff came to St. Louis to participate in a two-day meeting that involved independent environmental health experts. At the end of the meeting, the participants, including the lay public, developed a follow-up agenda. As a consequence, new CDC-funded warnings and training in the community were directed at everyone from physicians to those fishing in the river.[87] Secondly, the CAG won a commitment from the CDC to test the infant blood-spot methodology for assessment of fetal exposure to DDT and PBB. A core cost that the CAG once hoped to fund through the NIEHS health grant now was paid from regular CDC funds in the exceptional facilities in Atlanta.[88]

THE BATTLES WITH GLOBAL LOBBYISTS

In the midst of implementing the results from the 2005 health conference, a CAG member noticed an editorial in the *Wall Street Journal* entitled "DDT Saves Lives." As with Velsicol's approach in 1962, powerful interests could not resist tying their defense of DDT to attacks on the ghost of Rachel Carson. The *Journal* asserted: "The perception—going back to Rachel Carson—that DDT spraying is dangerous has long since been debunked."[89] Later in the month, the *Journal* published three letters to the editor praising the newspaper for its courage in printing the editorial, under a heading "DDT Hysteria Has Killed Millions of People." The letter writers were Harold M. Koenig, Jason Urbach, and Richard Tren.

Koenig, former surgeon general of the Air Force, would be one of two retired Defense Department health experts who would be prominent in the pro-DDT campaigns of the era. The other was Donald Roberts. In 2005, Koenig was a partner in Martin, Blanck and Associates, a Washington health-lobbying firm. However, in his letter to the *Journal*, he did not reference his role at the lobbying firm, but instead identified himself as the director of the Annapolis Center for Science Based Public Policy. Funded by ExxonMobil and the National Association of Manufacturers,[90] the Annapolis Center described itself as "a national, non-profit educational organization that supports and promotes responsible energy, environmental, health and safety policy-making through the use of sound science. . . . The Center is committed to ensuring that public policy decisions are based on scientific facts and reasoning."[91] As would be typical of the pro-DDT advocates, the center took on a number of other deregulation issues—for example, renewable-energy programs.[92]

Jason Urbach and Richard Tren, like many associated with the pro-DDT campaign, had free-market economics backgrounds. Urbach had worked for the South Africa Free Market Foundation, and Enterprise Africa. Tren, who led Africa Fighting Malaria, previously had written pro-tobacco and climate-skeptic essays linked to the Competitive Enterprise Institute.[93] Among Africa Fighting Malaria's (AFM) supporters were mining giant BHP Billiton, the Earhart Foundation of Ann Arbor, De Beers diamond company, the Anglo-American Chairman's Fund, and the fortune of Hans Rausing, perhaps the richest Englishman. AFM obtained public-relations help from free-market organizations such as the White House Writers Group.

Among Tren's works was an attack on Doctors Without Borders entitled "Doctors Without Principles."[94] Related to POPs, the primary role of Tren had been multiple critiques of health professionals working in the developing world, including staff of the World Health Organization (WHO). For example, in 2003, Tren wrote to the editor of the *Daily Telegraph*:

> In developing countries, DDT is still being used very effectively in malaria control and saves thousands of lives every day. Any restriction on its use will cost lives and hamper development. It is absurd . . . to call for the same environmental standards around the world when the risks faced from communicable diseases, pests and poor sanitation are vastly different. . . . Eco-imperialism will only lead to increased death and misery in Africa and should be stopped in its tracks.[95]

The attacks on DDT regulation shared several characteristics with the pro-tobacco, climate-skeptic, and related free-market campaigns. First, they attacked the policy process as anti-science and controlled by powerful "funders," a core feature of their work. Henry Miller, who worked on the pro-DDT campaigns through a series of what should be called "junk science" think tanks, attacked both the Gates Foundation and the UN for resisting DDT spraying, saying in 2010, "For someone as smart as Gates . . . some of the foundation's strategies are baffling. . . . Policies based on science and data have a short half-life at the UN."[96] In a pro-DDT book published by Africa Fighting Malaria in 2010, Tren and his coauthor wrote, "As an ideological movement, environmentalism has been astonishingly powerful, controls vast resources, and is not answerable to any government. . . . In the United States, the EPA, National Institute of Environmental Health Science and Agency for Toxic Substances and Disease Registry are functional components of this empire."[97] Second, they directly questioned the integrity of academic scientists and the public institutions that base policy on peer-reviewed science. In the same book, Tren praised J. G. Edwards, one of his allies, for a critique of the journal *Science*. "The record supports Edwards' claim that *Science* refused to publish articles favorable to DDT."[98] Another ally of Tren's, S. Fred Singer, also a tobacco advocate and climate skeptic, regularly attacked peer-reviewed science. What appeared to a general reader as both cautious and specialized became for modern laissez-faire proponents fodder for a strategy to undermine public respect for scientific consensus.[99]

The final characteristic common to DDT advocates, both in the late 1960s and in the early twenty-first century, was a fixation on Rachel Carson. Koenig, in his 2005 letter to the *Wall Street Journal* captured this approach: "Many of the world's leading scientists predicted 30 years ago that Carson's crusade against DDT would allow some of the world's deadliest diseases to return with dreadful consequences. The truth of that prediction is now upon us . . . right on schedule."[100] The attacks on Carson could take a strange turn for what billed itself as "sound science." In the 2010 AFM book, the authors attacked Carson for "not writing about personal experience . . . at no point did she claim to have personally witnessed the killing of birds from use of DDT."[101]

Like so much anti-science propaganda of the era, at the core of the pro-DDT strategy was conversion of the empirical testing of hypotheses into a source for debates about the results of testing.[102] Chris Mooney said of the American Petroleum Institute's campaign against global warming, "Victory will be achieved [when] recognition of uncertainties becomes part of the 'conventional wisdom.'"[103] Henry Pollack explained how these efforts build upon public misunderstanding of "scientific theory" and scientific "uncertainty." Pollack stressed the need for media intelligence in handling lobbying efforts to turn the meaning of theory on its head.[104] Yet, Africa

Fighting Malaria realized that the media's search for controversy worked to their advantage. Soon, policy change seemed to prove the wisdom of their strategy.

FROM PROPAGANDA TO POLICY CHANGE

In addition to the attacks on Carson and the science used in DDT policy-making, one of the repeated claims of the pro-DDT forces was that they were on the side of "locals against outside do-gooders."[105] That distortion, as well as news in 2006 that the World Health Organization was reversing its DDT policy, made the Pine River locals decide they had to confront—as "locals"—those advocating bad policy. Already concerned with the *Wall Street Journal*'s position, National Public Radio's (NPR) treatment of the issue appalled the community advisory group. NPR not only praised WHO's leadership for its policy change, but did so by attacking Rachel Carson:

> The World Health Organization today announced a major policy change. It's actively backing the controversial pesticide DDT as a way to control malaria. . . . While DDT repels or kills mosquitoes that carry the malaria parasite, it doesn't get much good press. In 1962, environmentalist Rachel Carson wrote a book, *Silent Spring*, about how it persists in the environment and affects not just insects but the whole food chain. . . .
>
> In the early 1960s, several developing countries had nearly wiped out malaria. After they stopped using DDT, malaria came raging back and other control methods have had only modest success.
>
> Which is why Arata Kochi, head of the WHO's anti-malaria campaign, has made the move to bring back DDT. His major effort at a news conference Friday in Washington, D.C., was not so much to announce the change, but to deflect potential opposition from environmental groups.[106]

The account briefly noted the resignation of a number of WHO staff in protest over Kochi's policy change, but the pro-DDT propaganda campaign worked so well that NPR quickly returned to the theme of saving little African children.

Rather than lament the power and influence of Africa Fighting Malaria, the CAG chose to confront them. The CAG would show Africa Fighting Malaria's true colors, not as defender of locals, but as a classic Washington special-interest lobby. Consequently, in the fall of 2007, the CAG contacted a large number of global malaria and DDT experts to determine if they would participate in an international conference on DDT. When the CAG received a number of encouraging responses from widely respected scholars, it decided to commit some of the resources from an earlier environmental settlement to the event. Alma College's Center for Responsible Leadership, under the direction of John Leipzig, agreed to cosponsor the conference with the College's Public Affairs Institute. The CAG named the conference for Eugene Kenaga, one of the founders of the Society for Environmental Toxicology and Chemistry (SETAC), who had died in 2007. Eventually the CAG won the official sponsorship of SETAC and the International Society for Environmental Toxicology. It recruited medical, public-health, and environmental experts from as far as South Africa, and from some of the leading universities in the United States. While the CAG paid travel expenses, the scholars donated their time.

As if CAG sponsorship were not enough to reflect a new local community role in shaping policy, the group designed a format for the conference that implemented their evolving theory of lay involvement. Half the attendees would be people from the region, who would both hear and interact with the global experts. As the CAG finalized arrangements for the conference, a controversy erupted related to an ATSDR study of "areas of concern" in the Great Lakes basin. Giving the lie to the Africa Fighting Malaria claim that agencies such as the ATSDR were under the control of well-funded anti-insecticide lobbyists, the CAG learned that ATSDR had blocked release of the study that identified the Pine River as one area of concern. The agency also had reassigned the scientist responsible for the report, Christopher DeRosa, to a vacant office with no duties. For several months, the CAG had tried to get ATSDR to send a representative to the Kenaga International DDT Conference, only to be rebuffed. Now, David Carpenter at Albany University urged the CAG to invite DeRosa privately. By this point, DeRosa's name was in the national media, having told the ATSDR director, Howard Frumkin, that the delay in release of the study had "the appearance of censorship of science and the distribution of factual information regarding the health status of vulnerable communities."[107] Two Michigan members of Congress, Bart Stupak and John Dingell, demanded that the CDC release both the report and information on its treatment of DeRosa. They warned, "ATSDR's apparent withholding of this report raises grave questions about the integrity of scientific research at CDC and ATSDR, as well as treatment of its scientists."[108]

Without support from ATSDR, DeRosa became the keynote speaker—traveling on vacation time, and with the college and CAG paying for his ticket. On the evening before the conference, with Dr. DeRosa on his way to Michigan, ATSDR released the embargoed Great Lakes report. Doris Cellarius of the Sierra Club captured the mood of all, saying, "This is totally wonderful. Dr. DeRosa is a scientist with the greatest integrity and has been standing up for the best science possible."[109]

DeRosa's keynote directly challenged the contempt of groups such as Africa Fighting Malaria for both the "precautionary principle" and Rachel Carson. He summarized his thoughts, saying:

> I began to focus in my graduate career on DDT alternatives, DDT substitutes. A very elegant article by Ian Nisbet turned me on to that idea and I did early research on the effects of carbon made pesticides on reproduction in wildlife populations. Then I started to think about the confluence of some of the remarks that have been shared with us by the different speakers [at the DDT Conference], and the fact that one of the themes we have to recognize is that there are limits of our knowledge base that is circumscribed by the questions we choose to ask and they change over time. That is why Bernardino Ramazzini in the 16th [*sic* 17th] century, the father of occupational medicine, invoked the precautionary principle, said it is better to prevent than to cure. So we banned DDT after the epiphany of *Silent Spring* and we thought perhaps that was the end of the problem, but that was the end of the beginning of the problem as we have heard today.[110]

At the conclusion of the Kenaga Conference, the participants requested that the experts draft a consensus document, summarizing all that had been learned. Drafted initially by Brenda Eskenazi as she returned to the University of California, it was edited and signed by fifteen experts, including the three South African medical and environmental scholars who had participated: Maria Bornman and Christiaan de Jager from the University of Pretoria, and Henk Bouwman from North-West University. Other signatories came from Cornell, Creighton, Indiana, Rockefeller, and Wisconsin, as well as the hosts from Alma College. The "Statement's" conclusion exemplified the unofficial slogan of the conference, "Think Locally, Act Globally." After reviewing hundreds of studies, it concluded:

> Current evidence on DDT exposure to human populations and on its potential health effects support the Stockholm Convention on Persistent Organic Pollutants which emphasizes that DDT should be used with caution, only when needed, and when no other effective, safe and affordable alternatives are locally available. . . . Given the paucity of data in populations who are currently potentially exposed to high levels of DDT, we urge the global community to monitor

exposure to DDT and to evaluate its potential health impacts both in malaria endemic regions of the world and in locations where DDT use has been historically high such as the Pine River Superfund site.[111]

The official announcement of publication in the NIEHS's journal, *Environmental Health Perspectives*, received national coverage. *Scientific American* summarized the findings: "The scientists from the United States and South Africa said the insecticide, banned decades ago in most of the world, should only be used as a last resort in combating malaria."[112] Institutions that sent experts to the conference—such as the Sprecher Institute for Cancer Research at Cornell, and the School of Public Health at the University of California at Berkeley—celebrated the publication that resulted. Its title, the "Pine River Statement," epitomized its link to the CAG, its community, and the world.

In the fall of 2008, the CAG learned that the draft Pine River Statement had been discussed at the "Stakeholder Meeting of the UN Environmental Program on the Stockholm Convention," attended by two of the Kenaga Conference experts, Hank Bouwman and Riana Bornman.[113] This discussion took place in front of Richard Tren, who attended as an NGO (nongovernmental organization) representative. He would attack the "Pine River Statement" in his pro-DDT book, *The Excellent Powder*, which was published by a vanity press in 2010. Following up on the commitment started with the Kenaga Conference, the CAG sent its public-health chair to the Washington press conference for Tren's book.

LESSONS

Whatever the long-term impact of the Pine River Statement, its publication represented a model for other isolated communities facing well-funded global threats to their sustainability. St. Louis residents and the CAG experimented with many strategies and tactics in order to be heard and taken seriously by government agencies and polluters. Their experiences provided other communities with several lessons in civic empowerment:

1. Defend the rights of the community to control its structure and processes as the community defines its expectations. Do not allow outside special interests to set the agenda or decide with whom negotiations will take place.
2. Expect and be prepared to use the law to achieve justice. Confront legal challenges, even from the highest levels of government, if those are compromised

by special interest influence. Understand that while the legal system may be biased against poorly funded communities, the door is never completely closed to presenting a community case.

3. Do your research and be prepared to challenge experts, even the most well-trained and respected, as in the community's battle with NIEHS. Realize, however, that the community must find some professional help to have an opportunity to influence experts.

4. Neither ignore nor fear global special interests that may seem so far away as to be irrelevant or beyond local influence. Dismiss slogans such as "Think globally, act locally." Certainly know and think about the global forces and movements that may threaten the community, but do not retreat into only local actions. With modern globalization, such a retreat opens the door to being hopelessly outmaneuvered by forces seemingly beyond one's control.

Obviously, these lessons are neither easy to apply nor, if applied, will necessarily lead to solutions to every community's problems with contamination, unemployment, or other forms of exploitation. Former workers in Toledo and Clarksville, and residents near the various Velsicol pollution sites outside St. Louis, and certainly laborers at Fruit of the Loom facilities in Honduras, face nearly insurmountable barriers to asserting their interests. Yet, just as clearly, St. Louis encountered nearly impossible obstacles, including from international lobbyists. While not fully defeating lobbyists, they found a method to battle them on the global stage. By contrast, the workers at Doehler-Jarvis and Acme Boot failed to influence their circumstances. If the spirit of St. Louis had prevailed within the UAW in Toledo, they could have taken control of Doehler and likely survived. The customers wanted the plant to survive; only union timidity and a focus on short-term risks doomed long-term recovery and preservation of jobs in the city. By contrast, the CAG in St. Louis exemplified hope. All similar communities must recall the words of Paulo Coehlo, the great Brazilian author: "When you really want something to happen, the whole universe conspires so that your wish comes true."[114]

CHAPTER 8

Consequences and Controls

THE HISTORY NARRATED IN THIS INQUIRY EXPOSES CORPORATE LEADER-ship failures and some institutional pathologies that require clarification and study. It recounts more than the rise and fall of a chemical, boot, or underwear manufacturer. It also details social, intellectual, political, and environmental problems, as well as the assumptions that underpin much modern behavior and the structures of many contemporary institutions. As exemplified in the three generations of Fruit of the Loom directors, its predecessors, and its subsidiaries, some practices have become so outrageous and some ideologies so contradicted by reality that they clamor for reform. On the positive side, this investigation uncovers a few models of community organization, law, and general citizen empowerment to check the worst institutional conduct and pathologies.

This work makes no claim that these firms are the best subject for such research. However, it does indicate that they are very good representatives of the problem. While primary attention has been directed to Velsicol, it is crucial to note that its managerial and other institutional failures were widespread. The negative consequences of poor corporate leadership ran the gamut from consumer and community exploitation to shareholder deceit. Because of the firm's many locations, its actions repeatedly led observers to reflect on the unsustainable behavior and to respond. Southern mayors sought to buy its jobs. Environmentalists plugged the firm's pipes.[1] Human-rights activists condemned its sweatshops. Workers lamented their layoffs. By contrast, market analysts blindly praised Fruit's financial manipulations. A president even played the saxophone for its owner. That so many support-ers could act so foolishly until the firm's collapse, and that so many could suffer loss of jobs, human rights, and environmental justice as a result of Fruit's activities, make it a subject of unquestioned importance if modern American leadership and its role in the world are to be understood and shaped for good, rather than ill. The insight that arises from the study of Fruit and its subsidiaries can be separated into four broad areas: economic, technological, cultural, and civic.

However, first it is essential to emphasize that this is not an attack on regular economic activity or on the general social and political structures of the United States. The U.S. has been blessed to develop a system that places limited public controls on individual belief and activity. This freedom fosters creativity and innovation, and resourcefulness must include mistakes, even big ones. Businesses will fail. Policies will not work. At its best, the process permits criticisms of those flaws and the gaining of insight from them. Rejecting a hereditary ruling class at its founding, the country chose not to protect unduly those who err from the consequences of their own folly. Businesses go bankrupt. Unproductive and incompetent workers get fired. Voters throw the rascals out.

The cases here raise fundamental questions about the efficacy of the idealized U.S. system. These are not studies based on simple failures of judgment, including negligence to prepare properly for the accidents of life. The focus here is on the systematic avoidance of the consequences of the gross pursuit of self-interest and the most incompetent leadership. It involved tolerating pollution so egregious that generations are forbidden to utilize common natural resources. It included terminating over 50,000 loyal First World workers and replacing them with people whose human rights were systematically denied. It entailed the repeated dereliction of duty by those with a fiduciary responsibility for worker and investor assets. Further, it revealed the blindness of the media, and of academic and civic leaders responsible for providing both intellectual critiques and legal sanctions for antisocial behavior. Unbelievably, at times, this elite replaced censure of egregious acts with celebration of what Hollis Kelley, the former West Point worker, labeled "community rape."

The failure to criticize the conduct that caused these negative consequences raises the most pressing problem identified in this study. Intellectually, while virtually all Americans value individual liberty, the excesses of the free-market ideology, with its emphasis upon hyper-individualism and consumerism, have undermined the ability of many communities to protect themselves from economic fates or to define the long-term public interest. Although once popular only within a segment of the American elite, in the last quarter of the twentieth century, many leaders in the United States accept neoliberalism as the hegemonic ideology of the world.[2] That ideology celebrates the antisocial consequences of unregulated economic behavior as the only logical norm. It vigorously attacks not merely Marxism, but the social-democratic model of Western Europe and corporatist schemes that evolved in the early twentieth century to protect the most vulnerable populations and ecosystems from the worst forms of exploitation. Within the United States, neoliberal ideology rejects the mixed system developed empirically through trial and error during the first half of the century.[3]

The mixed system in the United States provided some of the same protections for the working class found in social democracies, while protecting the interests of traditional or bureaucratic elites. It allowed the United States to become the global leader in environmental protection by the mid-1970s, for a time.[4] Since the Great Depression, pragmatically developed policies were some of the hallmarks of U.S. policy experts, recognizing what has been called the limits of rationality or policy incrementalism.[5] Without reducing private business control over the core terrain of management, and often permitting major roles for the private market in fields that other democracies had made public, the mixed system minimized ideology. It focused on what worked and supported regulation of what did not. Also, it endorsed working with nations throughout the world with alternative systems, not demanding that the Canadians abolish public ownership of health insurance or that the French relinquish government control of railroads. Toward the end of the twentieth century, however, this flexibility eroded. As the mixed system produced substantive protections of the environment and workers, with new bureaucracies such as the EPA and NIOSH charged with enforcing caution, elites reacted. Across the political spectrum, they demanded a return to laissez-faire fundamentalism or neoliberalism. Through global trading schemes, the United States sought to make every nation follow the one true free-market ideology.[6]

The irony of this development is that growing toleration of visible diversity at home has distracted potential critics. They have disregarded the fact that older traditions tolerated intellectual diversity, not physical differences. Historically, Americans welcomed people who rejected the core ideas of the majority. They accepted the Amish, who refused to use electricity; the Christian Scientists, who did not sanction intrusive medical care; and the Shakers, who lived communally. At times, the United States had fundamental problems with tolerance—for instance, of extreme groups like the Klan and others. Nonetheless, it won fame for protecting the right to differ, going so far as to allow the Amish exemption from compulsory schooling laws.[7] However, while the government has remained reluctant to press conformity, free-market ideologues have challenged the right of other nations to continue traditional community practices, such as the preservation of communal land ownership in Mexican *ejidos*.[8] In the country that led the world in a constitutional commitment to the law of nations, they have believed reference to any standards but those of the United States to be dangerous.[9] Applying their deregulation doctrine to the employment of human beings has justified domestic and international labor-rights abuse.[10] Stripped of any responsibility to the civil community, the fundamentalist pursuit of laissez faire has condoned looting the natural world and public and private treasuries, and when necessary, clashing with the world's other fundamentalists. Consequently, in the new millennium, in order to contain this hegemonic ideology, it seems necessary to begin with manageable examples. Few better places to start exist than the rise and collapse of the companies studied here.

ECONOMIC LESSONS

Two categories of economic lessons arise from this study. Some pertain to corporate leadership. Others relate to national economic policy, particularly trade policy. While business management could be added, its mistakes seem obvious and not in need of review. Each company studied was dreadfully managed in the closing years. In general, this mismanagement arose from a common mistake on the part of top leadership to replace technically competent managers, whether boot makers at Acme or die casters at Doehler, with marketing or financial manipulators. The issue of mismanagement did not originate with mid-level officials but came from the highest levels of corporate direction, including investors, the board, and the chief executives, who should have set priorities among the various stakeholders of each firm. At each level of leadership, the experiences at these firms have raised worrisome concerns about currently accepted practice.

The first problem is the investors. Increasingly, American firms, especially the larger ones such as Fruit, are dominated by institutional investors—pension funds, mutual funds, endowments and other trusts, insurance companies, and banks. These institutional investors on average control about 60 percent of the largest companies, reaching a high of 85 percent at Coca-Cola in 2000. As reflected in the lawsuit filed by New England Health Care Employees Pension Fund against Fruit, these institutional investors can become involved in management.[11] However, why did they wait so long? Clearly, Fruit of the Loom had fundamental difficulties throughout the 1990s under Farley's leadership, and yet the firm continued to compensate Farley at extraordinary levels. The information was available to the public, and institutional investors could have demanded changes from the company before its collapse. The experience at Fruit illustrates Lawrence Mitchell's argument about irresponsibility on the part of institutional investors: "The promise of institutional stockholder activism has been distinctly less significant than was originally hoped."[12]

Mitchell refers to the expectation that engaged institutional investors would reduce corporate interest on short-term returns and increase concern for institutional sustainability. However, at the heart of the Anglo-American corporate model is the belief that the sole purpose of management is to increase shareholder value.[13] As Thomas Berry wrote, "[Corporations] have obtained the rights of individual citizens without assuming responsibility in proportion to their influence on public concerns." Even investors from institutions with altruistic purposes share this irresponsible goal. Mitchell found that TIAA-CREF, the largest investment fund for the nation's educators, avoided intervening on issues such as those that ruined Fruit.[14] He concluded, "There is no question that American institutional investors promote the Anglo-American model with a vengeance."[15]

As Robert Monks and Nell Minow discovered in their study of corporate power, institutional investors face the classic "free rider" problem. If they act alone to reform a corporation, other investors reap the benefits with none of the costs.[16] Hence, the New England Health Care Pension Fund only intervened when it was too late, when it had nothing to lose and hoped to win damages from Farley. Unfortunately, neither TIAA-CREF nor the New England Health Care Pension Fund, two institutional investors composed of professionals who normally would be attentive to their social responsibility, questioned why many Kentuckians were losing their jobs, or why labor rights were being exploited in Honduras. Only some larger public pension plans utilized their power to monitor the most egregious corporate mismanagement.[17] TIAA-CREF finally reversed its usual position in 2006 by withdrawing investments in Coca-Cola over workers' rights concerns.[18] Yet, at the very time of this hopeful change, corporate leaders renewed a trend to "privatize" corporations to avoid stockholder oversight. The evolution of structural changes in corporate finance, such as hedge funds, greatly influenced corporate decisions in institutions fully exempt from public oversight.[19]

While an alternative model of corporate governance in Western Europe and Japan includes solicitude for workers and other stakeholders, the Anglo-American model ignores all but the investors.[20] Clearly, this is a major blind spot in corporate governance.[21] It also corrupts non-U.S. investors who put money into U.S. firms with the same limited expectations as domestic investors.[22] As a result, institutional investors and their clients, the vast majority of the global middle class, bear as much blame for "community rape" as the leaders of firms such as Fruit of the Loom. They exhibit no concern with workers in any country.[23] Likewise, they lack interest in other stakeholders in the communities of their facilities and their environment. These investors are only interested in actions that influence short-term change in share value of their endowments and pension funds. The short-term focus of these investors also undermines broadening the concept of value to include long-term wealth generation. Only short-term value increase is "wealth."[24]

The reality, however, appears worse. Corporate leaders not only abandoned communities, workers, human rights, and the environment for the "good" of the investors. In fact, there seems little evidence at firms such as Fruit of the Loom that management served the shareholders at all. However, shareholder interest has been the rationalization for the antisocial conduct of modern corporate leaders.[25] Classic theory assumes that shareholders remove a director who does not successfully increase the value of their holdings.[26] However, the evidence from Fruit of the Loom illustrates that incompetent leaders survived long after their failure was apparent. Furthermore, the failed executives continued to claim for themselves a rate of compensation out of proportion to their success.[27]

As demonstrated at Fruit, the compensation of leaders was not a detail best left to the private judgment of a board, since board members were not attentive to protecting the interests of either shareholders or communities. The study of Fruit of the Loom revealed that stipends for leaders were acceptable tactics for transferring significant corporate resources from the firm to the management friends of the board. After Fruit lost massive amounts of money, it persisted in paying eight-figure outlays to its CEO. Further, the compensation packages usually contained language automatically providing enormous benefits if management achieved goals, such as a merger or an acquisition that enhanced short-term value. As with the Fruit transfer to the Cayman Islands, however, more of the benefits of the move went to the CEO than to other shareholders, and Fruit of the Loom was not unique.[28]

These automatic bonuses to executives who achieve certain predetermined goals are one more example of a board surrendering its leadership role. Of course, top administrators should work to improve a company, but the board needs to guarantee that through its oversight of the corporation. Likewise, an automatic bonus is unnecessary considering the excessive pay given to executives.[29] In 2004, John Kenneth Galbraith summarized this process well:

> This fraud has accepted ceremonial aspects. One is a board of directors selected by management, fully subordinated to management but heard as the voice of

the shareholders. It includes men and the necessary presence of one or two women who need only a passing knowledge of the enterprise; with rare exceptions, they are reliably acquiescent. Given a fee and some food, the directors are routinely informed by management on what has been decided or is already known. Approval is assumed, including for management compensation—compensation set by management for itself.[30]

Proof of the absurdity of this relationship between the board and its CEO was Farley's defense of the loan that the board granted him to buy out Leon Black. Totally in his own self-interest and not that of the corporation, Farley announced to outside critics that he had only accepted it because the board asked him to.

Contemporary compensation policies originated in what Robert Frank and Philip Cook have called the rise of the "Winner-Take-All Society." They correlate the rise of compensation to the decreasing loyalty of executives to their employer. They suggest, "If every company were to promote from within, there would be no reason to pay the most talented senior officers what they were worth, because they would have no place else to go. This remains common in Germany and Japan, where CEOs are still promoted almost exclusively from within."[31] Of course another key benefit of promotion from within is the technical competence of managers. At the companies studied here, the era of promotion from within created exceptionally profitable and innovative leaders, while more recent management from outside often resulted in the removal of resources, not the development of additional "seed corn." Fruit of the Loom's desperate return of John Holland to management in the closing days before bankruptcy in order to save the collapsing firm is proof of the flaw in current trends.

Management from without and its technical incompetence resulting from the winner-take-all attitude relate directly to the debate about the impact of trade policy on firms.[32] While offshore lower-cost production may have compelled the abandonment of some domestic manufacturing, the experiences at Fruit and its subsidiaries raise doubts about the applicability of that justification. Companies such as Doehler-Jarvis and Velsicol declined despite no competitive threats to their domestic production. Doehler collapsed in 1998 from financial looting by leaders fixated on their personal survival. Imports played no role in the demise of Doehler.

Regarding textiles and apparel, it is not clear that Acme, WestPoint, or Fruit of the Loom itself needed to cut all or part of their domestic production. Geoffrey Underhill in his multinational research on the textile industry found that the emphasis on labor cost differences—the old comparative-advantage argument—ignored what he called competitive advantage growing from location, skilled management, technology, and communication resources.[33] Research on other industries impacted by trade has also illustrated that improvement of management and quality allows companies to survive and prosper. Blaming workers and/or government policy, such as tax rates, ignores the great competitive advantages of production in safe and secure locations, with literate and critical workers who monitor quality.[34]

In two other ways, the Acme and Fruit of the Loom experiences undermine arguments that wage differentials alone caused shifts in production. First, both firms had a reputation for quality and affordability, which could have offset, with more attention to consumers, much of the competitive price advantage from low-wage countries. The decline in quality and added delivery problems sustained by Fruit after shifting sewing operations offshore demonstrates that these were not trivial difficulties. In many ways, they precipitated the bankruptcy as market share declined. Detailed studies of the costs of production show that U.S. manufacturing had minor cost disadvantages compared to those in some developing countries.[35]

These qualifications serve to validate Underhill's interpretation of the data and the actual performance of companies. He concluded, "Explanations focusing on economic structural factors and omitting the crucial agency of firms is misleading."[36] Lazy or incompetent leadership seeks the economic cushion provided by cheap labor. Executives compensated at the high rates common in the United States should be able to deliver competitive production with a work force paid wages and provided with rights comparable to those of the top supervisors. Fruit of the Loom and Acme abandoned the United States because they ceased investing in new technology. Given the choice of utilizing available funds to keep the company competitive or fattening their excessive compensation packages, the leaders, with complete control, opted for the latter. Whether or not it was illegal, these firms were victims of Galbraith's definition of fraud. To confuse that assessment by blaming their collapse on the wages of American textile workers is disingenuous.

Secondly, the comparative-advantage argument also weakens when conditions in the new manufacturing centers are considered. Here, questions need to be asked about the motives, working conditions, and consequences for the new host communities. It must not be forgotten that local and national investors primarily funded the early development of the textile industry in Britain, New England, or the U.S. South. Clearly some northern firms and investors facilitated the growth of the southern textile and apparel industry. Yet, there also were the Laniers, who built West Point from the ruins of the Civil War without much Yankee assistance.

It is essential for advocates of economic development to note that the Honduran and Salvadoran plants of Fruit of the Loom were examples of a new economic animal.[37] They were not owned by locals who might use profits and technological insight from their operation to enhance national institutions or skills in their country. By contrast, the investors in the long-abandoned textile plants of New England left the region with an institutional legacy that allowed it to prosper despite its poor geography as a center of global intellectual and technological innovation. Some names of its institutions, such as Brown University, came from textile investors. Under current practices, Honduras will receive no such long-term benefit. Worse than an absentee owner, firms like Fruit often do not own or manage the production facilities for their wares. They have retained subcontractors or shelter corporations, often from outside the country, to further isolate the global manufacturer from any direct responsibilities.[38]

In order to understand a process such as trade, it is essential that motives be clear. In the case of Central American apparel production, the reason for the manufacturing shift must be taken for what it was. Production did not move to the region because it contained a large supply of raw materials. It did not transfer because of the region's technological advantage or its large pool of skilled workers. It did not move because local investors decided to attack the international apparel trade by pooling scarce national capital to create economies of scale tied to an abundant, available work force. It did relocate because of that abundant work force; however, the investors were foreigners—either North Americans or, more often, Asians—who found the region ripe for exploitation. Labor was desperate, and governments were eager to collaborate in suppressing the workers' claim of agency. In a sinister manipulation of human freedom, some introduced fundamentalist religion to instill docility in employees.

As Michael Mortimore stated, this form of development "instead of deepening national industrialization . . . truncates it . . . instead of giving birth to national champion companies that evolve into global competitors it threatens their very existence."[39]

Human Rights Watch determined that such development suppressed rights rather than enhanced them. Any account that describes this growth as other than human exploitation is at best mistaken, at worst devious and deceptive. It was driven by an effort in the service of greed to reverse generations of global progress in social protections. Only when understood in this way does it become rational for the general public to protest against trade liberalization in places such as Central America. It also clarifies why an elite uses all means possible to sneak through or "Fast Track" trade agreements.[40]

ENVIRONMENTAL AND TECHNOLOGICAL LESSONS

Likewise, a study of these firms reveals insight into neoliberal approaches to technological innovation and environmental stewardship. While evident in Fruit of the Loom's various Superfund sites, the company's policy for several generations has been to exploit all possible profit from the use of resources, with only minimal regard to the wishes and rights of other users of those assets. As with labor exploitation, care is necessary when reviewing environmental and technological issues to be certain that practices and facts are precisely defined. For example, scientists warned at the first use of DDT that overuse would both kill many valuable species and lead to resistance in targeted pests. These correct assessments not only were ignored in the name of profit, but also have been aggressively challenged for a half century in a series of ruthless campaigns to discredit rather than learn from them. An example was Velsicol's creation of the so-called International Stewardship Centre to lobby against global pesticide regulation.[41]

Throughout most of the century, as it fought one study after another that raised concerns with its production and use of chemicals, Velsicol displayed complete disregard for what has been called "the precautionary principle." Its repeated accidents, whether with cattle-feed contamination or Phosvel poisoning, directly sprang from an absence of precaution, as did more recent efforts to ignore indications of endocrine disruption arising from exposure to its products.[42] Its current success at avoiding most of the cleanup costs for the wastes dumped over decades of profitable operation point to the urgent need for basic accounting reform to require assessment of the total cost of the externalities arising from production. Independent boards of directors responsible to all stakeholders should oversee accounting for these costs.[43]

Another fundamental flaw of modern management, despite much rhetoric to the contrary, is its failure to promote knowledge of technology. While the popular culture celebrates new technological developments, management at the end of the century displayed remarkably little interest in funding it. The leaders who created firms such as Doehler, Fruit of the Loom, and Acme were technological geniuses; their successors at the time of collapse could make neither a transmission, a T-shirt, nor a boot. Two Latin Americanists—Richard Tardanico and Mark Rosenberg—have tied this attitude to the current shared dilemma of the American South and Central America.[44] As reflected in the migration of Fruit of the Loom jobs from poor Louisiana and Kentucky towns to poorer Salvadoran or Honduran villages, it has built an economy on balancing the adverse consequences of poor and self-serving management with low expenditures on technology, infrastructure, and quality education. Combined with these underinvestments has been toleration of growing income inequalities that preclude development of local financial resources. Even the most minimal sense of justice should reject the claim by Fruit of the Loom executives who earned $10,000 per hour that the company had to abandon the U.S. South because $10-per-hour wages were uncompetitive. Stephen Cummings has called this the "Dixification" of our world.[45]

In order to protect firms, their workers, and communities from this form of exploitation and foolishness, new accounting and oversight rules are needed. Governments can tax compensation levels that rise above global averages in order to create more equity. Companion rules that favor real technological innovation and research should provide guidance to prevent firms from consuming their seed corn. If the United States really favors development in the Third World, a valuable Caribbean Basin Initiative would provide tax and trade rewards only for production-enhancing skilled work and technology transfers.

As the experience of St. Louis, Michigan, teaches, if national environmental policy is not to be compromised, local communities should be empowered as the final check on environmental abuse. Clearly some development will leave scars on the landscape and waste resources. However, St. Louis demonstrates that local people have some sense of the acceptable tradeoff. More importantly, we have learned much about proper limits to empower people to insist on the limits being observed. As Cummings points out, the real growth sector for the future will likely be one that finds ways to produce with hardly any use of nonrenewable resources.[46] The company and host community that perfect one or more technologies that advance in that direction will prosper. To do so probably requires exceptional investment in education, as well as the infrastructure to support innovation.

CULTURAL LESSONS

To generate technological and environmental innovation requires training a new type of leader. Education must cease to endorse enslavement to the market. Required as well is a significant cultural change, focused on overcoming the disconnect between formal traditional beliefs and routine behaviors. Given high levels of individual religious identification, Americans would seem to be especially susceptible to connecting ethical and civic beliefs to their daily business conduct. Yet, the personal emphasis in much of American religious tradition might explain why the Anglo-American model permits a contradiction between private beliefs and actual economic and leadership deeds.[47]

The modern managers of Fruit of the Loom and its affiliated companies displayed a basic inability to accept any responsibility to stakeholders. A formal principle of American and international law defines labor not as a commodity, but as entitled to special recognition and treatment beyond the usual market relationship of buying and selling at the lowest cost possible.[48] A century ago, Richard Ely, a founder of the American Economic Association, and John Ryan, a Catholic priest, captured this concept in the term "living wage."[49] For many years, their work had the support of a majority of Americans who believed in reimbursing labor at a sufficient level so that they could afford the products they made. Henry Ford supported this approach in his decision to pay his workers well enough that they could buy one of his cars.[50] By the mid-twentieth century, business in the United States expanded "Fordism" to include guarantees for health insurance and pensions as well as a living wage.

This changed before the new millennium; "post-Fordism" became a common term.[51] Increasingly, companies have opted out, essentially breaking the social contract with their own employees by transferring social insurance costs either to the employee or onto the backs of all taxpayers.[52] While some business writers have found it perplexing that criticisms have mounted despite companies producing unprecedented levels of worker and consumer comforts, those criticisms have arisen in response to the callousness of leaders who have simultaneously excused failures and looted firms.[53] The repeated collapse of pension and health-benefit plans seems to parallel wider problems throughout the economy. Unfortunately, the 2006 Pension Protection Act only attempted to solve the pension problems of firms such as Fruit of the Loom by encouraging businesses to open individual retirement accounts for their workers in place of the once common guaranteed pension plans.[54] American business leaders need to modify their approach to community.

There was a general trend to oppose any actual or implied obligations to stakeholders, even after firms received subsidies or infrastructure support.[55] Michael Piore believes that inclusion of all stakeholders would be for the long-term good of free enterprise. Companies are so accustomed to using what he labels "strong-armed tactics," like wage concessions and tax breaks to continue operations, that they neglect technological innovation, as at Fruit of the Loom and Doehler.[56] While the textbook version of good leadership assumes that employee productivity and innovation result from treating workers humanely, Fruit did the opposite and hurt social progress.[57] In contrast to the faith among neoliberals in the "virtue" of greed, real conservative writers, such as Michael Novak, have emphasized the community and ethical responsibilities of business.[58] In international economic policy, Nobel laureate Joseph Stiglitz has pointed out the urgent need for fair trade, not just free-trade policies.[59]

Essential to effecting this change in conduct is an ethical renewal that identifies the business looting of natural and man-made resources as injustice. Despite theoretical condemnation of greed by the nation's civil religion, and references against it in the faith communities to which most Americans belong, few seem to apply this teaching to actual business or civic practices.[60] How can the public explain the transfer of excessive wealth from the corporation to the private accounts of its leaders without use of the word "greed"?[61] With every revelation that the income gap between the richest and the working class was multiplying, often accompanied by factory closures, remarkably few community ethical leaders, such as the clergy, castigated the behavior.[62]

The reason for this flaw seems to arise from the timeless problem of applying theory to life. As Lawrence Mitchell said from the perspective of the law, the difficulty is that "Radical individualism and rampant selfishness [contradicts] a lesson we are all taught as children and all teach our children . . . lay off the delight of McDonald's in favor of the asceticism of fruit. . . . But although it's a lesson we all know, we easily forget it. And we most easily forget it not when we are acting alone but together, in the faceless, soulless, thoughtless markets of American corporate capitalist society."[63]

Robert Wuthnow summarized data to reinforce Mitchell. One opinion survey has showed that "while 86 percent of weekly churchgoers say greed is a sin, only 16 percent of them say they were ever taught that wanting a lot of money is wrong, and in fact 79 percent of them say they wish they had more money than they do."[64]

The first step in applying ethical standards to economic behavior is getting Americans, especially the elite, to understand and acknowledge their power. As John Hall and Charles Lindholm found, "But all—the weak and the strong alike—disagree about who really has influence; none see themselves as having power; all see power as unfairly grabbed by others, and all see 'themselves as victims.'"[65] While this powerlessness may have been true of workers subject to layoff because of downsizing and globalization, it was not true of the people at the top. Unfortunately, once downsizing began, it discouraged actions essential to the survival of a firm, including innovation.[66] W. Edwards Deming found that the worst barriers to corporate success are "fear, mistrust, and isolation," yet they are tools often used by modern business leaders to solidify their authority.[67] Not only do these tools limit business success, they undermine support for freedom, an essential feature of good civil society. As Mitchell says, "Freedom [is] to act as a natural human being, situated in a society and in a moral and an ethical environment that allows all of us to make our choices and that holds us to account for the choices we make."[68]

The widespread ideology of free trade has been the most direct threat to this freedom. Its pervasiveness is illustrated by its endorsement across the truncated American political spectrum, from Ronald Reagan to Bill Clinton.[69] Its followers have imbibed the contradiction epitomized by Farley at West Point, who simultaneously exerted a powerful influence while claiming to be controlled by fate. This inconsistency was a farce, as global business leaders asserted their power, yet pleaded their helplessness to protect people, cultures, or resources from exploitation by the highest bidder. While interference in the market represents a sin as grave as any in public life, according to their principles, they need to admit that the greater sin is a prominent leader abandoning a community. They have to acknowledge as well that citizens were correct in maintaining that certain items could not be sold, and that setting a price for the sale of public resources by civic deliberation was not a "non-tariff barrier."[70] The market requires ethics.

Denying community power over the market is fundamentally denying that "economic systems are products of artifice, not nature."[71] To assure that labor is not to be sold to the lowest bidder is not establishing a non-tariff barrier; it is imposing justice. To decide to protect a community's environment for future generations is not an offense against the law of nations. In fact, it may be preventing a transgression. Yet, modern ideology discourages the insertion of values into market forces. It demands that the leader submit to ruthless laws that bring efficiency and set prices purely by supply and demand. Nevertheless, the compensation packages for these leaders betrayed the fraud they perpetuated. If the market set Bill Farley's wages, it was a strange one over which he had direct control. By contrast, when he had to close a Kentucky mill or withdraw funds from new investment to pay for his compensation, he claimed he was helpless to resist market pressures.

CIVIC LESSONS

The reality studied here suggests that the legacy of these firms built by several generations of investors and workers, including inherited natural resources and such priceless possessions as the human rights of replacement workers in maquiladoras, was sacrificed to "market forces" by powerful and increasingly affluent speculators. Except perhaps for Joseph Regenstein, these people did not use their wealth to build permanent cultural institutions. They were not civic leaders like Ben Heineman, who volunteered his time and resources to a struggle for human rights, fairer treatment of the poor, and wise transportation policy. The contributions of Regenstein and Heineman provided some counterbalance for the offenses their firms committed against communities. It may be puzzling for some to visit the wetlands exhibit at the Brookfield Zoo, funded with Regenstein money, and observe warnings not to dump contaminants into water. However, the legacy of more recent managers and owners has left no such puzzle. Their greed bestowed no benefit on the common good.

Galbraith considered the greatest human achievements not products of money but of creativity. He specifically mentioned Renaissance Florence and Venice.[72] Naturally, he recognized that Florence was a product of the munificence of generations of skillful and ruthless merchants who returned to the city with the wealth of the world. On a lesser scale, Regenstein did that for Chicago, unlike more recent leaders—although Bill Farley did give his alma maters significant grants. Having started as a leader in Chicago, Kentucky, and other states with his factories, as he became a global capitalist he seemed uninterested in Bowling Green or the Cayman Islands. He became a stateless person in a world where everyone needs a state.

Thomas Friedman, who wrote effusively of the wonders of globalization, nonetheless pointed out the continuing need for a state that supports and protects its citizens.[73] Of course, he failed to admit that the globalization he celebrated caused many leaders to abandon their birthplaces. With their excessive wealth, these elite figures forgot, as Gavin Wright argued, that everyone is part of a national economy that is the sole government capable of protecting the public welfare.[74] Yet, global executives persist in rejecting any role for publicly controlled institutions in guiding the economy, while adopting self-serving but shortsighted resistance to the only positive obligations of our citizenship, such as taxes.[75]

Society witnessed their obsessive opposition to precautionary environmental regulation or taxes. Subsequently, they successfully procured the abandonment of taxes dedicated to Superfund cleanups. This elite claimed that worker health and safety protection undermined competition. They asserted that the nation, although richer, no longer could fund pension obligations that were fully supported when it was poorer. Yet, they adamantly maintained their obligation to fulfill fiduciary responsibilities to their shareholders—at least, the part about paying themselves massive incentives to raise share values.

This is the practice that Cummings has labeled "Dixification." It cursed the American South with low-wage jobs that could easily move to Honduras.[76] It led to governments so short of funds that they had to cut support for public-education programs that promised future opportunity for children. The new global economy would not finance former levels of public assistance to higher education, and sought to privatize university research. The utter foolishness of these policies for the vast majority has seldom penetrated the fog created by free-market opposition to public involvement in innovation, technology, and leadership.

This study has demonstrated that citizen resistance to aggressive public intervention in the global economy and domestic market behavior should cease. It needs to stop because good public policy demands it and ethics requires it. A community that allows anyone to pollute its river in exchange for a few years' growth is foolish. Americans act immorally if they continue to permit leaders acting in their name and with their resources to exploit poor indigenous Hondurans in order to obtain cheap underwear.[77] Consumers need to assume responsibility. Retailers who market goods using deceitful double-entendres like Wal-Mart's "bring it home to the USA" must refrain from duplicity and truly promote domestic manufacturing. Investors, especially institutional ones, must accept responsibility for their conduct at home and abroad. Most importantly, a good society should create policies to enforce this behavior if investors, consumers, retailers, and managers of global corporations fail to act.

Likewise, structures must be strengthened that protect both government regulatory agencies and institutions of learning from lobbyists who seek to distort research to further their own agendas. The half-century campaign to frustrate policy grounded in the scientific consensus on DDT is a good example of both the lengths to which certain economic interests will go, and the susceptibility of public institutions to special-interest pressures. Great universities such as the University of Wisconsin, respected media such as the *New York Times*, and agencies responsible for consumer protection, including state, federal, and international food- and health-monitoring agencies, have been threatened or corrupted by special-interest lobbying.[78] Obviously, these institutions must be free to serve as independent monitors of behavior in order to help control egregious actions.[79] A better combination of secure public funding, more distance between researchers and grant makers, and vigorous public oversight might prevent such corruption.

For the benefit of business as well as communities, society should insist on restoring respect for civic regulation of individual public conduct. Such restrictions need not act as an impediment to innovation, but could supply a spur to make companies solve community-designated problems. Some people will complain about new rules, but good regulation established by law is a process for defining community values. Since DDT clearly harms some creatures in the natural world, it is fully appropriate for a civil society to decide it wishes to restrict DDT use. It is the responsibility of those who wish to profit from solving the problems addressed by DDT to invent a safer production method or alternative product. Civil societies have every right to demand sustainable actions from their citizens, institutions, and businesses, and that prerogative should not be undermined by ideological calls for unregulated markets.[80]

THREE SUCCESSES

John Kenneth Galbraith reminds us of a beautiful alternative to neoliberal civic ideology: Florence. The history of that city, with a millennium of prosperity and culture, provides a stark alternative to the neoliberal abuse of people, their civil communities, and the natural world. As chronicled in numerous histories, Florence has modeled a form of sustainable civic life that challenges the simplistic destructiveness implicit in modern free-trade ideology.[81] In contrast to communities such as St. Louis, Michigan, that witnessed modern carpetbaggers decimating its natural resources in a single generation, Florence, notwithstanding the problems it shares with any city, is a long-term enterprise. And, in spite of the twentieth-century popularity of the Communist Party in Tuscany, Florence remains an example of vibrant capitalism.[82] However, vibrancy is the operative word. Florence's capitalism created and has sustained a remarkable heritage and beauty, rather than the utilitarianism of modern America.

Marvin Becker, one of the great historians of Florence, described the developments that fostered Florentine sustainability. He identified the mid-fourteenth century as a period in which the city transitioned from a form of capitalism similar to that of modern America to a sustainable type. His explanation is worthy of reflection. Just before 1350, Florence faced a crisis that resulted from the inadequacy of both aristocratic rule and despotism. The city's leaders and the corporate bodies they controlled had requested private immunities in order to bring prosperity. They sought freedom to break the law. As Becker phrased it, they received "judicial dispensation and annulment of sentences."[83] Much like modern America, as their privileges multiplied, "confidence in government, with its reliance upon exhortation, admonition, and political magnanimity, ebbed."[84] No one trusted calls for sacrifice or service from a politician "on the take."

In this critical period, Florentines reformed the relationship of the citizen to the state, and thereby gave birth to a vibrant capitalism in which the city flourished, despite scarce resources, and emerged as the richest city in the world. The key after 1343 was the creation of a powerful state "that repressed recalcitrant patrician, obdurate magnate, and overly expansive corporate bodies. . . . Successive regimes replaced private immunities with public law. . . . Democratized rule that exalted the power and majesty of communal law came into being. . . . [The good citizen] was expected to demonstrate an intense concern for the life and welfare of the [city]."[85] Elsewhere, Becker characterized the era as one that saw the rise of "the public world." While the city would still face many crises, including invasions, the rise of new autocrats, plagues, and all manner of natural disasters, the commitment to a "public world" harnessed the energies of capitalists to work for the common good. As Becker declared, "The great oligarchs . . . were now virtually stockholders in a giant corporation called the state."[86] Ben Heineman would have supported that stock ownership. With the extreme individualism popular after the 1970s, later leaders of Farley's generation could not. They possessed no commitment either to the public or to a civic space defined by national borders.

Other examples of a vigorous polis consciously united by business leaders and their community to enhance the common good exist in the United States. A recent study of St. Paul, Minnesota, maintains that leaders in the late nineteenth century, especially railroad tycoon James J. Hill and archbishop John Ireland, helped sustain their community by minimizing the labor-industry conflicts that raged in other cities. They achieved peace with labor by seeking business commitments to minimal labor rights, not by suppressing worker demands. They achieved a city marked by stability in business, housing, and employment. Not as brash as its twin city, which based its economy on the success of a number of large multinational firms, St. Paul realized a more modest, peaceful, and sustainable development grounded in a "civic compact" forged by responsible business leaders and organized labor. Together, they reminded the entire elite of the "mutual obligations owed to fellow citizens."[87]

Perhaps revealing the longtime bias of laissez-faire ideology, a reporter sent by *Fortune* to report on labor unrest in Minneapolis in the 1930s described St. Paul as a dying commercial backwater, because it permitted organized labor to thrive.[88] Of course, the city survived and remained peaceful and prosperous into the next century. The reporter viewed progress as rapid growth and the suppression of traditional rights. Neoliberal leaders tended to appreciate neither the past nor specific communities. Therefore, it was inappropriate for them to claim the title "conservative." St. Paul, oddly, was both one of the most leftist labor citadels and one of the most "conservative" cities in America. The two features were not inconsistent; both reflected a rejection of the radical individualism inherent in neoliberalism.

Like leaders in Florence, those in St. Paul had a sense of place and an obligation to make the city work. Those who think only in terms of global perspectives can ignore the terrible consequences for specific communities and their citizens.[89] Global elites can focus comfortably on their personal enrichment and assume the world is better off. As Mary Wingerd concluded in her study of St. Paul:

> The political lesson is evident. The priorities of everyday life lie at the heart of political choice. We cannot understand the past, nor can we effectively use it to make a better future unless we take place-based consciousness seriously—in all its cultural complexity. . . . This reminder is particularly timely today, with national policy-making increasingly divorced from the realm of everyday life and political apathy a growing infection. In the search to reverse this trend, a reconnection between local and national concerns seems imperative.[90]

The global neoliberals committed to no place, only to an ideology, can ignore the jobs lost in Clarksville, Tennessee, or Toledo in order to accumulate a fortune.[91] Likewise, rationalizations can justify dumping tons of pesticide wastes in a river in Michigan or a rural aquifer in Tennessee as the necessary cost of profiting from the genius of Julius Hyman. Globalized thinking overlooks arguments against its self-serving behavior. It paralyzes community reaction, ironically labeling it as selfish, while it tramples workers' rights and loots natural resources in the name of economic growth.

At the end of her study of St. Paul, Wingerd referenced the work of Jackson Lears, who described a link between those focused on the local and those concerned with future sustainability. Lears discovered that places such as St. Paul that had preserved "loving memories of the past could spark rebellion against the present in the service of future generations."[92] This observation could be applied equally to St. Louis, Michigan. Hardly the size of Florence or St. Paul, this community nonetheless modeled similar civic empowerment, confronting threats from irresponsible professionals and demanding accountability for the community's future.

If modern neoliberal ideologues are to be prevented from abuses of economic and technological power in the name of the global economy and scientific truth, an alternative source of knowledge must be recognized. As Frank Fisher has described it, and as the St. Louis CAG has shown, those with local knowledge must be empowered. It is not to claim that the locals always are right. As Mary Wingerd has said, "Place based cultural defensiveness also has a darker side, with the capacity to nurture a catalogue of evils— nativism, racism, fascism, ethnic cleansing—as critics of localism often have rightly noted."[93] However, those same vices have been inherent in modern neoliberalism, which seeks to impose the economic system that benefits an elite on the world. They have been intolerant of the concerns of residents of rural Michigan, the neighbors of the North Hollywood dump, and the workers brought to the free-trade zones in Honduras.

The empowerment of local knowledge is not a panacea. However, it is an imperative if a fair policy debate is to occur. Also essential is a renewed appreciation of the need for vigorous public policy. Superfund's history provides an excellent example. The program has not worked to fully clean up many places. There have been mistakes, as in St. Louis in the early 1980s. However, the purpose of public policy is not perfection. We are just muddling through, to use Charles Lindblom's phrase.[94] But the majesty of public law, including Superfund, is that it captures a sense of the civil society that must clean up its messes and assess major penalties on those who cause them. At its best, Superfund is a reminder that the national policy is to use resources with respect for the benefit of community descendants, not for short-term self-interest. Because they are in touch with a place and because, as parents and grandparents, they are in contact with the future, the citizens of St. Louis fought the former polluter and the government experts for what Becker called the "judicial dispensation" they unjustly gave Velsicol in 1982. As the defenders of democracy in Florence in 1343, they battle to "[tame] the [American] magnate and exercise . . . political pressure [to compel them] to conform to the norms of law."[95]

This study has shown that society needs to celebrate the role of the citizen in overcoming two modern challenges to democratic policymaking and civic integrity. First is the ideology of expertise that insists public development of policy should be replaced by policy controlled by experts. In the case of the environment, professionals insist that citizens are not to take precautions; they must yield to the results of risk assessments conducted by experts funded by powerful interests. Second is the ideology that asserts leadership is powerless before economics, which limits the possibility of responsiveness and responsibility. In July 2005, Scott Hummel, the Michigan state senator who represented St. Louis, used the common neoliberal excuse of powerlessness, saying that the community should realize it must trade some environmental protection for jobs.[96] He and his fellow neoliberal leaders exhorted citizens to abandon any hope of defining or protecting the public good when it conflicts with current market norms.[97]

These twin assaults can be controlled if civic institutions are restored. When Senator Hummel excused inaction in St. Louis, a CAG member, Verna Hollenbeck, responded firmly: "The only way to have a strong economy is to have a clean environment. . . . To not take care of our environment is to not take care of our children, our most vulnerable citizens, our future."[98] Verna Hollenbeck's voice was enhanced in that she had a civic structure to empower her, the Pine River Task Force. Unfortunately, continuous employer attacks on organized labor have weakened its ability, especially on a multinational level, to provide a similar civic structure to give expression to fired and exploited workers. If the culture could celebrate labor's power rather than demean it, the society might overcome the ideology of economic helplessness. Despite their greater wealth, shareholders or participants in various institutional investment structures, such as pension plans and endowments, have little more influence than labor.[99] The hope for investors, communities, or workers harmed by the conduct of leaders described in this study is to copy the citizens of Florence in 1343 and "assert the primacy of the public world over private interest."[100] The tendency in the United States for citizens to exit failing enterprises rather than compel reform must end.[101] Citizens need to be inspired and civic life renewed by the cases in this review, where a handful of people, as in St. Louis, Michigan, said "No!" to leaders who justified community rape in the name of expertise and economics.

Notes

PREFACE

1. They presented their findings first at the Michigan Public Management Institute in May 1992, receiving special attention in the public administration newsletter, *PA Times* (July 1992), 8; later their study was converted into one in a series of reports on environmentally devastated communities: see *St. Louis, Michigan: The Fallout from the Michigan PBB Crisis* (Livingston, KY: Appalachia Science in the Public Interest Publications, 1993).

INTRODUCTION

1. For a good review of this, see Lou Grumet, "Commentary: A Financial Frankenstein: Deconstructing the Argument for the Return of Glass-Steagall," *Accounting Today*, 5 January 2009, 6.
2. On the triple bottom line, see Rob Gray and Jan Bebbington, *Accounting for the Environment*, 2nd ed. (London: Sage Publications, 2001), 254, 315. More generally on sustainable capitalism, see Paul Hawken, *The Ecology of Commerce* (New York: HarperCollins, 1993); and Paul Hawken, Amory Lovins, and L. Hunter Lovins, *Natural Capitalism: Creating the Next Industrial Revolution* (Boston: Little, Brown and Co., 1999).
3. Exemplifying the need for this understanding was the report of Gene Dobaro, acting controller general of the United States, *A Framework for Crafting and Assessing Proposals to Modernize the Outdated U.S. Financial Regulatory System*, U.S. General Accountability Office Report 09–216, 8 January 2009, which when dealing with AIG only focused on its mortgage insurance mistakes, not seeing them as part of a larger pattern of speculative failure.
4. For example, Michael Milken, who arranged Fruit of the Loom's disastrous expansion in the 1980s, also funded the Milken Institute in Los Angeles, which with officials in the Office of the Comptroller of the Currency lobbied for the repeal of the Glass Steagall banking regulation regime that contributed to the

financial crisis in 2008; see James R. Barth, R. Dan Brunbaugh Jr., and James A. Wilson, "Policy Watch: The Repeal of Glass-Steagall and the Advent of Broad Banking," *Journal of Economic Perspectives* 14 (Winter 2000): 191–204.

5. For example, under the 2008 bank rescue process, Wells Fargo and Company received $25 billion to buy (bail out) Wachovia Corp. The merger was finalized at a 23 December 2008 vote of shareholders, with the merger effective 31 December 2008. On 15 December 2008, Wachovia paid dividends to all shareholders. See Wachovia Corp. Notice of Special Meeting of Shareholders to be held December 23, 2008, letter dated 21 November 2008, from Jane Sherburne, Secretary.

6. Thomas Berry, *The Great Work: Our Way into the Future* (New York: Bell Tower, 1999), 59.

7. Jacob Hacker, *The Great Risk Shift: The Assault on American Jobs, Families, Health Care and Retirement* (New York: Oxford University Press, 2006), 61–85.

8. More generally on this process see Benjamin Barber, *Consumed: How Markets Corrupt Children, Infantilize Adults, and Swallow Citizens Whole* (New York: W.W. Norton, 2007), 116–165.

9. For example, even John P. Diggins, *Ronald Reagan: Fate, Freedom, and the Making of History* (New York: Norton, 2008), 304–310, 324, while generally praising Reagan's accomplishments, holds that his environmental record as president was his major failure.

10. Walter J. Stone, *Republic at Risk: Self-Interest in American Politics* (Pacific Grove, CA: Brooks Cole, 1990), 10–46, has a good summary of self-interest theory.

11. *Norma Rae* (Century City, CA: 20th Century Fox, 1979), film based upon the book by Henry P. Leifermann, *Crystal Lee: A Woman of Inheritance* (New York: Macmillan, 1975).

12. Timothy J. Minchin, *"Don't Sleep with Stevens!" The J.P. Stevens Campaign and the Struggle to Organize the South, 1963–1980* (Gainesville: University Press of Florida, 2005), 166–178.

13. David Greising, "Bill Farley Could Lose His Shirt—and His Underwear," *Business Week*, 11 March 1991.

14. Robert Kuttner, *The Squandering of America: How the Failure of Our Politics Undermines Our Prosperity* (New York: Knopf, 2007), 97–98, reviews some of the conglomerate origins of the behavior of the 1980s; elsewhere Kuttner focuses upon the argument that the Reagan administration ushered in the conditions that allowed for junk bonds.

15. For a good review of Superfund, see Daniel Mazmanian and David Morell, *Beyond Superfailure: America's Toxics Policy for the 1990s* (Boulder, CO: Westview Press, 1992). For a more recent study, see Katherine N. Probst and

David M. Konisky, *Superfund's Future: What Will It Cost?* (Washington, DC: Resources for the Future, 2001).

16. John Albertine, "This Covert Recovery—Commentary: Burdensome Rules Hurting Economic Growth," CBS *Market Watch*, 6 June 2003 (accessed 2 July 2003). Albertine said, "Unfortunately, we had eight years of bone-headed Clinton administration policies that put barriers in the path of entrepreneurs and average working Americans who were trying to grow the real output of this economy."

17. *Detroit News*, 21 August 1997. The *Detroit Free Press*, 8 May 1993, 3A–4A, got some of the story straight, since it interviewed farmers' families who disputed the reports of "no lasting harm."

18. Of course some say the United States has no ideology; however, Richard Hofstadter, the great U.S. historian, once said, "It has been our fate as a nation not to have ideologies but to be one"; quoted in Michael Kazin, "The Right's Unsung Prophet," *The Nation*, 20 February 1989, 242.

19. See, for example, the studies of American exceptionalism, such as John W. Kingdon, *America the Unusual* (New York: Worth Publishing, 1999); Seymour M. Lipset, *American Exceptionalism: A Double Edged Sword* (New York: W.W. Norton, 1996); and for a collection of thoughts on the topic, Byron E. Shafer, ed., *Is America Different? A New Look at American Exceptionalism* (New York: Oxford University Press, 1991).

20. Joseph Schumpeter, *Capitalism, Socialism and Democracy*, 3rd ed. (New York: Harper and Row, 1950), 81–86.

21. Robert Fogel, *Without Consent or Contract: The Rise and Fall of American Slavery* (New York: W.W. Norton, 1989), 244–246, discusses this difficulty.

22. Frank Tugwell and Andrew S. McElwaine, "The Challenge of the Environmental City: A Pittsburgh Case Study," in *Toward Sustainable Communities: Transition and Transformations in Environmental Policy*, ed. Daniel Mazmanian and Michael Kraft (Cambridge, MA: MIT Press, 1999), 195–197.

23. For example, *Lucas v. South Carolina Coastal Council*, 505 U.S. 1003 (1992).

24. Albert O. Hirschman, *Exit, Voice, and Loyalty* (Cambridge, MA: Harvard University Press, 1970), 106–119.

25. These higher principles could arise in the defense of civil liberties, as critiqued by writers such as Mary Ann Glendon, *Rights Talk the Impoverishment of Political Discourse* (New York: Free Press, 1991), or in libertarian philosophies, such as objectivism, which influenced leaders such as long-term federal reserve chair Alan Greenspan.

26. For a good discussion of fear and individualism, see Christopher Lasch, *The Culture of Narcissism: American Life in an Age of Diminishing Expectations* (New York: W.W. Norton, 1978).

27. Gary S. Becker and Kevin M. Murphy, "The Upside of Income Inequality," *The American*, May/June 2007, 20–23. The same issue of *The American* had an article raising concerns that the fastest-growing economies now were in countries that suppressed political freedom and human rights; see Kevin Hassett, "Economic Freedom, Political Freedom," *The American*, May/June 2007, 96–97.

28. John E. Ikerd, *Sustainable Capitalism: A Matter of Common Sense* (Bloomfield, CT: Kumarian Press, 2005), 6; quotation from R. M. Ryan and E. L. Deci, "On Happiness and Human Potentials: A Review of Research on Hedonic and Eudaimonic Well-Being," *Annual Review of Psychology* 52 (2001): 141–166.

29. For a discussion of income trends, see Thomas Piketty and Emanuel Saez, "The Evolution of Top Incomes: A Historical and International Perspective," *American Economic Review, Papers and Proceedings* 96 (2006): 200–205. On voter participation, see information from the Vanishing Voter Project at Harvard; Thomas E. Patterson, *The Vanishing Voter: Public Involvement in an Age of Uncertainty* (New York: Knopf, 2002). On changes in churches, see Thomas Reeves, *The Empty Church: Does Organized Religion Matter Anymore?* (New York: Free Press, 1998).

30. Kenneth D. Wald, *Religion and Politics in the United States*, 4th ed. (Lanham, MD: Rowman and Littlefield, 2003), 14–16; see also comments in Seymour Lipset, *Continental Divide: The Values and Institutions of the United States and Canada* (New York: Routledge, 1990), 74–89.

31. For the work of the master, see Edward L. Bernays, *The Engineering of Consent* (Norman: University of Oklahoma Press, 1955), 3–25; for context, Larry Tye, *The Father of Spin: Edward Bernays and the Birth of Public Relations* (New York: Crown, 1998).

32. Alexis de Tocqueville, *Democracy in America*, vol. 1, ed. Phillips Bradley (New York: Alfred A. Knopf, 1956), 299.

33. Tocqueville, *Democracy in America*, vol. 2, 123.

34. Robert N. Bellah et al., *Habits of the Heart: Individualism and Commitment in American Life* (New York: Harper and Row, 1985), especially 142–163.

35. One of the early studies of these modern trends was Phillip Slater, *The Pursuit of Loneliness* (Boston: Beacon Press, 1970), 117–118, which warned that individualism undermined his utopian hope to overcome the excessive fears and loneliness of Americans: "I am not arguing that individualism need be totally extirpated in order to make community possible, but new-culture enterprises often collapse because of a dogmatic unwillingness to subordinate the whim of the individual to the needs of the group." Repeatedly, communitarians have shared his concerns, such as Amitai Etzioni, *The Spirit of Community: Rights,*

Responsibilities, and the Communitarian Agenda (New York: Crown Publishers, 1993), 26; Jean Bethke Elshtain, *Democracy on Trial* (New York: Basic Books, 1995), 11–12. Alan Ehrenhalt, *The Lost City: The Forgotten Virtues of Community in America* (New York: Basic Books, 1995), 210–211, reflected on the evolving U.S. relationship between individualism and middle-class conformity.

36. K. William Kapp, *The Social Costs of Private Enterprise* (Cambridge, MA: Harvard University Press, 1950), vii.

37. Peter F. Drucker, *Innovation and Entrepreneurship: Practice and Principles* (New York: Harper and Row, 1985), vii.

38. The technical assistance grant (TAG) process began under the Superfund Amendments and Reauthorization Act of 17 October 1986. The community advisory group (CAG) process grew out of the environmental justice movement and a separate process, often called Round 1, of efforts to improve Superfund, both launched by William Reilly, EPA administrator under President George H. W. Bush. The final procedures for establishing CAGs were written under the Clinton administration, Office of Solid Waste and Emergency Response (OSWER) Directive 9230.0–28, December 1995.

39. On 19 June 2008, the task force and Alma College won the Jimmy and Rosalynn Carter Partnership Award for Campus Community Collaboration, *(Alma, MI) Morning Sun*, 20 June 2008, 1.

40. The hearings were: U.S. Department of Justice, Public Meeting Re: Settlement between the Department of Justice and TPI, held at the Heritage Center for the Performing Arts, Alma College, Alma, Michigan, 10 May 2000; also U.S. Department of Justice Public Meeting Concerning the St. Louis Facility (Velsicol Chemical Superfund Site), Gratiot Senior Center, St. Louis, Michigan, 19 June 2002.

41. Meanwhile, conflicts with regulators escalated as community leaders in St. Louis battled for a health study and further remediation at related sites; see letters from Pine River Superfund Citizen Task Force to Senator Carl Levin, 29 August 2001, and to Dr. Anne Sassaman, Director, Division of Extramural Research and Training, National Institutes of Health, 4 September 2001; letter from Senator Levin to Ruth Kirschstein, Acting Director, National Institutes of Health, 17 September 2001; letter from Pine River Task Force to Frederick Tyson, Program Administrator, Division of Extramural Research, National Institute of Environmental Health Sciences, 22 October 2001; letter from Pine River Task Force to Assistant Administrator, Agency for Toxic Substances Disease Registry, 30 April 2004—all in Health Grant 2001 file, in Pine River Files, Alma College Archives, Alma, MI. These resulted in a Centers for Disease Control community health consultation at Alma College,

in Alma, Michigan, on 19–20 January 2005; see Health Grant 2001 file, in Pine River Files, Alma College Archives.

42. John Elkington, *Cannibals with Forks: The Triple Bottom Line of the Twenty-First Century Business* (Oxford: Capstone, 1997), 69?98.

43. *New York Times*, 5 June 2007, F1 and F6; see also earlier editorial, 5 October 2006, A28.

44. Suzanne Snedeker, "View of the Pine River and Beyond: The Legacy of DDT Use and Health Effects," *The Ribbon* 13 (Spring 2008). Dr. Snedeker was a participant in the conference and coauthored the conference consensus statement.

45. Nicholas Sambanis, "Using Case Studies to Expand Economic Models of Civil War," *Perspectives on Politics* 2 (June 2004): 261.

46. On Carter-era deregulation efforts, see, for example, L. S. Keyes, *Regulatory Reform in Air Cargo Transportation* (Washington, DC: American Enterprise Institute, 1980), or AEI Legislative Analyses, *Proposals for Railroad Regulatory Reform* (Washington, DC: American Enterprise Institute, 1980). For a typical neoliberal assessment, see "How Deregulation Increased Efficiency and Reduced Fares," *Business Week*, 18 August 1980, 80–81. On environmental policy deregulation, see Paul R. Portney, ed., *Natural Resources and the Environment: The Reagan Approach* (Washington, DC: Urban Institute Press, 1984).

47. Nancy Cartwright, *The Dappled World: A Study of the Boundaries of Science* (New York: Cambridge University Press, 1999), 151.

48. Pierre Duhem, *The Aim and Structure of Physical Theory*, trans. Philip P. Wiener (New York: Atheneum, 1962), 151.

49. Kim Lane Scheppele, "Cultures of Facts," *Perspectives on Politics* 1 (June 2003): 364.

50. On growing inequality in the era, see Piketty and Saez, "The Evolution of Top Incomes." For an earlier version of the income debate, see Kevin Phillips, *The Politics of Rich and Poor: Wealth and the American Electorate in the Reagan Aftermath* (New York: Random House, 1992), 154–209. And for the opposing view, see Becker and Murphy, "The Upside of Income Inequality," and also George Gilder, *Wealth and Poverty* (New York: Basic Books, 1980), 96–101.

51. See, for example, Charles Bowden, *Juarez: The Laboratory of Our Future* (New York: Aperture, 1998).

52. David H. Rosenbloom, "Public Administrative Theory and the Separation of Powers," *Public Administration Review* 43 (May/June 1983): 219–226.

53. David H. Rosenbloom, *Public Administration: Understanding Management, Politics, and Law in the Public Sector*, 2nd ed. (New York: Random House, 1989), 14–28, has a more thorough review application of the three approaches.

54. For a list of the sites, see *In re: Fruit of the Loom, Inc., et al.*, United States Bankruptcy Court for the District of Delaware, No. 99–4497 Settlement Agreement, attachment A, A&I Facility List.

55. This number is either the total of those employed in 1979, or, for firms added later, employment at the time of acquisition, using industrial directory information from each state.

56. See, for example, the video *Sweating for a T-Shirt* (San Francisco: Global Exchange, 1998).

57. Daniel A. Mazmanian and Michael E. Kraft, "The Three Epochs of the Environmental Movement," in *Toward Sustainable Communities: Transition and Transformations in Environmental Policy*, ed. Mazmanian and Kraft (Cambridge, MA: MIT Press, 2001), 17, 18.

58. Grant McConnell, *Private Power and American Democracy* (New York: Vintage Books, 1966), 336–368; Theodore Lowi, *The End of Liberalism* (New York: W.W. Norton, 1969), 287–293. The continued importance of this interpretation was reflected in the publication in 2009 of a 40th anniversary edition of *The End of Liberalism*.

59. Cass R. Sunstein, *Risk and Reason: Safety, Law, and the Environment* (New York: Cambridge University Press, 2002), 19–27, has a good summary of his position.

60. Cass R. Sunstein, "Beyond the Precautionary Principle," *University of Pennsylvania Law Review* 151 (January 2003): 1003–1010; Indur M. Goklany, *The Precautionary Principle: A Critical Appraisal of Environmental Risk Assessment* (Washington, DC: Cato Institute, 2001), 85–88; Jonathan B. Wiener, "Precaution in a Multi-Risk World," in *Human and Ecological Risk Assessment: Theory and Practice*, ed. Dennis D. Paustenbach (New York: J. Wiley, 2002).

61. For example, on the uncertainty of calculating natural-resource damages, see Jason J. Czarnezki and Adrianne K. Zahner, "The Utility of Non-Use Values in Natural Resource Damage Assessments," *Boston College Environmental Affairs Law Review* 32 (2005): 509–526.

62. James C. Scott, *Seeing Like a State: How Certain Schemes to Improve the Human Condition Have Failed* (New Haven: Yale University Press, 1998), 4–6, 89–97.

63. On Milliken's general environmental record and PBB, see Dave Dempsey, *William G. Milliken: Michigan's Passionate Moderate* (Ann Arbor: University of Michigan Press, 2006), 97–101.

64. Letter from William Rustem, Special Assistant to the Governor of Michigan, to Arnold Bransdorfer, County Commissioner, Gratiot County, Michigan, 27 March 1978, Pine River files, Alma College Archives.

65. Paul Oskar Kristeller, *Renaissance Thought: The Classic, Scholastic, and Humanist Strains* (New York: Harper and Row, 1961), 139, wrote of the need

to apply its lessons to the work of neoliberal experts, whom he called positivists. "The startling events of our own time have shaken our confidence in the sufficiency, if not the truth, of positivism. We wonder whether its principles are broad enough to explain our experience and to guide our endeavors. We have become more modest about our achievements and hence more willing to learn from the past."

66. Frank Fischer, *Citizens, Experts, and the Environment: The Politics of Local Knowledge* (Durham, NC: Duke University Press, 2000), 1.

67. Max Weber, "Science as a Vocation," in *From Max Weber: Essays in Sociology*, ed. and trans. H. H. Girth and C. Wright Mills (New York: Oxford, 1946), 145–147, 154–155, on 143, after reviewing how scientific experts have claimed to explain all things, concludes quoting Tolstoy, "Science is meaningless because it gives no answer to our question, the only question important for us: 'What shall we do and how shall we live?'"

CHAPTER 1. CORPORATE LEADERSHIP PROBLEMS

1. Sid Cato, "World's 10 Worst Annual Reports for 1998," at http://www.sidcato.com/topten98.htm (accessed 23 June 2005).

2. Filed 9 August 2000, with Donlin, Recano and Co., Agent for U.S. Bankruptcy Court, *In re: Fruit of the Loom, Inc., et al.*, No. 99-04497 (PJW).

3. David Harvey, *A Brief History of Neoliberalism* (New York: Oxford University Press, 2005), 11–12, uses these two terms as in this study.

4. *New York Times*, 29 June 2005, A1 and C4.

5. On evolution of Fruit of the Loom trademark, see *New York Times*, 14 July 1974, 138.

6. For a general discussion, see Edward Lorenz, *Defining Global Justice: The History of U.S. International Labor Standards Policy* (Notre Dame, IN: University of Notre Dame Press, 2001), 105–119.

7. *Moody's Manual of Investments, 1928* (New York: Moody's Investors Service, Inc., 1928), 1337.

8. See, for example, Louis Galambos, *Competition and Cooperation* (Baltimore: Johns Hopkins University Press, 1966); James Hodges, *New Deal Labor Policy and the Southern Textile Industry* (Knoxville: University of Tennessee Press, 1986); and case studies such as Louise Lamphere, *From Working Daughters to Working Mothers: Immigrant Women in a New England Industrial Community* (Ithaca, NY: Cornell University Press, 1987).

9. April S. Dougal and Susan W. Brown, "Fruit of the Loom, Inc.," in *International Directory of Company Histories* (New York: St. James Press, 1999), 164–167.

10. Marc S. Reisch, "From Coal Tar to Crafting a Wealth of Diversity," *Chemical and Engineering News*, 12 January 1998, said, "The Great Depression did not slow chemical industry development."

11. *New York Times*, 29 May 1965, 30.

12. Interview with Deacon Dunbar, 12 May 1999.

13. Michigan Chemical Company, *Annual Report, 1947.*

14. Michigan Chemical Corp., *Pestmaster Progress*, 17 May 1946, in St. Louis (MI) Historical Society.

15. Ibid.

16. Cecil K. Drinker et al., "The Problem of Possible Systemic Effects from Certain Chlorinated Hydrocarbons," *Journal of Industrial Hygiene and Toxicology* 19 (September 1937): 283.

17. *New Yorker*, 26 May 1945, 18.

18. John K. Terres, "Science and Technology: Dynamite in DDT," *New Republic*, March 1946, 415.

19. *New York Times*, 16 March 1947, "Sunday Magazine," 19.

20. Terres, "Science and Technology," 416.

21. O. Garth Fitzhugh and Arthur A. Nelson, "The Chronic Oral Toxicity of DDT (2,2-bis p-chlorophenyl-1,1,1-tri-chloroethane)," *Journal of Pharmacology and Experimental Therapeutics* 89 (January 1947): 18–30. See also the following early studies: Ray F. Smith et al., "Secretion of DDT in Milk of Dairy Cows Fed Low Residue Alfalfa," *Journal of Economic Entomology* 41 (1948): 759–763.

22. Arthur A. Nelson and Geoffrey Woodard, "Severe Adrenal Cortical Atrophy (Cytotoxic) and Hepatic Damage Produced in Dogs by Feeding DDD or DDE," *A.M.A. Archives of Pathology* 48 (October 1949): 392.

23. G. C. Carmen et al., "Absorption of DDT and Parathion by Fruits," *Abstracts, 115th Meeting American Chemical Society* (1949): 30A.

24. U.S. Congress, House of Representatives, House Select Committee to Investigate Use of Chemicals in Food Products, part 1, 275. This Delaney Committee greatly influenced Rachel Carson; see Linda Lear, *Rachel Carson: Witness for Nature* (New York: Henry Holt, 1997), 321.

25. Wolfgang F. Von Oettingen, *The Halogenated Aliphatic, Olefinic, Cyclic, Aromatic, and Aliphatic-Aromatic Hydrocarbons: Including the Halogenated Insecticides, Their Toxicity and Potential Dangers* (Washington, DC: U.S. Dept. of Health, Education, and Welfare, Public Health Service Publication No. 414, 1955), see especially 341–367.

26. On the impact of this legislation, see Donna U. Vogt, "The Delaney Clause Effects on Pesticide Policy," Congressional Research Service Report 95–514 (13 July 1995).

27. Frank Graham Jr., *Since Silent Spring* (Greenwich, CT: Fawcett, 1970), 35–47, discusses first warnings in 1945.

28. Michigan Chemical, *Annual Report, 1960*, 12.

29. *New York Times*, 30 March 1958, F7.

30. *New York Times*, 3 December 1965, 60.

31. Marvin came to the company after serving with Hercules Powder for more than thirty years, and was president of Michigan Chemical from 1953–1963; on his appointment, see *New York Times*, 1 July 1953, 45.

32. "Michigan Chemical Corporation," *Chemical and Engineering News*, 8 April 1957, 46.

33. Ulrich Beck, *Risk Society: Towards a New Modernity* (Newbury Park, CA: Sage Publications, 1992), discusses the problems of professionals policing themselves.

34. On the Van Lenneps, see *New York Times*, 13 January 1955, 1; and *Lexington Herald-Leader*, 18 August 2000, which describes their role for many years as owners of Castleton Farms, a major breeder of thoroughbreds.

35. Various Michigan Chemical annual reports explain expansions: Pine Bluff joint plant 1947–1948; a second Michigan plant in Manistee in 1953; El Dorado, Arkansas, in 1956; Port St. Joe, Florida, in 1958; Cedar Rapids in 1961; Napoleon, Ohio, in 1963; and Boise in 1965; annual reports in St Louis (MI) Historical Society.

36. *New York Times*, 7 February 1946, 22, has Morris's obituary.

37. City of Saginaw (MI), *City Council Minutes*, 3 December 1935, 407.

38. *St. Louis (MI) Leader*, 4 September 1941, 1–2.

39. On May 1, 1955, when in response to such criticisms the state Water Resources Commission ordered the City of St. Louis to stop polluting the river, the city blamed industry; see *Gratiot County (MI) Herald*, 4 April 2002, 13.

40. Michigan Department of Natural Resources, "A Biological Survey of the Pine River above Alma to M-30 to Determine Effects of Pollution, May 31, 1955."

41. Philip Shabecoff, *Earth Rising: American Environmentalism in the 21st Century* (Washington, DC: Island Press, 2000), 122; on the early EPA, see Benjamin Kline, *First along the River: A Brief History of the U.S. Environmental Movement*, 2nd ed. (San Francisco: Acada Books, 2000), 92–93; on Nixon and Muskie, see J. Brooks Flippen, *Nixon and the Environment* (Albuquerque: University of New Mexico Press, 2006), 75, 85–93.

42. Letter from Senator Edmund S. Muskie to Ralph Purdy, Executive Secretary, Michigan Water Resources Commission [date stamp illegible, apparently December 1970].

43. *American Men and Women of Science*, 12th ed. (New York: R.R. Bowker, 1972), 3:2937.

44. Most of this history is taken from court documents, primarily from the opinion of the court in *Julius Hyman v. Joseph Regenstein*, U.S. Court of Appeals Fifth Circuit, 258 F. 2d 502, 23 July 1958.

45. *United States Envelope Co. et al. v. Transo Paper Co. et al.*, U.S. District Court for Connecticut, 221 F. 79, 1 March 1915.

46. *Velsicol Corporation v. Julius Hyman*, Appellate Court of Illinois, First District, 338 Ill. App. 52; 87 N.E. 2d 35, 20 June 1949.

47. *Pure Oil Co. v. Hyman et al.*, U.S. Court of Appeals Seventh Circuit, 95 F. 2d 22, 29 January 1938.

48. *Velsicol v. Hyman*, Appellate Court of Illinois, 11.

49. *Julius Hyman v. Velsicol Corporation et al.*, Appellate Court of Illinois 342 Ill. App. 489, 31 January 1951, 3.

50. *Hyman v. Regenstein* at the U.S. Court of Appeals included Helen Regenstein, Betty Hartman, and Joseph Jr.

51. Muller won the prize for applications of DDT; the address was delivered on 4 December 1948.

52. For details on Hyman's legacy at Rocky Mountain, see Jack Doyle, *Riding the Dragon: Royal Dutch Shell and the Fossil Fire* (Boston: Environmental Health Fund, 2002), 39–68; also U.S. Environmental Protection Agency, Region 8, Rocky Mountain Arsenal site description, available at http://www.epa.gov/region08/superfund/co/rkymtnarsenal. Hyman used his fortune to establish a small lab in Berkeley, California, which locals still believe is a place of contamination; interview of Ron Penndorf, Berkeley, CA, 29 June 2005.

53. J. Lisa Jorgenson, "Aldrin and Dieldrin: A Review of Research on Their Production, Environmental Deposition and Fate, Bioaccumulation, Toxicology, and Epidemiology in the United States," *Environmental Health Perspectives* 109 (March 2001): 114.

54. Walter C. Alvarez, MD and Samuel Hyman, MD, "Absence of Toxic Manifestations in Workers Exposed to Chlordane," *AMA Archives of Industrial Hygiene and Occupational Medicine* 8 (November 1953): 480–483. Two years earlier, research by Frank Princi, MD and George H. Spurbeck, MD, "A Study of Workers Exposed to the Insecticides Chlordan [*sic*], Aldrin, and Dieldrin," *AMA Archives of Industrial Hygiene and Occupational Medicine* 3 (January 1951): 64–72, was openly funded by "a grant from Julius Hyman and Company, Denver."

55. See "History of Rutgers' Department of Entomology," available at http://www.rei.rutgers.edu-insects/indust.hym, which shows Hyman funded fellows from 1951 to 1955.

56. Generally on this issue see Derek Bok, *Universities and the Marketplace: The Commercialization of Higher Education* (Princeton, NJ: Princeton University

Press, 2003); and focused on medical research, such as that on chlordane, see Sheldon Krimsky, *Science and the Private Interest: Has the Lure of Profits Corrupted Biomedical Research* (Lanham, MD: Rowman and Littlefield, 2003). For a general overview of the issue, see Nathan Rosenberg and Richard R. Nelson, "American Universities and Technical Advance in Industry," *Research Policy* 23 (1994): 323–348. Gerald Markowitz and David Rosner, *Deceit and Denial: The Deadly Politics of Industrial Pollution* (Berkeley: University of California Press, 2002), 5, optimistically see a new generation of scientists supported by foundations and independent of industry who are willing to serve the community; their view reinforces the point here about the 1960s.

57. *Julius Hyman and Co., et al. v. Velsicol Corporation*, Supreme Court of the United States 342 U.S. 87; 72 S. Ct. 113, 5 November 1951; and *Hyman v. Velsicol Corp.*, Supreme Court of the United States 339 U.S. 966; 70 S. Ct. 1002, 29 May 1950.

58. On Thurmond Arnold, see Wilson D. Miscamble, "Thurmond Arnold Goes to Washington: A Look at Antitrust Policy in the Later New Deal," *Business History Review* 56 (Spring 1982): 1–15. On Pepper's liberalism, see Tracy E. Danese, *Claude Pepper and Ed Ball: Politics, Purpose, and Power* (Gainesville: University Press of Florida, 2000).

59. James MacGregor Burns, *Roosevelt: The Lion and the Fox* (New York: Harcourt, Brace and World, 1956), 198–202, had an early discussion of the origins of the broker state; see also his bibliography, 510–511. Theodore Lowi, "Four Systems of Policy, Politics, and Choice," *Public Administration Review* 32 (July-August 1972): 303, discusses the difficulty of achieving good regulatory policy, even during the New Deal, when the president had special support versus interest groups.

60. *New York Times*, 25 July 1955, 29.

61. *Chemical and Engineering News*, 8 April 1957, 73.

62. B. Davidow and J. L. Radomski, "Isolation of an Epoxide Metabolite from Fat Tissues of Dogs Fed Heptachlor," *Journal of Pharmacology and Experimental Therapeutics* 107 (March 1953): 259–265; and later Norman Gannon and J. H. Bigger, "The Conversion of Aldrin and Heptachlor to Their Epoxides in Soil," *Journal of Economic Entomology* 51 (February 1958): 1–2.

63. C. S. Myers, "Endrin and Related Pesticides: A Review," *Pennsylvania Department of Health Research Report Number 45* (1958); and Harold Jacobziner and H. W. Raybin, "Poisoning by Insecticide (Endrin)," *New York State Journal of Medicine* 59 (15 May 1959): 2017–2022.

64. The *New York Times* ran display ads in the spring, for example in 1958 on 30 March, 12; 13 April, 22; 20 April, 5; and 18 May, 24; and in 1959 on 22 March, 32; 5 April, 10; and 12 April, 20.

65. *New York Times*, 16 March 1947, "Sunday Magazine," 19.

66. Both are available online at the Prelinger Archives, http://www.archive.org/movies/details-db.php?collection=prelinger&collectionid (accessed 11 July 2003).

67. On the fire-ant program, see Graham, *Since Silent Spring*, 24–29, 226–227; and Lear, *Rachel Carson*, 312–315, 332–336, 340–344.

68. Robert W. Akers, "Tell the Truth—But Tell It!" *Chemical and Engineering News*, 15 December 1958, 78–80.

69. Letter from Rachel Carson to DeWitt Wallace, 27 January 1958, in box 44, file 821, Rachel Carson Papers, Yale Collection in American Literature, Beinecke Rare Book and Manuscript Library, Yale University.

70. Robert S. Strother, "Backfire in the War against Insects," *Reader's Digest*, June 1959, 64–69.

71. "Scientific Background for Food and Drug Administration action against Aminotriazole in Cranberries," Food and Drug Administration, U.S. Department of Health, Education, and Welfare, 17 November 1959; see letter from Dr. David Rutstein, *New York Times*, 16 November 1959.

72. See Benson's *Freedom to Farm* (Garden City, NY: Doubleday, 1960); on his agricultural policy, see Edward L. Schapsmeier and Frederick H. Schapsmeier, *Ezra Taft Benson and the Politics of Agriculture* (Danville, IL: Interstate Printers, 1975).

73. *New York Times*, 20 January 1960, 17.

74. The case twice went before the federal court as *Murphy et al. v. Benson et al.*, U.S. District Court for the Eastern District of New York, Civ. No. 17610, 151 F. Supp. 24 May 1957, and as 164 F. Supp. 120, 23 June 1958; it was appealed to the U.S. Court of Appeals for the Second Circuit, No. 270, Docket 25448, 270 F. 2d 418, decided 1 October 1959.

75. Rachel Carson, *Silent Spring* (Boston: Houghton Mifflin, 1962), 106–113.

76. George J. Wallace Papers, Michigan State University Archives, UA 17–50, folder 16.

77. *Toledo Blade*, 10 February 1960, 27.

78. In addition, they sponsored hearings in the Michigan Senate on pesticide safety, bringing in witnesses who spoke approvingly of U.S. Agriculture Secretary Benson: "Chemicals enable us to produce the great variety of foods people want in the tremendous quantities needed," in Statement of Dr. Wayland Hayes, Jr., Chief, Toxicology Section, U.S. Public Health Service, 20 September 1960, before the Michigan Senate Hearings on Pesticides, in Michigan Archives, Lansing, Michigan, Michigan Department of Agriculture Records, record group 69–17, box 47–51.

79. Kenneth S. Lowe, "George Wallace and the Fight against DDT," *Michigan Out-of-Doors*, June 1989, 56–58. For an interesting account, the former chair of the entomology department, Gordon Guyer, who became president of

MSU in the 1990s, delivered a commencement award to Dr. Wallace in 1993, admitting his mistake in 1960 and calling Wallace a hero and role model; see Sue Nichols, "Wallace's Stand on DDT Bears Remembering," *MSU News Bulletin*, 5 August 1993, 6.

80. Rachel Carson, "Silent Spring," *New Yorker*, 16 June 1962, 35–99; 23 June 1962, 31–89; and 30 June 1962, 35–67.

81. *Chemical and Engineering News*, 13 August 1962, 24.

82. Ibid., 25.

83. Ibid.

84. The university agriculture establishment did attack Carson; see, for example, I. L. Baldwin, "Chemicals and Pests," *Science* 137 (28 September 1962): 1042–1043. Baldwin was an agriculture professor at the University of Wisconsin; his book review of *Silent Spring* was recalled approvingly by John Tierney in his attack on Carson forty-five years later; *New York Times*, 5 June 2007, F1 and F6.

85. *Chemical and Engineering News*, 30 July 1962, 5.

86. Ibid.

87. *Chemical and Engineering News*, 23 July 1962, 5.

88. M. Therese Southgate, MD, "Review of Silent Spring," *Journal of the American Medical Association* 182 (10 November 1962): 704.

89. Paul Brooks, *The House of Life: Rachel Carson at Work* (Boston: Houghton Mifflin, 1972), 293.

90. "Biology—Pesticides: The Price of Progress," *Time*, 28 September 1962, 45–48.

91. Brooks, *The House of Life*.

92. Memo from the *New Yorker* to Rachel Carson, 17 July 1962, with the heading "Confidential excerpt for Miss Carson's information only. Taken from letter received by staff member," in Carson Papers, box 85, folder 1496; according to the *New Yorker* memo, Martinson's letter was dated 7 July 1962.

93. Ibid.

94. Letter from Louis McLean to William Spaulding, President, Houghton Mifflin, 2 August 1962, in Carson Papers, box 89, folder 1575.

95. This was from 18 November 1961, 749.

96. Terres, "Science and Technology," 416.

97. Letter from Louis McLean to William Spaulding, President, Houghton Mifflin, 2 August 1962, in Carson Papers, box 89, folder 1575.

98. The events of August 6 are reviewed in letter from Paul Brooks to Marie Rodell, 6 August 1962, in Carson Papers, box 89, file 1575.

99. Discussed in letter from Carson to Brooks, 8 August 1962, box 89, file 1575, in Carson Papers; the sources she relied upon in this letter were van Oettingen, *Clinical Memorandum on Economic Poisons*, U.S. Public Health Service

Publication 476 (1956). See also James B. DeWitt, "Chronic Toxicity to Quail and Pheasants of Some Chlorinated Insecticides," *Journal of Agricultural and Food Chemistry* 4 (1956): 863–866; and E. C. Klostermeyer and C. B. Skotland, *Pesticide Chemicals as a Factor in Hop Die-out, Washington Agricultural Experiment Stations Circular 362* (1959); and material supplied by Justine Ward of USDA. There is also a letter in the Carson papers from Brooks to Jeptha Wade, Houghton Mifflin's attorney, dated 9 August 1962, urging prompt attention to this matter and sharing Carson's letter of August 8.

100. Draft letter from Brooks, apparently never sent, in Carson Papers, box 89, file 1575.
101. Letter from William Spaulding to Velsicol, attention Louis McLean, 10 August 1962, in Carson Papers, box 89, file 1575.
102. Letter from McLean to Spaulding, 14 August 1962, in Carson Papers, box 89, file 1575.
103. McLean to Spaulding, 14 August 1962, 4.
104. Memo from Brooks to Lovell Thompson, 22 August 1962, in Carson Papers, box 89, file 1575.
105. Letter from Brooks to McLean, 22 August 1962, in Carson Papers, box 89, file 1575.
106. Unsent draft of early August, from Brooks to Velsicol, in Carson Papers, box 89, file 1575.
107. Draft letter from Brooks to National Agricultural Chemicals Association (NACA), undated, in Carson Papers, box 89, file 1574; while only a draft is in the Houghton Mifflin files, the response from the NACA refers to "Your letter of September 21."
108. Letter from Brinkley to Brooks, 1 October 1962, in Carson Papers, box 89, file 1574.
109. See, for a classic discussion, the work of William Riker, *The Theory of Political Coalitions* (New Haven: Yale University Press, 1962), and his *The Art of Political Manipulation* (New Haven: Yale University Press, 1986); for a good discussion, see Kenneth A. Shepsie, "Losers in Politics (and How They Sometimes Become Winners): William Riker's Heresthetic," *Perspectives on Politics* 1 (June 2003): 307–315.
110. Frontispiece to Carson, *Silent Spring.*
111. Letter from E. B. White to Carson, 24 August 1962, in Carson Papers, box 89.
112. On Kronick, see Jaques Cattell, ed., *American Men of Science: The Physical and Biological Sciences*, 10th ed. (Tempe, AZ: Jaques Cattell Press, 1960), 2:2258.
113. "Pesticide Sales Pick-Up," *Chemical and Engineering News*, 2 July 1967, 22.

114. Letter to the editor from Paul Kronick of Haverford College, *Chemical and Engineering News*, 16 July 1962, 5.

115. On Caeser, see Cattell, *American Men of Science*, 1:556.

116. Letter to the editor from G. V. Caeser, *Chemical and Engineering News*, 16 July 1962, 5.

117. On Muffat, see Cattell, *American Men of Science*, 3:2824.

118. Letter to the editor from Peter Muffat, *Chemical and Engineering News*, 20 August 1962, 5.

119. On Blau, see Cattell, *American Men of Science*, 1:346.

120. Letter to the editor from Edmund Blau, *Chemical and Engineering News*, 3 September 1962, 4–5.

121. Lear, *Rachel Carson*, 419, has the text of the question and answer; even though the president referred to "Miss Carson's book," it would not be published for another month. See also Graham, *Since Silent Spring*, 50–51.

122. Lear, *Rachel Carson*, 422, discusses the Kennedy seminar.

123. On Freeman, see Lear, *Rachel Carson*, 414, 424, 447; and Mary A. McCay, *Rachel Carson* (New York: Twayne Publishers, 1993), 81–82.

124. McLean remained a hero to pro-pesticide lobbyists into the next century; see Donald Roberts and Richard Tren, *The Excellent Powder: DDT's Political and Scientific History* (Indianapolis: Dog Ear Publishing, 2010), 118–120; this book will be discussed further in chapter 7.

125. *Washington Post*, 14 March 1971, A4.

126. Jamie L. Whitten, *That We May Live* (Princeton, NJ: D. Van Nostrand, 1966).

127. *Library Journal* 91 (July 1966): 3454. The *Washington Post*, 14 March 1971, A4, discussed the strategy and Whitten's link to Velsicol and industrial agriculture.

128. Richard H. Goodwin, "The Future Role of the Biologist in Protecting Our Natural Resources," *Bioscience* 17 (March 1967): 161–162.

129. Ibid., 163.

130. Louis A. McLean, "Pesticides and the Environment," *Bioscience* 17 (September 1967): 613, italics in original.

131. Ibid., 615.

132. Ibid., 616.

133. Ibid. In his "References," McLean identified the medical authorities cited in the longest quotation in this paragraph as follows: F. J. Stare, "Nutritional Quackery," *The New Physician* (June 1966); V. W. Bernard, "Why People Become the Victims of Medical Quackery," *American Journal of Public Health* 55 (1965): 1142–1147; and J. Marmor, V. W. Bernard, and P. Ottenberg, "Psychodynamics of Group Opposition to Health Programs," *American Journal of Orthopsychiatry* 30 (1960): 330–345.

134. For a discussion of the Wisconsin case and McLean, see Thomas R. Dunlap, *DDT: Scientists, Citizens, and Public Policy* (Princeton, NJ: Princeton University Press, 1981), 158–175, 181.

135. Sunstein, *Risk and Reason*, 14, wrote of the reasons for his book, "One analysis concludes: 'The health loss from the [DDT] ban has been much greater than the health gain . . . completely banning DDT did more harm than good.' This is a controversial conclusion, but at least it raises some questions about Carson's analysis and about 1970s environmentalism in general—the same sorts of questions that I will be pressing throughout this book." His quotation came from another conservative critic of Carson, Aaron Wildavsky, *But Is It True? A Citizen's Guide to Environmental Health and Safety* (Cambridge, MA: Harvard University Press, 1995).

136. *New York Times*, 24 April 1964, 35, sets the date endrin was announced as the cause of fish kills as March 19; for a thorough review of the fish-kill issue, see Pete Daniel, *Toxic Drift: Pesticides and Health in the Post–World War II South* (Baton Rouge: Louisiana State University Press, 2005), 84–100.

137. *New York Times*, 23 April 1964, 41.

138. Graham, *Since Silent Spring*, 94–108.

139. For quantity in the sewers, see *New York Times*, 17 January 1965, 47.

140. On Dirksen, see *New York Times*, 17 June 1964, 38. See also the description of Dirksen as a tool of Chicago-area businesses; Graham, *Since Silent Spring*, 107.

141. *New York Times*, 17 December 1964, 26, listed the seventeen who had given large amounts to both parties in 1964; as with any good effort to buy influence, Regenstein matched his donation to the Republicans with an equal contribution to the Presidents Club for Johnson-Humphrey.

142. See *New York Times*, 24 April 1964, 67, for a copy of the advertisement.

143. For the officers of Velsicol in 1963–64, see the *Illinois Manufacturers Directory*.

144. *New York Times*, 17 January 1965, 47. On Loeb and Memphis politics in the era, see David M. Tucker, *Memphis since Crum: Bossism, Blacks, and Civic Reformers, 1948–1968* (Knoxville: University of Tennessee Press, 1980), 81–82; and Marcus D. Pohlmann and Michael P. Kirby, *Racial Politics at the Crossroads* (Knoxville: University of Tennessee Press, 1996), 17–20. Both point out Loeb was not interested in winning black support; since much of the concerns with Velsicol primarily arose in Afro-American communities, it is not surprising he sided with the company.

145. See *New York Times*, 17 January 1965, 471. Tucker, *Memphis since Crum*, 111–112, describes Ingram's election as a revolution.

146. See *New York Times*, 26 April 1964, 60.

147. *New York Times*, 23 April 1964, 41.

148. *New York Times*, 26 April 1964, 60.

149. It was in response to the fish-kill hearings that Velsicol's lobbyist, Samuel Bledsoe, worked to have Congressman Jamie Whitten publish the book defending pesticides.

150. *New York Times*, 24 April 1964, 35.

151. *New York Times*, 28 April 1964, 24.

152. *New York Times*, 27 June 1964, 9.

153. Quoted in Graham, *Since Silent Spring*, 108.

154. *New York Times*, 29 July 1964, 68.

155. *New York Times*, 19 May 1965, 55.

156. See Michigan Chemical, *Annual Report, 1964* and Proxy Statement of March 12, 1964, to be voted on at annual meeting on April 14, 1964. Eggum would be the longest-serving link between Michigan Chemical and the Regensteins, since he served on the board of the Regenstein Foundation to the end of the century; see *The Foundation Directory, 1999*, 602.

157. *New York Times*, 14 February 1964, 46.

158. *Chicago Sun-Times*, 7 March 1999, 72.

159. *Crain's Chicago Business*, 12 May 1986, 77.

160. *Chicago Tribune*, 6 March 1999, 17.

161. *Chicago Sun-Times*, 17 April 2002, 8.

162. *Chicago Sun-Times*, 13 May 1993, 5.

163. Graham, *Since Silent Spring*, 108.

164. See Walter Rosenbaum, *The Politics of Environmental Concern* (New York: Praeger, 1973), 122–124. William A. Shutkin, *The Land That Could Be: Environmentalism and Democracy in the Twenty-First Century* (Cambridge, MA: MIT Press, 2001), 99–101, discusses Muskie's links to environmental organizations and creation of EPA; see also Sunstein, 15–18.

165. On bipartisan support for EPA, see David Vogel, *Fluctuating Fortunes: The Political Power of Business in America* (New York: Basic Books, 1989), 69, discussing lack of intense business opposition to EPA. On support from business for creating a single regulatory agency and freeing business from the uncertainties of case-by-case settlement of environmental complaints, see Rosemary O'Leary, "Environmental Policy in the Courts," in *Environmental Policy: New Directions for the Twenty-First Century*, 5th ed., edited by Norman J. Vig and Michael E. Kraft, 152–153 (Washington, DC: CQ Press, 2003). Also on EPA founding, see Jack Lewis, "Looking Backward: A Historical Perspective on Environmental Regulations," *EPA Journal* 14 (March 1988): 42–46; and John Quarles, *Cleaning Up America: An Insider's View of the EPA* (Boston: Houghton Mifflin, 1976).

166. Ray C. Anderson, *Mid-Course Correction—Toward a Sustainable Enterprise: The Interface Model* (White River Jct., VT: Chelsea Green, 1998), 7.

CHAPTER 2. NEW CORPORATIONS AND NEW REGULATIONS

1. *New York Times*, 29 May 1965, 30, contained the first rumors of the purchase, which were denied by Velsicol's John Kirk; the purchase was reported in *New York Times*, 15 June 1965, 57.

2. H. Roger Grant, *The North Western: A History of the Chicago and North Western Railway System* (DeKalb: Northern Illinois University Press, 1996), 213–214, shares this interpretation of Heineman's motives for diversification.

3. The Michigan Chemical proxy statement of 12 March 1964 told the proportion of Velsicol owned by the Regenstein interests; in Pine River Files, Alma College Archives, Alma, MI.

4. Pete Daniel, *Lost Revolutions: The South in the 1950s* (Chapel Hill: University of North Carolina Press, 2000), 28.

5. For a good discussion of the significance of 1964, see David Steigerwald, *The Sixties and the End of Modern America* (New York: St. Martin's Press, 1995), 272–277; for a discussion of the change from a liberal perspective, see Richard Hofstadter, "The Pseudo-Conservative Revolt," in *The Radical Right*, ed. Daniel Bell (Garden City, NY: Doubleday, 1964); and for a conservative view, Russell Kirk, "New Direction in the U.S. Right?" *Time*, 7 August 1966.

6. For a basic discussion of the early incorporation process, see Mansel G. Blackford and K. Austin Kerr, *Business Enterprise in American History*, 2nd ed. (Boston: Houghton Mifflin, 1990), 142–145.

7. For a contemporary report on Heineman, see *New York Times*, 9 June 1943, 21.

8. For a fascinating account of OPM, see John Kenneth Galbraith's op-ed piece at the time of Henderson's death reviewing their common work; *Washington Post*, 2 November 1986, C7.

9. Heineman lived on the south side of Chicago in the upscale but integrated neighborhood around the University of Chicago called Hyde Park; see *New York Times*, 18 August 1966, 30; and Michael Kilian, Connie Fletcher, F. Richard Ciccone, *Who Runs Chicago?* (New York: St. Martin's Press, 1979), 351–352.

10. *New York Times*, 16 May 1956, 54.

11. Jim Scribbins, *The 400 Story* (Park Forest, IL: PTJ Publishing, 1982), 118, 143, 196–204, and Grant, *The North Western*, 205–207, describe the service cutbacks under Heineman; also see *New York Times*, 30 November 1958, 39.

12. *New York Times*, 25 February 1965, 41. On 4 June the railroad announced profits were up; *New York Times*, 5 June 1965, 36.

13. *New York Times*, 3 March 1959, 1; see also *Harper's* 232 (January 1966): 65–68; *Life*, 4 December 1964, 61–62; *Reader's Digest*, May 1963, 175–180.

In general on Heineman's style of learning as he managed the C&NW, see Pamela S. Barr, J. L. Stimpert, and Anne S. Huff, "Cognitive Change, Strategic Action, and Organizational Renewal," *Strategic Management Journal* 13 (Summer 1992): 15–36.

14. *New York Times*, 2 May 1966, 16.

15. On agenda, see *New York Times*, 25 May 1966, 28.

16. *New York Times*, 26 August 1966, 17; 27 August, 1. On results of report, see *New York Times*, 8 March 1968, 28; Kathleen Connolly, "The Chicago Open-Housing Contest," in *Martin Luther King, Jr. and the Civil Rights Movement*, ed. David Garrow (New York: Carlson Publishing, 1989), 70–79.

17. *New York Times*, 27 August 1966, 1; Kathleen Connolly, "The Chicago Open-Housing Contest," 70–79.

18. Michigan Chemical had worked on a special rare-earth research project with Davison Chemical, W. R. Grace, Mallinckrodt, and a few others; see *New York Times*, 20 March 1958, F7.

19. *New York Times*, 1 July 1965, 39; see also "a convenient reference to the products of Michigan Chemical Corporation" in the St. Louis (MI) Historical Society, which reports the company produced lanthanum, cerium, praseodymium, neodymium, samarium, gadolinium, terbium, ytterbium, dysprosium, holmium, erbium, thulium, lutetium, as well as europium and yttrium.

20. *Michigan Chemical Company Annual Report, 1965*, 4.

21. For a long profile of Hathaway, see *New York Times*, 2 October 1966, 157.

22. *New York Times*, 2 October 1966, 157.

23. *Chemical Week*, 27 September 1969, 63.

24. For text of the full letter from Baer to W. F. Clark, 17 July 1902, during the anthracite strike, see Mary A. Merrick, RSM, "A Case in Practical Democracy: Settlement of the Anthracite Coal Strike of 1902" (Ph.D. dissertation, University of Notre Dame, 1942), 57.

25. On the transition at Philadelphia and Reading, see William M. Adler, *Mollie's Job: A Story of Life and Work on the Global Assembly Line* (New York: Scribner, 2000), 169–170.

26. *Moody's Industrial Manual, 1968*, 2058–2059; *Moody's Industrial Manual, 1959*, 2102–2103; *Moody's Industrial Manual, 1957*, 202–203.

27. "Fable for Our Times," *Forbes*, 1 March 1973, 21.

28. Grant, *The North Western*, 216–217; Eugene M. Lewis, *12,000 Days on the North Western Line* (North Riverside, IL: Chicago and North Western Historical Society, 2005), 659–666, has a more cynical view of Heineman and employee ownership.

29. *Northwest Industries Annual Report, 1968*, 1.

30. Anthony Borden, "Ben Heineman's In-House Revolution," *The American Lawyer*, September 1989, 100.

31. *Northwest Industries Annual Report, 1970*, 5–6.

32. *Northwest Industries Annual Report, 1971*, 17.

33. *Northwest Industries Annual Report, 1974*, 10.

34. *Chemical Week*, 25 September 1974, 47, had a profile of Girard that did not mention any problems at Michigan Chemical.

35. Inter-Office Memorandum, from W. W. Thorne to Salt Plant Managers, 14 June 1973, in U.S. Congress, House of Representatives, Committee on Public Works and Transportation, *PBB (polybrominated biphenyls) Pollution Problem in Michigan: Hearings before the Subcommittee on Water Resources*, 96th Cong., 1st sess., 1979, supplement L (3).

36. U.S. Department of Health, Education, and Welfare, NIOSH, *Health Hazard Evaluation Determination Report HE 77–73–610, Velsicol Chemical Corp*, August 1979, 21–23.

37. *Saginaw (MI) News*, 15 August 1976, D3.

38. Ibid. Scholtz would become an environmental activist in town and was elected mayor in 2009.

39. *Saginaw (MI) News*, 15 August 1976, D2–3.

40. Michigan Water Resources Commission, "A Biological Survey of the Pine River above Alma to M-30 to Determine the Effect of Pollution," 31 May 1955, 1, in Pine River Files, Alma College Archives, Alma, MI.

41. Michigan Water Resources Commission, "Report of Industrial Survey, Michigan Chemical Company, St. Louis, Michigan," 28 August–1 September 1967, in Pine River Files, Alma College Archives, Alma, MI.

42. Memo from John R. Byerlay, Geologist, Michigan Geological Survey, to John Bohunsky, Regional Engineer, Water Management Bureau, Department of Natural Resources, 10 September 1969, in Pine River Files, Alma College Archives, Alma, MI.

43. State of Michigan, Water Resources Commission, *Water Resource Conditions and Uses in the Tittabawasee River Basin* (Lansing, MI: Water Resources Commission, 1960), 116.

44. Department of Natural Resources, "File Resume Michigan Chemical Company, St. Louis, Michigan," in Pine River Files, Alma College Archives, Alma, MI.

45. Ibid., 65.

46. Department of Natural Resources, Michigan Water Resources Commission, Division of Water Quality Control, "Biological Survey of the Pine River Vicinity of Alma and St. Louis, 1967 and 1970," 7.

47. Letter from John Rademacher, Vice President Environmental Health and Regulatory Affairs, Velsicol, to Dale Bryson, Deputy Director, Enforcement, Div., US EPA Region 5, 11 April 1980, 2–5. These numbers were found in 2005 to have been significantly less than reality: see letter from Bruce L.

Jorgensen, Chief Decommissioning Branch, Nuclear Regulatory Commission, to John E. Hock, Senior Project Consultant, Civil and Environmental Consultants, Inc., 20 December 2001, which found, "*Not all radioactive material identified*" (underlined in original); Environ, "Breckenridge Disposal Site Project Status Update Meeting," presented at the Pine River Superfund Citizen Task Force meeting, 15 June 2005, St. Louis, MI, in Pine River Files, Alma College Archives, Alma, MI.

48. Joyce Egginton, *The Poisoning of Michigan* (New York: W.W. Norton & Co., 1980), 87–90 (this book was reissued by Michigan State University Press in 2009).

49. Egginton, *The Poisoning of Michigan*, 87–90; A. G. Aftosmis et al., "Toxicology of Brominated Biphenyls: I. Oral Toxicity and Embryotoxicity," paper presented at Society Toxicology Meeting, Williamsburg, VA., 5–9 March 1972, pp. 1–2, in Michigan House of Representatives, Special Committee Examining the Effects of PBB on Michigan Chemical Corporation Workers, *The Workers: Effects of PBB on Michigan Chemical Corporation Workers*, December 1978, appendix H.

50. Edwin Chen, *PBB: An American Tragedy* (Englewood Cliffs, NJ: Prentice Hall, Inc., 1979), 35–37; Egginton, *The Poisoning of Michigan*, 92; Aftosmis et al., "Toxicology of Brominated Biphenyls: I.," 7.

51. In addition to the Aftosmis study, Perry Gehring reported on PBB at a November 1972 polymer conference; Chen, *PBB: An American Tragedy*, 36.

52. Egginton, *The Poisoning of Michigan*, 35 and 88–90.

53. Michigan House, Special Committee Examining . . . , "Summary of Study Done for Michigan Chemical by Hill Top Research, Inc., of Ohio on the Health Effects of Brominated Biphenyls," in Michigan House of Representatives, *Special Committee Examining the Effects of PBB*, appendix G.

54. Willis Frederick Dunbar, *Michigan through the Centuries* (New York: Lewis Historical Publishing Co., 1955), 2:574–585; 4:579–580.

55. State of Michigan, Joint Legislative Committee on Water Resources Planning, *Study on Needs for Water Pollution Control Works*, 31 December 1966, 1:1–13; and Nicholas V. Olds, "Conflicts: Federal v. State Pollution Laws and Administration," *Papers and Talks Presented at the 40th Annual Technical Conference Michigan Water Pollution Control Association*, Boyne Falls, Michigan, 7–9 June 1965, 1–24.

56. E. H. Gault and C. N. Davisson, *The Saginaw Valley Problem* (Ann Arbor: Bureau of Business Research, School of Business Administration, University of Michigan, 1945), 31–32.

57. State of Michigan, Water Resources Commission, *First Annual Report: 1950–52*, 29–32, only noted the presence in St. Louis of an old sugar-beet processing company as a polluter.

58. "Lt. Governor's Message to the 75th Michigan Legislature: January 9, 1969," in State of Michigan, *Addresses and Special Messages of Governor William G. Milliken, 1969–1982*, 1–2.

59. For a review of these changes, see Judith A. Layzer, *The Environmental Case: Translating Values into Policy* (Washington, DC: CQ Press, 2002), 25–47; also Helen M. Ingram and Dean E. Mann, "Environmental Policy from Innovation to Implementation," in *Nationalizing Government: Public Policies in America*, ed. Theodore Lowi and Alan Stone (Washington, DC: Congressional Quarterly, 1984), 131–162.

60. Attachment I, 3, to Memorandum of Agreement between the Michigan Water Resources Commission and Region V, United States Environmental Protection Agency, in State of Michigan, Michigan Water Resources Commission, *Michigan's Wastewater Discharge Permit Program*.

61. The subsequent contamination of large numbers of dairy cattle and most of the population of Michigan received widespread attention, including numerous press accounts, four books, a British television documentary, several domestic television dramatizations, legislative and congressional hearings, and civil and criminal lawsuits. Typical magazine articles include Edwin Chen, "Michigan: If Something Odd Happens . . . ," *Atlantic*, August 1977, 12–18; Ted J. Rakstis, "The Poisoning of Michigan," *Reader's Digest*, September 1979, 104–108; Mike Cary and Jon Thompson, "Your Money and Your Life," *Seven Days*, December 8, 1978, 21–24. Books include Egginton, *The Poisoning of Michigan*; Chen, *PBB: An American Tragedy*; Frederic Halbert and Sandra Halbert, *Bitter Harvest* (Grand Rapids, MI: W. B. Eerdmans Publishing Co., 1978); and Lois Touzeau, *Our Side of the Story* (New York: Vantage Press, 1985); also chapters in the following books: Ellen E. Grzech, "PBB," in *Who's Poisoning America: Corporate Polluters and Their Victims in the Chemical Age* (San Francisco: Sierra Club Books, 1981), 60–84; and Thomas H. Corbett, *Cancer and Chemicals* (Chicago: Nelson-Hall, 1977), 121–145. Movies include *The Poisoning of Michigan* (Thames Television, 1977); and *Bitter Harvest* (Charles Fries Productions, 1981).

62. Egginton, *The Poisoning of Michigan*, 95–100, as well as various records in the appendices of various public hearings describe these problems.

63. For copies of relevant documents, see box 693, Staff Files, Milliken Papers, Bentley Historical Library, Ann Arbor, MI; also see Egginton, *The Poisoning of Michigan*, including specifically 51–52, 56, 71, 75, 112, 141, 155–156.

64. See Michigan Chemical File in box 820, Staff Files, Kathy Starika, Milliken Papers.

65. Memo 29 June 1976 from Governor Milliken to legislature related to balancing health standards with interests of farmers; Milliken said, "I believe that we should take the precautionary step of a partial lowering of the levels

in a manner that will not only provide greater assurances of public health protection and consumer confidence, but also will protect the interests of affected farmers" [all in capital letters and with blue underlining]. See also Michigan Department of Agriculture, "Proposed Public Relations Program to Restore Confidence in Michigan Agricultural Products," 24 June 1977, in Michigan Archives, Records of Department of Agriculture, record group 88–229, box 1.

66. Ibid., 131–139, 155–166; and Chen, *PBB: An American Tragedy*, 56–57; *The Poisoning of Michigan* by Thames Television has many interviews related to state failure to act quickly.

67. Egginton, *The Poisoning of Michigan*, 31, 36, 224–229.

68. William D. Rowe, "Identification of Risk," in *Risk and Reason: Risk Assessment in Relation to Environmental Mutagens and Carcinogens*, ed. Per Oftedal and Anton Brogger (New York: Alan R. Liss, Inc., 1986), 18–21, contrasts normal operations, abnormal operations, and accidents; see also Charles Perrow, *Normal Accidents: Living with High-Risk Technologies* (New York: Basic Books, Inc., 1984), 101–122 and 304–352.

69. On the theory of capture, see Anthony Downs, *Inside Bureaucracy* (Boston: Little, Brown and Co., 1967).

70. For a contemporary review of regulation, see Leonard Dworsky, *Pollution* (New York: Chelsea House Publishers, 1971), 765–851.

71. Samuel P. Hays, *Beauty, Health, and Permanence: Environmental Politics in the United States, 1955–1985* (New York: Cambridge University Press, 1987), 60, refers to the lost momentum in environmental reform after 1972; for a general discussion of points here, see Edward P. Weber, *Pluralism by the Rules: Conflict and Cooperation in Environmental Regulation* (Washington, DC: Georgetown University Press, 1998).

72. On the growth of contracting out to consultants, see William Danhof, *Government Contracting* (Washington, DC: Brookings Institution, 1968); Danhof saw this growth beginning in the 1960s. On the consequences of the phenomenon, see Susan R. Bernstein, *Managing Contracted Services in the Nonprofit Agency: Administrative, Ethical, and Political Issues* (Philadelphia: Temple University Press, 1991).

73. Theodore Lowi, *The End of Liberalism* (New York: W.W. Norton, 1969), 310–314, predicted the weakness of "interest group liberalism's" regulatory regime.

74. On the concept of subgovernments, see Joseph A. Pika, "Interest Groups and the Executive: Presidential Intervention," in *Interest Group Politics*, ed. Allan Cigler and Burdett Loomis (Washington, DC: CQ Press, 1983), 303.

75. Examples of contemporary risk-assessment literature include Baruch Fischhoff et al., "Handling Hazards," *Environment* 20 (September 1978):

16–24; Baruch Fischhoff, Paul Slovic, and Sarah Uchtenstein, "How Safe Is Safe Enough? A Psychometric Study of Attitudes towards Technological Risks and Benefits," *Policy Sciences* 9 (1978): 127–151; Burke K. Zimmerman, "Risk-Benefit Analysis," *Trial*, February 1978, 43–47; Robert W. Kates, *Risk Assessment of Environmental Hazards* (Chichester, UK: John Wiley and Sons, 1978), 98–100; William W. Lowrance, *Of Acceptable Risk* (Los Altos, CA: William Kaufmann, Inc., 1976), 174–175.

76. W. Kip Viscusi, *Risk by Choice* (Cambridge, MA: Harvard University Press, 1983), 167–168; also see Nicholas A. Ashford, *Crisis in the Workplace: Occupational Disease and Injury* (Cambridge, MA: MIT Press, 1976), 18–22; on limits of knowledge see Anders Grimvall and Rolf Ejvegard, "The Dynamics of Scientific Uncertainty and Its Implications for the Use of Conservative Procedures in Risk Analysis," in Oftedal and Brogger, *Risk and Reason*, 23–29.

77. Memo, 12 December 1975, from Frankland to Governor Milliken, box 625, file PBB 1975, Milliken Papers.

78. Corbett, *Cancer and Chemicals*, 133, describes meetings on PBB incident where "The major concern seemed to be the economic loss. . . . No one mentioned the possible harm to the human population"; also *Detroit News*, 19 July 1978, B8, reported Northwest Industries claimed it was only a stockholder in Velsicol, not a party to lawsuits. Chen, *PBB: An American Tragedy*, 272–292, describes the litigation following a farmer's loss of his herd after low-level PBB contamination; *Gratiot County (MI) Record-Leader*, 29 November 1974, described Farm Bureau's $277 million suit against Michigan Chemical. As late as 1990 lawsuits were being filed in the deaths of farmers; see for example, *Saginaw (MI) News*, 4 February 1990, A3. See Louis Touzeau, *Our Side of the Story* (New York: Vantage Press, 1985) for point of view of plant management.

79. Oil Chemical and Atomic Workers Grievance Report, February 21, 1974 (photocopy), and Management Replies, 21 February 1974, and 8 May 1974 (photocopies), in Michigan House, *Special Committee Examining the Effects of PBB*, appendix B. While management rejected the grievance of 21 February on 8 May 1974, after information on the cattle-feed disaster began to appear they agreed to grant shower time and issue overalls to be laundered by the company to workers in the PBB operations; Egginton, *The Poisoning of Michigan*, 91, says shower time was granted on 21 February 1974, but documents above confirm it was on 8 May. While some union workers believe the cattle-feed accident began during the October 1973 strike, most evidence indicates it began in early May 1973; see Chen, *PBB: An American Tragedy*, 42–43; Egginton, *The Poisoning of Michigan*, 99.

80. Steve Spalding, "Velsicol Workers to Talk to State House Committee, *(Alma, MI) Morning Sun*, 17 January 1978, 1; this statement was reiterated by

comments of Ron Orwig (former officer of local 7–224, Oil, Chemical and Atomic Workers Union), interview at Alma, Michigan, 15 April 1992. For unemployment data, see David L. Verway, ed., *Michigan Statistical Abstract* (East Lansing: Division of Research, Graduate School of Business Administration, Michigan State University, 1980), 159.

81. Arthur M. Weimer, "An Economic History of Alma, Michigan" (Ph.D. dissertation, University of Chicago, 1934), 36–56 and 99–110, reviews deforestation in the region around St. Louis; see also Federal Writer's Program, *Michigan: A Guide to the Wolverine State* (New York: Oxford University Press, 1941), 446. Willard D. Tucker, *Gratiot County, Michigan, Historical, Biographical, Statistical* (Saginaw, MI: Seeman and Peters, 1913), 921–937, describes an 1880s chemical business on the Michigan Chemical site that extracted bromines; this failed after a few years (935). On Michigan lumber industry generally, see William Cronon, *Nature's Metropolis* (New York: W.W. Norton, 1991), 151–180.

82. U.S. Congress, House of Representatives, Committee on Public Works and Transportation, *PBB (polybrominated biphenyls) Pollution Problem in Michigan: Hearings before the Subcommittee on Water Resources*, 96th Cong., 1st sess., 1979, 18.

83. Michigan Department of Labor, Bureau of Employment and Training, Employment Policy Division, *Analysis of the Employment Impact of the Velsicol Chemical Company Plant Closing*, September 1978.

84. General Release, 20 September 1976; see also news stories in *Gongwer News Service, Michigan Report No. 178*, 14 September 1976; *Detroit News*, 15 September 1976.

85. *Lansing State Journal*, 24 May 1977. The agreement would be referenced as late as 2008 to explain why the state could not recover damages from Velsicol in spite of the $100 million being spent in the previous decade; see e-mail from John Scherbarth, Michigan Attorney General's office to Ed Lorenz, Legal Chair Pine River Task Force, 24 October 2008, in Pine River Files, Alma College Archives, Alma, MI.

86. For a good summary of this problem, see Henry N. Pollack, *Uncertain Science . . . Uncertain World* (New York: Cambridge University Press, 2003), which makes the point that the tentativeness of scientific evidence is confused with uncertainty by the media and politicians, allowing interests opposed to caution to undermine actions based on scientific consensus.

87. In January 1976, the governor wrote to a number of experts, ranging from major university health researchers to a variety of chemical companies, asking for knowledge of PBB health threats. While many responded with warnings about either actual known problems or potential ones, such as I. A. Bernstein

at the University of Michigan and James Aftosmis at DuPont, the governor
included Michigan State and Michigan Chemical in the contact list, and both
responded with less concern with PBB; see letters in box 645, PBB Department of Agriculture File, Milliken Papers.

88. Michael R. Reich, "Environmental Politics and Science: The Case of PBB
Contamination in Michigan," *American Journal of Public Health* 73 (March
1983): 307–311; examples of conflicting PBB stories in the local press are
the following headlines in *(Alma, MI) Morning Sun:* "PBB Damaging to
Children," 5 October 1978; "PBB Stays, but May Not Hurt," 24 July 1980;
"Federal Study Shows PBB is Carcinogenic," 25 June 1981; "PBB Fights
Cancer," 11 December 1982. On the press perspective on environmental cri-
ses, see Joanne Omang, "Perception of Risk: A Journalist's Perspective," in
The Analysis of Actual versus Perceived Risks, ed. Vincent T. Covello et al. (New
York: Plenum Press, 1983), 267–271.

89. On the concept of issue networks, see Hugh Heclo, "Issue Networks and the
Executive Establishment," in *The New American Political System*, ed. Anthony
King (Washington, DC: American Enterprise Institute, 1978).

90. After the PBB crisis, some Michigan DNR officials claimed DNR had no pre-
accident information on Michigan Chemical's impact; see U.S. Congress,
Senate Committee on Commerce, Science, and Transportation, *Toxic Substances Polybrominated Biphenyl (pbb) Contamination in Michigan: Hearings
before the Subcommittee on Science, Technology, and Space*, 95th Cong., 1st
sess., 1977, 704.

91. The Commission of Agriculture and the Natural Resources Commission had
their origins in legislation in 1921 creating the departments of Agriculture
and Conservation; on the history of administrative reforms of the 1920s,
see Frank B. Woodford, *Alex J. Groesbeck: Portrait of a Public Man* (Detroit:
Wayne State University Press, 1962), 124–127. Willis F. Dunbar, *Michigan: A
History of the Wolverine State* (Grand Rapids, MI: William Eerdmans Publishing Company, 1965), 545–546, does not believe the reforms were consistent
with the "Progressive movement," but they were consistent with the view
of progressivism's links to efficiency. See also Samuel Haber, *Efficiency and
Uplift: Scientific Management in the Progressive Era, 1890–1920* (Chicago:
University of Chicago Press, 1973).

92. For a classic statement of this goal, see Morris Llewellyn Cooke, "Scientific
Management of the Public Business," *American Political Science Review* 9
(August 1915): 488–495.

93. For a discussion of the national agricultural cozy triangle, see William P.
Browne, "Policy and Interests: Instability and Change in a Classic Issue
Subsystem," in *Interest Group Politics*, 2nd ed., edited by Allan J. Cigler and

Burdett A. Loomis (Washington, DC: CQ Press, 1986), 183–201; on environmental cozy triangles, see Christopher J. Bosso, *Pesticides and Politics: The Life Cycle of a Public Issue* (Pittsburgh, PA: University of Pittsburgh Press, 1987), 3–8; on the general concept of cozy triangles, see Douglass Cater, *Power in Washington* (New York: Random House, 1964), 26–48.

94. Egginton, *The Poisoning of Michigan*, 223–226; Chen, *PBB: An American Tragedy*, 183–199.

95. Produced by Thames Television, this ran on 4 October 1977.

96. U.S. Congress, Senate, Committee on Commerce, Science, and Transportation, *Toxic Substances Control Act: Hearings before the Subcommittee on Science, Technology, and Space*, 95th Cong., 1st sess., 1977, 1781–1783; Walter Rosenbaum, *Environmental Politics and Policy*, 2nd ed. (Washington, DC: Congressional Quarterly, 1991), 113, 220–224; and Editorial Research Reports, *Environmental Issues: Prospects and Problems* (Washington, DC: Congressional Quarterly, 1982), 105–111.

97. For example: *Detroit News*, 10 August 1978, A22, editorial on governor's vulnerability; 28 September 1978, A1, on PBB found in most state residents; 19 October 1978, A1 and A19, on *Lou Grant Show*.

98. On Fitzgerald campaign, see *Detroit News*, 6 October 1978, B1; 10 October 1978, B1. Saul Friedman, "George Romney's Boys are Slowing Down," *The New Republic*, 4 November 1978, 21–22, incorrectly relates governor's problems to "PCBs" not PBBs, as did the *New York Times*, 8 November 1978, 20. On Albosta, see Michael Barone and Grant Ujifusa, *The Almanac of American Politics: 1984* (Washington, DC: The National Journal, 1983), 596.

99. Chen, *PBB: An American Tragedy*, 295; *Detroit News*, 21 October A13. The *Detroit Free Press*, 21 October 1978, A5, said of Halbert appointment, "A Milliken spokesman said the appointment was neither opportunistic nor an effort to defuse PBB as a political issue."

100. *Detroit Free Press*, 19 October 1978, A1 and A19, reported the governor's office said he did not ask CBS to delay the show, rather the CBS affiliate in Detroit asked that it not be shown on the eve of the election; *Detroit News*, 20 October 1978, B4, reported Democratic protests over delay in *Lou Grant Show*.

101. *Detroit News*, 2 September 1978, B21.

102. Chen, *PBB: An American Tragedy*, 293–296; Egginton, *The Poisoning of Michigan*, 313–315. For election results, see Michigan Department of Management and Budget, *Michigan Manual, 1979–80* (Lansing: Michigan Department of Management and Budget, 1979), 610, 651, 762. On post-election reforms of Milliken, see Bryan Coyer and Don Schwerin, "Bureaucratic Regulation and

Farmer Protest in the Michigan PBB Contamination Case," *Rural Sociology* 46 (Winter 1981): 718–721.

103. *(Alma, MI) Morning Sun*, 8 April 1978.

104. U.S. House, Committee on Public Works, *PBB*, iii, included eighteen witnesses giving oral testimony, of whom four simply were county residents, seven county or city officials, one union officer, three state legislators, and three state or federal officials.

105. For example, at the U.S. Senate, Committee on Commerce, Science, and Transportation, Hearings on *Toxic Substances*, see p. iii, held in Lansing on 31 March 1977, of 21 witnesses: 14 were federal or state officials, 3 represented Velsicol and the Farm Bureau (the cattle feed supplier), 3 were state legislators, and only 1 was an unaffiliated citizen.

106. For a conservative analysis of free-market environmentalism, see John R. E. Bliese, *The Greening of Conservative America* (Boulder, CO: Westview Press, 2001), 1–6, 57–59, where he discusses, for example, the Foundation for Research on Economics and the Environment.

107. At its most extreme, this position was advocated by William Ophuls, *Ecology and the Politics of Scarcity* (San Francisco: W. H. Freeman, 1977), 145; and Robert Heilbroner, *An Inquiry into the Human Prospect* (New York: W.W. Norton, 1974), 161.

108. John Passmore, *Man's Responsibility for Nature* (London: Duckworth, 1974), 183.

109. *Saginaw (MI) News*, 15 August 1976, D1, interview with William T. Dennis.

110. Ibid., interview with Kenneth Barnum.

111. Ibid., interview with William Welch.

112. Ibid., D4, interview with Dr. C. J. Bender.

113. Ibid., interview with Eugene Nikkari.

114. Ibid., D1, interview with Alfred Bush, manager of branch of Michigan Chemical Bank.

115. Ibid., D2, interview with Thomas Overley, co-owner of T and G Bar and Grill.

116. Ibid., interview with Lawrence Shirely, former mayor and owner of Shirely Furniture Store.

117. Ibid., D3, interview with Mary Wrath, high school sophomore.

118. Ibid., interview with Tammy Billiau, high school freshman.

119. Ibid., D2, interview with Joseph Scholtz.

120. Ibid., interview with Richard Wrath, who worked for General Motors in Lansing.

121. Ibid., D3, interview with Jim Kelly.

122. Ibid., interview with Ralph Boyles. See also letter from Michigan Department of Management and Budget to Governor, 27 July 1978, which stated, "It was reported that the community has not pressed for relaxed enforcement efforts despite the importance of the plant to the local economy"; in box 820, R-1 Michigan Chemical, Milliken Papers.

123. *Saginaw (MI) News*, 15 August 1976, D3.

124. For a radical discussion of the concept of changing the role of schools and expertise, see Ivan Illich, *Deschooling Society* (New York: Marion Boyars, 1971), 1–24.

CHAPTER 3. DISEMPOWERING COMMUNITIES

1. Federal Election Commission records point out that Heineman in the 1980 campaign cycle gave $1,000 to the Carter-Mondale Campaign, $1,000 to Senator Gaylord Nelson, and $2,000 to the Democratic Congressional Campaign Committee. He gave no contributions to Republicans. See http://www .tray.com/cgi-win/x_allindiv.exe (accessed 15 September 2003).

2. Philip Shabecoff, *Earth Rising: American Environmentalism in the 21st Century* (Washington, DC: Island Press, 2000), 7–8, referring to the 1970s, said, "Some commentators have called it the 'golden age' of environmentalism . . . corporate America, which had been caught off guard by the militant environmentalism that emerged in the 1960s and 1970s, began to mount an effective resistance." Yet Samuel P. Hays, in *Beauty, Health, and Permanence: Environmental Politics in the United States, 1955–1985* (New York: Cambridge University Press, 1987), and in "From Conservation to Environment: Environmental Politics in the United States since World War II," *Environmental Review* 6 (Fall 1982): 14–29, sees only a brief era of environmental success, not a full decade.

3. A good collection of essays on Reagan's impact is in Norman J. Vig and Michael E. Kraft, eds., *Environmental Policy in the 1980s: Reagan's New Agenda* (Washington, DC: Congressional Quarterly, 1984); also for a response to Reagan, Susan J. Tolchin and Martin Tolchin, *Dismantling America: The Rush to Deregulate* (New York: Oxford University Press, 1985), 20–22, 39–40, 57–59, and 104–105.

4. *New York Times*, 22 July 1983, A1, summarizes the New Jersey Supreme Court's description of the site.

5. Steven P. Bann, "Case Digest: New Jersey Supreme Court," *New Jersey Law Journal*, 26 July 1993, 45, reviews the outcome of what was called the Ventron

case. Ventron was the successor to Velsicol at Berry's Creek; however, the case included Velsicol as a defendant.

6. *New York Times*, 24 May 1978, B27.

7. Bann, "Case Digest," 45.

8. Steven P. Bann, "Case Digests: State Digests, Appellate Division," *New Jersey Law Journal*, 5 February 1996, 68.

9. *Bergen (NJ) Record*, 17 August 1997, A1.

10. David Alpern, Dewey Gram, and James Bishop Jr., "The Phosvel Zombies," *Newsweek*, 13 December 1976, 38.

11. *Chemical Week*, 8 December 1976, 25.

12. Charles Xintaras et al., *NIOSH Health Survey of Velsicol Pesticide Workers: Occupational Exposure to Leptophos and Other Chemicals* (Cincinnati: Division of Biomedical and Behavioral Science, NIOSH, U.S. Dept. of Health, Education, and Welfare, 1978), 32–33. The rate for the population was between 6 and 14 per 100,000; the figure 0.014 is computed here from the higher expected rate per 100,000.

13. James H. Colopy, "Poisoning the Developing World: The Exportation of Unregistered and Severely Restricted Pesticides from the United States," *UCLA Journal of Environmental Law and Policy* 167 (1994–95): 180.

14. Alpern et al., "The Phosvel Zombies," 38; they also were discussed in the testimony of S. Jacob Scherr, attorney for the Natural Resources Defense Council, in House Committee on Government Operations, *U. S. Export of Banned Products*, 95th Cong., 2d sess., 11, 12, 13 July 1978, p. 35.

15. See various comments on origins of flammability standards in House Committee on Interstate and Foreign Commerce, Subcommittee on Oversight and Investigation, *Regulation of Cancer Causing Flame-Retardant Chemicals and Governmental Coordination of Testing of Toxic Chemicals*, 95th Cong., 1st sess., 11, 13, 16 May 1977.

16. Ibid., 89. Hoffman said that the company did standard toxicity tests, but admitted they were not advanced and that they trusted others, including government, to have tested the product.

17. *Chemical Week*, 12 October 1977, 24, reviewed the filing of this case against both Velsicol and Celanese for "misrepresenting their products."

18. For stories that found Velsicol more guilty and for other interpretations, see for example: *Washington Post*, 6 August 1978, A25; *Chemical Week*, 25 May 1977, 16; testimony of David Moulton of Public Citizen in House Committee on the Judiciary, Subcommittee on Administrative Law and Governmental Relations, *Reimbursement for Losses Incurred by Government Banning of Tris*, 95th Cong., 2d sess., 14, 15 June 1978, pp. 322–325.

19. For Carter's veto message, see House Committee on the Judiciary, Committee on Administrative Law and Governance, *Payment of Losses Incurred as a Result of the Ban on tris*, 9th Cong., 2d sess., 16 June 1982, 2.

20. On Carter, see Walter A. Rosenbaum, *Environmental Politics and Policy*, 5th ed. (Washington, DC: CQ Press, 2002), 141; on 156, he calls Carter "the most avowed environmentalist of all recent presidents." See also Hays, *Beauty*, 57–60; on Carter generally, Erwin Hargrove, *Jimmy Carter as President: Leadership and the Politics of the Public Good* (Baton Rouge: Louisiana State University Press, 1988).

21. For a critique of the bailout, see an op-ed column by Ralph Nader and Gene Kimmelman, "Bad Precedent," *New York Times*, 9 January 1983, F2.

22. *New York Times*, 13 December 1977, 89.

23. *Washington Post*, 13 December 1977, D7.

24. *Chemical and Engineering News*, 19 December 1977, 6.

25. *Velsicol Chemical Corp. v. Honorable James B. Parsons*, U.S. Court of Appeals for the Seventh Circuit, 561 F 2dg71: 1977, 29 July 1977, lists the attorneys from 1975 to 1977.

26. Denney served as an EPA attorney on one of the chlordane cases, *Environmental Defense Fund, Inc., v. epa et al.*, U.S. Court of Appeals for the District of Columbia Circuit, 179 U.S. App. D.C. 43; 548 F. 2d 998, 10 November 1976.

27. Pete Daniel, *Toxic Drift: Pesticides and Health in the Post–World War II South* (Baton Rouge: Louisiana State University Press, 2005), 95, says the Hardeman dump was opened specifically because of the reaction to the fish kills.

28. *Washington Post*, 4 September 1978, A1.

29. D. R. Rima et al., *Potential Contamination of the Hydrologic Environment from the Pesticide Waste Dump in Hardeman County, Tennessee*, U.S. Dept. of the Interior, U.S. Geological Survey, Water Resources Division, Administrative report to the Federal Water Pollution Control Administration, August 1967.

30. Craig L. Sprinkle, *Leachate Migration from a Pesticide Waste Disposal Site in Hardeman County, Tennessee*, U.S. Geological Survey, Water Resources Investigations 78–128, 1978; the report cover states this was prepared in conjunction with the Tennessee Department of Public Health.

31. *Woodrow Sterling et al. v. Velsicol*, U.S. Court of Appeals for the Sixth Circuit, 855 F 2d 1188, 29 August 1988, has the best summary of this complex litigation.

32. House Committee on Interstate and Foreign Commerce, Subcommittee on Oversight and Investigation, 96th Cong., 2d sess., *Hazardous Waste: Memphis, Tennessee Area*, 2 April 1980.

33. For a summary of events at the Hollywood Dump, see *Isabel v. Velsicol*, U.S. District Court for the Western District of Tennessee, Western Div., Case No. 04–2297, decided 20 June 2006.

34. *Wall Street Journal*, 13 February 1978, 1.

35. Ibid.

36. Memorandum from John Fisher, Enforcement Investigator, Michigan DNR, to Jack Bails, Chief, Environmental Enforcement Division, 10 December 1979, in Pine River Archives, Alma College, Alma, MI.

37. *Chemical Week*, 26 September 1979, 19, had noted his role in the "new" Velsicol campaign when it reported on his resignation.

38. W. David Gibson, "Velsicol: Still Striving to Live Down Old Image," *Chemical Week*, 14 May 1980, 29.

39. Ibid.

40. Gordon Davis, ed., *Who's Who in Engineering*, 9th ed. (Washington, DC: American Assoc. of Engineering Societies, 1995), 605.

41. Gibson, "Velsicol," 29.

42. While the text refers to Rademacher and several others, there were even more, such as David B. Graham from the U.S. Department of the Interior, and Charles Frommer, a respected New York State environmental-health expert.

43. Martindale-Hubbell Lawyer Locator at http://lawyers.martindale.com (accessed 31 March 1999).

44. *Chemical Week*, 25 August 1962, 57.

45. Gibson, "Velsicol," 29.

46. *Chemical Week*, 25 August 1962, 57.

47. See, for example, Michael Moskow, "Disruptions in Global Financial Markets: The Role of Public Policy," *Economic Perspectives [Third Quarter 2000]* (Chicago: Federal Reserve Bank of Chicago, 2000), 24.

48. Gibson, "Velsicol," 29.

49. "After a Big Overhaul, Velsicol Is Stepping Out," *Chemical Week*, 24 March 1982, 26.

50. Cass Sunstein, *Risk and Reason: Safety, Law, and the Environment* (New York: Cambridge University Press, 2002), 36–52, discusses the misunderstanding of tradeoffs such as between the health benefits of using DDT (reduced malaria deaths) and the costs of its use (possible cancers).

51. For an analysis fifteen years into the tracking of the cohort, see Harold Humphrey (project director), "Michigan Long Term PBB Study: The Largest and Longest Running Toxic Chemical Exposure Registry in the Nation," Michigan Department of Public Health, 1991, mimeo. While the cohort disproportionately included farm families, it was so large that it includes people from most residential and occupational categories.

52. For a review of all the contaminants of concern produced at Michigan Chemical, see U.S. Department of Health, Education, and Welfare, Centers for Disease Control, National Institute for Occupational Safety and Health, "Health Hazard Evaluation Determination Report HE 77–73–610, Velsicol Chemical Corporation," August 1979.

53. Henry A. Anderson et al., "Investigation of the Health Status of Michigan Chemical Corporation Employees," *Environmental Health Perspectives* 23 (April 1978): 187–191.

54. Elizabeth Seagull, "Developmental Abilities of Children Exposed to Polybrominated Biphenyls (PBB)," *American Journal of Public Health* 73 (March 1983): 281–285.

55. Edward M. Schwartz and William Rae, "Effect of Polybrominated Biphenyls (PBB) on Developmental Abilities of Children," *American Journal of Public Health* 73 (March 1983): 277–280.

56. Daniel W. Ebert, Janet Elashoff, Kenneth Wilcox, "Possible Effect of Neonatal Polybrominated Biphenyl Exposure on the Developmental Abilities of Children," *American Journal of Public Health* 73 (March 1983): 286–288.

57. Alden K. Henderson et al., "Breast Cancer among Women Exposed to Polybrominated Biphenyls," *Epidemiology* 6 (September 1965): 544–546.

58. Ashraful Hoque et al., "Cancer among a Michigan Cohort Exposed to Polybrominated Biphenyls in 1973," *Epidemiology* 9 (July 1998): 373–377. For a comment on the significance of these results, see the editorial "Does the Study of Environmental Disease Determinants Call for Skepticism or Open-Mindedness?" *Epidemiology* 9 (July 1998): 367–369.

59. Matt Clark, Jon Lowell, and Dan Shapiro, "Cancer in Clusters," *Newsweek*, 17 October 1977, 119.

60. Richard S. Schwartz, Jeffrey P. Callen, and Joseph Silva, "A Cluster of Hodgkin's Disease in a Small Community: Evidence for Environmental Factors," paper presented at American Federation for Clinical Research Meeting, New Orleans, 22 January 1976 [the cover of the paper says 1976, but the 6 is circled and there is a stamped date of Jan. 21, 1977], in the files of the Michigan Department of Natural Resources and Environment.

61. Ibid., 11.

62. Richard S. Schwartz, Jeffrey P. Callen, and Joseph Silva, "A Cluster of Hodgkin's Disease in a Small Community: Evidence for Environmental Factors," *American Journal of Epidemiology* 108 (1978): 19–26.

63. J. F. Plouffe et al., "Abnormal Lymphocyte Responses in Residents of a Town with a Cluster of Hodgkin's Disease," *Clinical Experimental Immunology* 35 (1979): 163–170.

64. Attachment to letter from Matthew Zach of CDC and John L. Isbister and George Van Amburg, of the Michigan Department of Public Health, to the Editor, *American Journal of Epidemiology* (3 November 1978), commenting on the draft of Silva's article, in Pine River Archives, Alma College, Alma, MI.

65. Matthew Zack, "Cancer Death Rates in Michigan Communities with and without Navy Bean Grain Elevators, 1970–1975," has a typed date of 7 July 1977 and a printed warning "not for publication"; in Pine River Archives, Alma College, Alma, MI.

66. Clark et al., "Cancer in Clusters," 119.

67. Heidi M. Blanck et al., "Age at Menarche and Tanner Stage in Girls Exposed *in utero* and Postnatally to Polybrominated Biphenyls," *Epidemiology* 11 (2000): 641–647.

68. On age at menarche, see W. C. Chumlea et al., "Age at Menarche and Racial Comparisons in US Girls," *Pediatrics* 111 (2003): 110–113, who find that "the median (mean) age at menarche for all US girls has not changed significantly in 30 years with a shift of only ~ 4 months in that period." Some general studies maintain that the age has been falling for the last two centuries; however, even if true, that would not impact the findings related to PBB, since all girls in the sample were born in the same era.

69. "Summary of Study Done for Michigan Chemical by Hill Top Research, Inc. of Ohio on the Health Effects of FireMaster BP-6," 22 May 1970, 10, mimeo, in Pine River Archives, Alma College, Alma, MI.

70. J. G. Aftosmis et al., "The Toxicology of Brominated Biphenyls: I. Oral Toxicity and Embryotoxicity," paper presented at the Society of Toxicology Meeting, Williamsburg, VA, 5–9 March 1972, 6–7; the subsequent paper mentioned in the quotation was presented at the same meeting, authored by the same authors, entitled "Toxicology of Brominated Biphenyls: II. Skin, Eye and Inhalation Toxicity and an Acute Test Method for Evaluating Hepatotoxicity and Accumulation in Body Fat," in Pine River Archives, Alma College, Alma, MI.

71. Grievance Number 74–4 of Oil Chemical and Atomic Workers International Union, Local 7–224, in Pine River Archives, Alma College, Alma, MI.

72. Michigan Chemical Corp., "Report on Operator Exposure to Firemaster BP-6 during Production," 12 June 1975, mimeo, in Pine River Archives, Alma College, Alma, MI.

73. Michigan Chemical Corp., "BP-6 Contamination Control," 19 November 1974, 7, mimeo, in Pine River Archives, Alma College, Alma, MI.

74. Michigan Chemical Corp., interoffice memo, L. H. Hahn to F. A. Daniher, 24 June 1974, in Pine River Archives, Alma College, Alma, MI.

75. For example, see House Committee on Interstate and Foreign Commerce, Subcommittee on Oversight and Investigations, *Adverse Effects of Polybrominated Biphenyls (pbb's)*, 95th Cong., 1st sess., 2 and 3 August 1977; and Senate Committee on Commerce, Science, and Transportation, Subcommittee on Science, Technology, and Space, *Toxic Substances Polybrominated Biphenyl (pbb) Contamination in Michigan*, 95th Cong., 1st sess., 31 March 1977.

76. U.S. Department of Health and Human Services, Public Health Service, Agency for Toxic Substances and Disease Registry, "Site Review and Update: Velsicol Chemical Mich.," 30 September 1993, 1–2, mimeo.

77. *EDF v. Director of Agriculture Department*, Court of Appeals of Michigan, 11 Mich. App. 693, 10 November 1967; the second case was an appeal from this one.

78. Hal Higdon, "Obituary for DDT (in Michigan)," *New York Times Magazine*, 6 July 1969, 6.

79. Mary S. Wolff et al., "Correlation of DDE and PBB Serum Levels in Farm Residents, Consumers, and Michigan Chemical Corporation Employees," *Environmental Health Perspectives* 23 (1978): 178.

80. Walter J. Rogan et al., "Polychlorinated Biphenyls (PCBs) and Dichlorodiphenyl Dichloroethene (DDE) in Human Milk: Effects on Growth, Morbidity, and Duration of Lactation," *American Journal of Public Health* 77 (October 1987): 1294–1297.

81. David H. Garabrant et al., "DDT and Related Compounds and Risk of Pancreatic Cancer," *Journal of the National Cancer Institute* 84 (1992): 764–771; see letter of comment from Nuria Malats, Francisco X. Real, and Miquel Porta, "DDT and Pancreatic Cancer," *Journal of the National Cancer Institute* 85 (February 1993): 329, and response from David Garabrant, Janetta Held, and David Hima, 328–329.

82. Mary S. Wolff et al., "Blood Levels of Organochlorine Residues and Risk of Breast Cancer," *Journal of the National Cancer Institute* 85 (April 1993): 648–652; John F. Acquavella, Belinda Ireland, and Jonathan Ramlow, "Organochlorines and Breast Cancer," *Journal of National Cancer Institute* 85 (November 1993): 1872–1875; and correspondence from Lisa Shames, Mark Munekata and Malcolm Pike, "Re: Blood Levels of Organochlorine Residues and Risk of Breast Cancer," *Journal of National Cancer Institute* 86 (November 1994): 1642–1643.

83. Anders Ekborn, Anders Wicklund-Glynn, Hans-Olov Adami, "DDT and Testicular Cancer," *The Lancet* 347 (24 February 1996): 553–554.

84. Matthew P. Longnecker et al., "Association between Maternal Serum Concentration of the DDT Metabolite DDE and Preterm and Small-for-Gestational Age Babies at Birth," *The Lancet* 358 (14 July 2001): 110–114.

85. Michigan Department of Public Health 1998, data for period 1990–1995.

86. John Hesse, DNR, Michigan Bureau of Water Management, "Water Pollution Aspects of Polybrominated Biphenyl Production: Results of Initial Surveys on the Pine River in the Vicinity of St. Louis, Michigan," presentation to the Governor's Great Lakes Regional Interdisciplinary Pesticide Council, 17 October 1974, copy in Pine River Archives, Alma College, Alma, MI.

87. Michigan Department of Natural Resources, Environmental Protection Branch Staff Report, "A Study of Polybrominated Biphenyl Uptake by Caged Fish Held in the Pine River after Termination of Polybrominated Biphenyl Production by the Michigan Chemical Corporation," 27 November–18 December 1974; copy in Pine River Archives, Alma College, Alma, MI.

88. This was a settlement with another polluter, Oxford Automotive, which clearly had contributed less contamination than Velsicol; see United States Bankruptcy Court, Eastern District of Michigan, Southern Division, *In Re: Oxford Automotive, Inc., et al*, Case No. 04–743777, Chapter 11 (Jointly Administered), Hon. Steven W. Rhodes, Order Authorizing Stipulation Resolving Joint Objection to Claim Number 358 of the Pine River Superfund Citizen Task Force, 19 September 2005.

89. Michigan DNR, Water Quality Division, "Biological Survey of the Pine River 1974 and 1978: Publ. No. 4833–5159," 15 June 1979, 2.

90. Ibid.

91. Ibid., 10.

92. John L. Hesse and Richard A. Powers, "Polybrominated Biphenyl (PBB) Contamination of the Pine River, Gratiot and Midland Counties, Michigan," *Environmental Health Perspectives* 23 (April 1978): 24–25.

93. On CERCLA origins, see James P. Lester and Ann Bowman, eds., *The Politics of Hazardous Waste Management* (Durham, NC: Duke University Press, 1983). For an early critique, Daniel Mazmanian and David Morell, *Beyond Superfailure: America's Toxics Policy for the 1990s* (Boulder, CO: Westview Press, 1992), 1–26.

94. A summary of the sites from a company perspective is in a letter and supporting documents submitted by John M. Rademacher, Vice President, Environmental Health and Regulatory Affairs, Velsicol, to Dale Bryson, Deputy Director, Enforcement Division, US EPA, 11 April 1980, 2–7. This document must have been presumed accurate, since it was part of the settlement process, and Velsicol was given protection from further liability for all sites it identified; in fact, the radioactive waste site contained much more waste than identified.

95. *Chemical and Engineering News*, 29 October 1979, 8.

96. *Schunk v. Northwest Industries*, and *Knight v. Northwest Industries*, settled 21 February 1991.

97. *The Poisoning of Michigan* (London: Thames Television, 1977). There was much concern among the governor's staff about this film and the possibility

of a second in the next election year; see Poisoning of Michigan files in William Long Staff Papers, box 778, William Milliken Papers, Bentley Historical Library, Ann Arbor, MI.

98. Frederick Halbert, *Bitter Harvest* (Grand Rapids, MI: W.B. Eerdmans, 1978); and *Bitter Harvest* (Hollywood: Charles Fries Productions, 1981).

99. The episode, called "Slaughter," was originally scheduled to run on 6 November 1978, the day before the gubernatorial election; however, it was pulled by CBS and ran on 27 November 1978. On the controversial decision to delay the show, a Detroit CBS official said, "It's too sensitive a political issue coming just before the election"; this made critics wonder what was the purpose of the media. See *Detroit News*, 19 October 1978, A1 and A19; and 20 October 1978, B4, which reported that Republican Party denied requesting postponement of episode until after the elections.

100. Letter from William R. Rustem to Arnold Bransdorfer, 27 March 1978, copy in Pine River Archives, Alma College, Alma, MI.

101. *New York Times*, 19 November 1982, A18; Shabecoff later became a major environmental writer of works such as *Earth Rising: American Environmentalism in the 21st Century* (Washington, DC: Island Press, 2000).

102. UPI story, 19 November 1982, http://web.Lexis-Nexis.com (accessed 28 April 1999).

103. UPI story, 20 November 1982, http://web.Lexis-Nexis.com (accessed 28 April 1999).

104. *New York Times*, 19 November 1982, A18.

105. Ibid., not in quotes; two days later the statement appeared as a quotation from Kaufman in *New York Times*, 21 November 1983, D7.

106. *United States and State of Michigan v. Velsicol*, U.S. District Court, Eastern District of Michigan, Northern Division, Civil Action No. 82–10303, 13. The *Christian Science Monitor*, 26 November 1982, 14, correctly identified the $38.5 million as "cash, goods, and services," and mentioned the prominent role played by former EPA employee Rademacher.

107. *Washington Post*, 13 December 1982, A13.

108. *Washington Post*, 11 January 1983, A2.

109. *Washington Post*, 5 March 1983, C1, and 9 March 1983, A24.

110. *(Alma, MI) Morning Sun*, 27 July 1983.

111. *(Alma, MI) Morning Sun*, 20 June 1986.

112. *(Alma, MI) Morning Sun*, 1 October 1985.

113. Oddly, in the haste to show progress, the EPA "delisted" the golf-course site without it ever being "listed" on the National Priorities List.

114. *(Alma, MI) Morning Sun*, 13 June 1985.

115. Michigan Dept. of Environmental Quality, Remedial Investigation Report for Operable Unit One Velsicol Chemical Corporation Superfund Site St. Louis, Gratiot County, Michigan (Okemos, MI: Weston Solutions, 2006), I: §2 31–34; §3 49–53.

116. *(Alma, MI) Morning Sun*, 6 November 1981.

117. *(Alma, MI) Morning Sun*, 1 December 1981.

118. *(Alma, MI) Morning Sun*, 2 February 1983.

119. Ibid.

120. *Saginaw (MI) News*, 8 January 1990.

121. *Saginaw (MI) News*, 29 January 1992, B4.

122. *(Alma, MI) Morning Sun*, 21 June 1982.

CHAPTER 4. NEW MANAGEMENT

1. *New York Times*, 14 July 1974, 138.

2. All plant data is taken from the *1979 Northwest Industries Annual Report*.

3. Ibid.

4. A good summary of the restructurings is in Thomas Derdak, ed., "Farley Northwest Industries, Inc.," *International Directory of Company Histories* (Chicago: St. James Press, 1988), 1:440–441.

5. Employment data from *Moody's Industrial Manual, 1982* (p. 4143) and *1985* (p. 4226).

6. *New York Times*, 10 September 1981, D6.

7. *New York Times*, 16 February 1982, D4.

8. *New York Times*, 5 May 1983, D18, reported on his recent sale of 250,000 shares, reducing his holdings to 150,000.

9. *New York Times*, 19 October 1983, D4.

10. *Wall Street Journal*, 12 December 1984, 1.

11. To see the Farley Field House, go to www.bowdoin.edu/athletics/facilities.shtml.

12. For a good review of Farley's background, see Kenneth Labich, "Bill Farley Has a Dream Machine," *Fortune*, 27 May 1985; also Julia Flynn Siler, "Taking Over in a One-Company Town," *New York Times*, 5 March 1989, F6.

13. Siler, "Taking Over in a One-Company Town"; and Labich, "Bill Farley Has a Dream Machine."

14. For an in-house perspective on Baumfolder post-Farley, see Baumfolder Corp. website, http://www.baumfolder.com/baumhistory.html (accessed 2

July 2003); on their engineering and quality leadership, see *American Printer*, 1 April 2003, 1.

15. Labich, "Bill Farley Has a Dream Machine."

16. See Connie Bruck, *The Predators' Ball* (New York: Penguin, 1988), 163; and James B. Stewart, *Den of Thieves* (New York: Simon and Schuster, 1991), 498, 504.

17. Ellyn Spragins, "Northwest Industries: The Acid Test for Bill Farley's Offbeat Style," *Business Week*, 9 September 1985, 68. Horatio Alger (1832–1899) was the writer of "rags-to-riches" stories, where poor heroes rose to great wealth by ingenuity and hard work; on Alger, see Richard Weiss, *The American Myth of Success: From Horatio Alger to Norman Vincent Peale* (New York: Basic Books, 1969).

18. Jennifer Merritt, "The ABCs of Failure," *Business Week*, 9 June 2003, 126.

19. Spragins, "Northwest Industries," 68, is the source for all quotes in this paragraph.

20. Labich, "Bill Farley Has a Dream Machine," 114.

21. Cindy Skrzycki, "The Buyout Man," *U.S. News and World Report*, 26 January 1987, 63.

22. William H. Miller, "Bill Farley: The Hands-on Conglomerator," *Industry Week*, 15 June 1987, 42.

23. Skrzycki, "The Buyout Man," 63.

24. *Washington Post*, 26 May 1988, D1.

25. *Crain's Chicago Business*, 6 June 1988, A44.

26. See announcement from Boston College at the time Farley gave $1.5 million to endow a chair in legal studies, at http://www.bc.edu/bc_org/rvp/pubaf/chronicle/v6/mr26/farley.html (accessed 8 July 2003).

27. See Horatio Alger Association of Distinguished Americans at http://www .horatioalger.com (accessed 8 July 2003).

28. Jack Egan, "Farley Gets a Workout," *U.S. News and World Report*, 12 March 1990, 57. Other references to Farley as Alger include one in Spragins, "Northwest Industries," 68; and Steve Weiner, "Some Things Are Easier Said Than Done," *Forbes*, 27 April 1987, 56.

29. Ibid.

30. Paul Merrion, "William Farley Winning Fight to Tame Northwest Ind. Debt," *Crain's Chicago Business*, 13 January 1986, 3.

31. PR Newswire, 14 January 1986, accessed on LexisNexis, 3 July 2003.

32. Information on Universal taken from websites of MagneTek and Universal Lighting.

33. *Chemical Week*, 24 July 1985, 5; *Chicago Tribune*, 9 September 1985, C2.

34. Accounts of the exact purchase price varied: see *Chicago Tribune*, 29 January 1986; *Financial Times*, 29 January 1986, 21; *New York Times*, 29 January 1986, D3; and *Chicago Tribune*, 22 March 1986, C6.

35. Ibid.; this text is stamped on every page of the agreement.

36. *Marcelle Crumbaugh v. Velsicol Chemical Corp.*, Court of Appeals of Michigan No. 212295, 29 August 2000.

37. PR Newswire, 16 September 1986, at http://web.lexis-nexis.com/universe (accessed 3 July 2003).

38. *Fortune* 29, September 1986, 10.

39. Weiner, "Some Things Are Easier Said Than Done," 56.

40. Mentioned in Michael Oneal and Dean Foust, "Bill Farley's $500 Million Needs a Home—Fast," *Business Week*, 27 June 1988, 35. In September 1989 only Fruit's cash was still mentioned as off limits; see David Greising and Dean Foust, "Bill Farley Is on Pins and Needles," *Business Week*, 18 September 1989, 58. Since Foust was involved in both stories, it would seem the raiding of Acme cash began between the two stories.

41. See *Moody's Industrial Manual, 1988*, pp. 1257–1260.

42. Oneal and Foust, "Bill Farley's $500 Million Needs a Home," 35.

43. See, for a discussion of the general problem of management control of corporate governance, Paul W. MacAvoy and Ira M. Millstein, *The Recurrent Crisis in Corporate Governance* (New York: Palgrave Macmillan, 2003), 11–31.

44. On Hall, see PR Newswire, 7 August 1986, http://www.lexis-nexis.com/universe (accessed 3 July 2003); on board, see *Moody's Industrial Manual, 1987*, p. 1287.

45. Weiner, "Some Things Are Easier Said Than Done," 56.

46. Miller, "Bill Farley," 42.

47. *New York Times*, 27 February 1988, 38.

48. Spragins, "Northwest Industries," 68.

49. Skrzycki, "The Buyout Man," 63.

50. Weiner, "Some Things Are Easier Said Than Done," 56.

51. Miller, "Bill Farley," 42.

52. Bill Saporito, "Look Who Wants to Be President," *Fortune*, 23 November 1987, 141.

53. *Washington Post*, 27 May 1988, D1.

54. *Crain's Chicago Business*, 30 April 1990, WW1.

55. Saporito, "Look Who Wants to Be President," 141.

56. Ibid.

57. "WestPoint Stevens," in Tina Grant, ed., *International Directory of Company Histories* (Detroit: St. James Press, 1997), 16:533–536; see also *Moody's Industrial Manual, 1960*, pp. 247–248, 1809; *1974*, p. 2331.

58. *Moody's Industrial Manual, 1985*, pp. 2704–2705; *Cluett Peabody Annual Report, 1984*, pp. 2–8; Joan Kiplinger, "Sanforized: Fabrics' Best Friend," *Fabrics.net*, July/August 2003, available from http://www.fabrics.net/print/joan803.asp.

59. For a list of licensees in the 1960s, see *Cluett, Peabody, and Co. 1968 Annual Report*, 19.

60. Grant, "WestPoint Stevens"; *Moody's Industrial Manual, 1968*, p. 1393; *1987*, p. 4365.

61. *Business Week*, 7 November 1988, 39.

62. *CBS Evening News*, 21 November 1991, available from CBS Archives.

63. Stephen Madden, "Georgia on His Mind," *Fortune* 30, January 1989, 191.

64. *CBS Evening News*, 21 November 1991.

65. Joseph Schumpeter, *Capitalism, Socialism and Democracy*, 3rd ed. (New York: Harper and Row, 1950), 81–86.

66. Madden, "Georgia on His Mind."

67. *CBS Evening News*, 16 November 1988, available from CBS Archives.

68. Interview with Wayne Clark, News Editor, *Valley Times-News* (Lanette, AL), 13 June 2005.

69. Greising and Foust, "Bill Farley Is on Pins and Needles," 58.

70. *CBS Evening News*, 21 November 1991.

71. *New York Times*, 27 June 1990, D5.

72. "A Fruit of the Loom Stock Harvesting," *Business Week*, 6 May 1991, 42.

73. Riva Atlas, "Doehler-Jarvis Sale Brings Good News at Last to Farley," *Mergers and Acquisitions Report*, 30 April 1990, 2; and Lisa Keefe, "ICM to Buy Farley Unit," *Crain's Chicago Business*, 23 April 1990, 1.

74. For a good example of the complexity of deciding on trends that define eras, see Richard L. McCormick, *The Party Period and Public Policy: American Politics from the Age of Jackson to the Progressive Era* (New York: Oxford University Press, 1986), 263–288.

75. John Kenneth Galbraith, *The New Industrial State* (Boston: Houghton Mifflin, 1967).

76. David Halberstam, *The Reckoning* (New York: Morrow, 1986), 673–679.

77. Mark Roe, *Weak Owners, Strong Managers* (Princeton, NJ: Princeton University Press, 1994).

78. Martin Feldstein, *American Economic Policy in the 1980s* (Chicago: University of Chicago Press, 1994).

79. *CBS Evening News*, 16 November 1988.

80. *CBS Evening News*, 21 November 1991.
81. Max Weber, *The Protestant Ethic and the Spirit of Capitalism*, trans. Talcott Parsons (London: G. Allen and Unwin, 1930), 48–54, 180–181, where he describes Benjamin Franklin and his progeny as the ideal type for the Protestant ethic stripped of links to orthodox Christianity.
82. Weiss, *The American Myth of Success*; Stephen Hunt, "Winning Ways: Globalization and the Impact of the Health and Wealth Gospel," *Journal of Contemporary Religion* 15 (October 2000): 331–347.
83. See, for example, Larry C. Farrell, *The Entrepreneurial Age: Awakening the Spirit of Enterprise in People, Companies, and Countries* (New York: Allworth Press, 2001), reflecting the modern triumph of the faith. Even the International Labor Organization published a book to teach the subject: Geoffrey G. Meredith, Robert E. Nelson, and Philip A. Neck, *The Practice of Entrepreneurship* (Geneva: International Labour Office, 1982); the book was not focused on entrepreneurship in a general sense, where it might have been of value to union leaders, it focused on business entrepreneurship.
84. *CBS Evening News*, 21 November 1991.
85. In addition to Bruck, *The Predators' Ball*, and Stewart, *Den of Thieves*, were books such as Michael Lewis, *Liar's Poker* (New York: Norton, 1989); Sarah Bartlett, *The Money Machine* (New York: Warner Books, 1992); George Anders, *The Merchants of Debt* (New York: Basic Books, 1992); and Brian Burrough and John Helyer, *Barbarians at the Gate* (New York: Harper and Row, 1990).
86. George P. Baker and George David Smith, *The New Financial Capitalists: Kohlberg Kravis Roberts and the Creation of Corporate Value* (New York: Cambridge University Press, 1998), 30.
87. Glen Yago, "High Yield or Junk," *National Review*, 31 August 1992, 40–45.
88. *CBS Evening News*, 21 November 1991.

CHAPTER 5. CREATING JUNK

1. Robert Kuttner, *The Squandering of America: How the Failure of Our Policies Undermines Our Prosperity* (New York: Alfred Knopf, 2007), 37–40.
2. William Adler, *Mollie's Job: A Story of Life and Work on the Global Assembly Line* (New York: Simon and Schuster, 2000); and *Booted Out: Exporting Jobs from the U.S.*, video (New York: Films for the Humanities, 1993).
3. Clearly many economists reject this view of the changes in the era. One of the best, with admirable historical perspective, is Daniel Cohen, *Globalization*

and Its Enemies (Cambridge, MA: MIT Press, 2006), who points out the comparatively modest levels of globalization today, and celebrates the benefits of weakening the state and freeing trade and movements of labor and capital.

4. For a good overview of trade's impact on labor rights, see Lance Compa, *Unfair Advantage: Workers' Freedom of Association in the United States under International Human Rights Standards* (Ithaca, NY: ILR Press, 2000). On interest-group lobbying that undermined labor protections under trade laws, see Gordon L. Clark, "Limits of Statutory Responses to Corporate Restructuring Illustrated with References to Plant Closing Legislation," *Economic Geography* 67 (January 1991): 22–41.

5. Many would hold that the gloom implicit in this paragraph is not backed by reality, which has been a growth in jobs and decline in unemployment. For a more worrisome review of these trends, which may give some support to the interpretation in this book, see the work of Susan Houseman at the W.E. Upjohn Institute, who has found the economic growth some have celebrated in recent years has not been as great as represented; see Michael Mandel, "The Real Cost of Offshoring," *Business Week*, 18 June 2007, 29–34.

6. On the theory, see Walter Braddock Hickman, *Corporate Bond Quality and Investor Experience* (Princeton, NJ: Princeton University Press, 1958).

7. Connie Bruck, *The Predators' Ball* (New York: Penguin Books, 1988), 106–107.

8. Biographical and early company history taken from story of Acme in *New York Times*, 6 September 1957, 21 and 27; and obituary of Jessel Cohn in *Clarksville (TN) Leaf-Chronicle*, 21 February 1961, 1; also description "Acme Boot Co., Inc." in Reference/Genealogy Dept., Clarksville-Montgomery County Public Library, Clarksville, Tennessee.

9. Ads ran in many publications, including *Detective Comics*, July 1982.

10. Photo in files of the Reference/Genealogy Dept., Clarksville-Montgomery County Public Library, Clarksville, Tennessee.

11. *New York Times*, 30 December 1993, D3.

12. James B. Stewart, *Den of Thieves* (New York: Simon and Schuster, 1991), 200–203, 222–225, is good on Beatrice.

13. *Footwear News*, 25 March 1991, 7.

14. Narrator comment in *Booted Out*, video (Nashville: Tennessee Public Television, 1992).

15. *Booted Out* incorporates footage of a Farley campaign speech.

16. *Footwear News*, 15 April 1991, 10.

17. *Memphis Commercial Appeal*, 25 April 1991, B3.

18. *Booted Out*; italics added to capture vocal emphasis in video.

19. *U.S. News and World Report*, 2 November 1992, 36.

20. *New York Times*, 28 February 1993, C37.

21. *Memphis Commercial Appeal*, 18 October 1992, C2.

22. *Atlanta Journal and Constitution*, 21 May 1993, A3.

23. *Sydney Morning Herald*, 19 December 1994, 14.

24. *Memphis Commercial Appeal*, 5 May 1993, B10.

25. *Miami Herald*, 13 June 1993, K3.

26. *Memphis Commercial Appeal*, 5 September 1993, C2.

27. *Sydney Morning Herald*, 19 December 1994, 14.

28. *Daily News Record*, 11 April 1996, 12, reported another presidential change at Acme.

29. *Omaha World Herald*, 10 March 2002, 1D.

30. Pension Benefit Guaranty Corporation, news release, 29 April 1998, at http://www.pbgc.gov/media/news-archive/news-releases/1998/pr98-23.html (accessed 3 July 2007).

31. *Clarksville (TN) Leaf Chronicle*, 28 August 1996; and 31 August 1996.

32. *Toledo Blade*, 19 October 1964.

33. Herman Doehler, *Die Casting* (New York: McGraw Hill, 1951).

34. For example, see *Toledo Blade*, 26 March 1950, and 22 March 1952.

35. *Toledo Blade*, 14 October 1969.

36. *Moody's Manual of Investments, 1952*, 180–181.

37. Ibid., 181; on perpetual agreement with UAW see *Toledo Times*, 10 October 1937.

38. *Toledo Blade*, 29 November 1950.

39. *Toledo Blade*, 22 May 1962, 27 June 1971, and 21 May 1974.

40. *Toledo Blade*, 12 January 1981.

41. *Toledo Blade*, 7 July 1981.

42. *Toledo Blade*, 11 October 1982.

43. Ibid.

44. *Toledo Blade*, 1 April 1984.

45. *Toledo Blade*, 5 May 1985.

46. *Toledo Blade*, 12 November 1984, reported Iwarsson had served as president of American-Lincoln and Abex Corporation, in Winchester, Virginia.

47. *Toledo Blade*, 5 May 1985.

48. *Toledo Blade*, 25 February 1986.

49. *Toledo Blade*, 4 March 1987.

50. *Toledo Blade*, 6 March 1987.

51. For a study of subsidies with a review of the literature, see Dennis A. Rondinelli and William J. Burpitt, "Do Government Incentives Attract and Retain International Investment? A Study of Foreign-owned Firms in North Carolina," *Policy Sciences* 33 (June 2000).

52. *Toledo Blade*, 6 March 1987.

53. *Toledo Blade*, 27 March 1987.

54. *Toledo Blade*, 5 May 1985.

55. *Toledo Blade*, 31 March 1987.

56. *Toledo Blade*, 7 April 1987.

57. *Toledo Blade*, 11 April 1987.

58. Tocqueville, *Democracy in America*, 1:59–97, on local government; and 2:106–110, on voluntary civic associations.

59. *Toledo Blade*, 6 June 1987.

60. *Toledo Blade*, 20 May 1987.

61. *Toledo Blade*, 13 May 1987.

62. *Toledo Blade*, 20 April 1989.

63. *Toledo Blade*, 12 December 1989.

64. *Crain's Chicago Business*, 23 April 1990, 1, has story about sale of Doehler-Jarvis to ICM Industries; see also *Mergers and Acquisitions Report*, 30 April 1990, 2.

65. *New York Times*, 14 August 1991, D5; the relationship of Farley and ICM was revealed in the settlement of a tax dispute in Alabama over the Sheffield Die Casting facility that Doehler briefly leased as part of its efforts to win concessions in Toledo. According to those records, Farley Industries was a general partner of ICM. See State of Alabama, Department of Revenue, Administrative Law Division, *Dept. of Revenue v. U.S. Die Casting and Development*, Docket No. L91-208, 24 November 1993, 3.

66. Haigh served on advisory committee for a study of technology and economic development sponsored by the Congressional Office of Technology Assessment; see U.S. Congress, Office of Technology Assessment, *Technology, Innovation, and Regional Economic Development: Encouraging High-Technology Development—Background Paper #2* (Washington, DC: U.S. Congress, OTA-BP-STA-25, February 1984).

67. *Toledo Blade*, 13 October 1991. On the short time horizons that dominate U.S. business planning, see Susan Christopherson, "Why Do National Labor Market Practices Continue to Diverge in the Global Economy? The 'Missing Link' of Investment Rules," *Economic Geography* 78 (January 2002): 1–20.

68. Harvard Industries Securities and Exchange Commission (SEC) 10-K 1995, 39.

69. *Lakeland (FL) Ledger*, 10 March 1995, D7.

70. Generally on the topic see John R. Gerdy, *Sports: The All-American Addiction* (Jackson: University of Mississippi Press, 2002), 21–39.

71. *USA Today*, 4 September 1992, 5C; for general review of the modern "yellow journalism," see Darrell West, *The Rise and Fall of the Media Establishment* (New York: St. Martin's Press, 2001).

72. *PR Newswire*, 23 June 1995, accessed through LexisNexis, 14 October 2003.

73. On the problem of the growth of public relations in public life, see Kevin Moloney, *Rethinking Public Relations: PR Propaganda and Democracy*, 2nd ed. (New York: Routledge, 2000), 75–118; also for a comparative perspective, Alex Carey, *Taking the Risk Out of Democracy: Corporate Propaganda versus Freedom and Liberty* (Urbana: University of Illinois Press, 1997), 153–172.

74. *PR Newswire*, 23 June 1995, accessed through LexisNexis, 14 October 2003.

75. *1995 Harvard Industries Annual Report*.

76. For a study of the tendency of private management to authoritarianism, see Michael Walzer, *Spheres of Justice: A Defense of Pluralism and Equality* (New York: Basic Books, 1983), 155–164, 291–303.

77. *1995 Harvard Industries Annual Report*.

78. On the need for checks upon executive power, see Linda deLeon, "Ethics and Entrepreneurship," *Policy Studies Journal* 24 (Autumn 1996): 495–511; and, while focused on public executives, John Kane and Haig Patapan, "In Search of Prudence: The Hidden Problem of Managerial Reform," *Public Administration Review* 66 (September/October 2006): 711–725.

79. *PR Newswire*, 13 August 1991, accessed through LexisNexis, 14 October 2003.

80. *Business Week*, 26 August 1991, 32.

81. *Chicago Daily Law Bulletin*, 17 January 2001, 1.

82. *Toledo Blade*, 23 August 1991.

83. *Toledo Blade*, 4 March 1995.

84. *Toledo Blade*, 11 January 1995.

85. *Toledo Blade*, 18 January 1995.

86. *Toledo Blade*, 13 January 1995. The classic study of unions falling victim to the "iron law of oligarchy" is Robert Michels, *Political Parties: A Sociological Study of the Oligarchical Tendencies of Modern Democracy*, trans. Eden and Cedar Paul (Glencoe, IL: The Free Press, 1962); for a study that finds unions resisting this trend, see Andrew W. Martin, "Organizational Structure, Authority and Protest: The Case of Union Organizing in the United States, 1990–2001," *Social Forces* 85 (March 2007): 1413–1436.

87. *Automotive News*, 29 May 1995; *PR Newswire*, 26 July 1995.

88. *PR Newswire*, 7 August 1995.

89. *1995 Harvard Industries Annual Report*.

90. *Toledo Blade*, 21 June 1998, B1.

91. Harvard Industries SEC 10-K 1995, 17.

92. *Toledo Blade*, 21 June 1998, B2.

93. *Toledo Blade*, 21 June 1998, B2.

94. Ibid.

95. Harvard Industries SEC 10-K 1995, 22.
96. *Junk Bond Reporter,* 18 November 1996, 1.
97. *Automotive News,* 10 August 1998.
98. Harvard Industries SEC 10K 1997, 7.
99. *PR Newswire,* 8 May 1997; see also *New York Times,* 9 May 1997, D4.
100. Harvard Industries SEC 10K 1997, 3 and 7; see also *Automotive News,* 24 November 1997, 6.
101. *AP News,* 26 August 2002.
102. The lessons are clear because the closure of Acme, Doehler, and Universal had little or nothing to do with trade policy. In the case of Doehler, see for example U.S. Department of Labor, Employment and Training Administration, Determination on Reconsideration of application by workers for benefits under the trade adjustment assistance program [TA-W-38, 550], at http://www.dol.gov/eta/regs/fedreg/notices/2001028983.htm (assessed 30 June 2004). As the *Blade's* business writer Gary Pakulski wrote on 21 June 1998, the weekend after the company shut down, "Unlike many industrial fatalities, Doehler's demise can't be blamed on declining sales or cheap foreign imports. Rather, say many observers, the company had too little cash, too much business, and too little consistent management savvy." The "too much business" reference is a second fact that needs emphasis. Doehler's products were in great demand until the day it closed. As the *Blade* said on 5 April 1007, the company had "so much business in Toledo that the company was forced to run seven days a week."
103. *Toledo Blade,* 3 October 1997.
104. *Automotive News,* 22 September 1997, 4.
105. *Toledo Blade,* 3 October 1997.
106. *FDU Magazine* (Fall/Winter 1998).
107. *Toledo Blade,* 12 February 1998, 21.
108. *Toledo Blade,* 10 April 2002.
109. In an ultimate irony, in February 2004 Naimoli filed suit against Anchor Glass for the loss of his pension, which the PBGC cut from $6,488 per month to $3,514; see *St. Petersburg Times,* 24 February 2004. Such developments caused Naimoli to lose some respect; see for example the *Tampa Tribune,* 16 July 2002, Sports 1, which mocked his financial management of the Tampa Devil Rays.
110. *Toledo Blade,* 21 June 1998, B1.
111. For wage differential data, see Marlene A. Lee and Mark Mather, "U.S. Labor Force Trends," *Population Bulletin* 63 (June 2008): 3–18.
112. For a discussion of Farley as one example of greed, see Harrison Rainie, Margaret Lofkus, and Mark Madden, "The State of Greed," *U.S. News and World Report,* 17 June 1996, 62.

113. For a study of the wider trends such as this, see Kim Bobo, *Wage Theft in America* (New York: New Press, 2009); this study and reports from UCLA's Institute for Research on Labor and Employment, "Broken Laws, Unprotected Workers: Violations of Employment and Labor Laws in American Cities," focus on treatment of only the poorest workers in the semi-formal economy.
114. *Toledo Blade*, 5 April 1997.
115. *Toledo Blade*, 4 December 1997.

CHAPTER 6. IMPORTING FRUIT

1. Focusing on workers suffering as a result of the leadership ideology of the "liberal democratic state," see Kathleen R. Arnold, *America's New Working Class: Race, Gender, and Ethnicity in a Biopolitical Age* (University Park: Pennsylvania State University Press, 2008).
2. Charles Maier, *Among Empires: American Ascendancy and Its Predecessors* (Cambridge, MA: Harvard University Press, 2006), 191–284.
3. Chris Hedges, *Empire of Illusion: The End of Literacy and the Triumph of Spectacle* (New York: Nation Books, 2009), 145–151.
4. "'Buy Industry' or 'You May not Get It,' Says Lafayette Mayor Who Learned the Hard Way," *Tennessee Town and City* 8 (July 1957), 6.
5. Pete Daniel, *Standing at the Crossroads: Southern Life in the Twentieth Century* (New York: Hill and Wang, 1986), 25 and 150, makes the case that the South was like the Third World in its dependence on outside investors, taming southerners "as dependable workers."
6. V. O. Key Jr., *Southern Politics in State and Nation* (New York: Alfred Knopf, 1950), 657–658, is the classic study of southern behavior. Peter Applebome, *Dixie Rising: How the South Is Shaping American Values, Politics, and Culture* (New York: Random House, 1996), 202, shows these attitudes continue down through the 1990s, quoting Bruce Raynor, southern organizer for UNITE, the current textile union, who said, "But Southern workers aren't anti-union. There's a tremendous anti-union culture built up in these towns; you have to fight the whole town, the politicians, the preachers, the funeral home director, the grocery store owner."
7. For a general study of subsidies and other support for low-wage industrialization, see James C. Cobb, *The Selling of the South: The Southern Crusade for Industrial Development, 1936–1990*, 2nd ed. (Urbana: University of Illinois Press, 1993), 98–105. More generally, see F. Ray Marshall, *Labor in the South* (Cambridge, MA: Harvard University Press, 1967), 276–331; and George B.

Tindall, *The Emergence of the New South, 1913–1945* (Baton Rouge: Louisiana State University Press, 1967), 513–525.

8. Lucy Randolph Mason, *To Win These Rights: A Personal Story of the CIO in the South* (Westport, CT: Greenwood Press, 1970), 183.

9. Alex Carey, *Taking the Risk Out of Democracy: Corporate Propaganda versus Freedom and Liberty* (Urbana: University of Illinois Press, 1997), 153, linked this trend to the evolution of the Protestant ethic through what he called a "moral revision."

10. For a study of this religious sanction for owners, see Liston Pope, *Millhands and Preachers: A Study of Gastonia* (New Haven: Yale University Press, 1942), 141–204.

11. On this transition, see David Burner, *The Politics of Provincialism: The Democratic Party in Transition, 1918–1932* (New York: W.W. Norton, 1967), who explained the small-town origins of southern thought; and, more recently, Christine L. Heyrman, *Southern Cross: The Beginnings of the Bible Belt* (Chapel Hill: University of North Carolina Press, 1998), 253–260, focuses on the historic origins of the transition.

12. *Park City Daily News* (Bowling Green, KY), 5 April 1940.

13. *Park City Daily News* (Bowling Green, KY), 22 January 1961; see also 17 May 1959.

14. *Park City Daily News* (Bowling Green, KY), 3 November 1975; see also in-house the *Union Herald*, December 1975.

15. *Park City Daily News* (Bowling Green, KY), 1 April 1980.

16. For a full description of headquarters, see Union Underwear, "Bowling Green, Kentucky: World Headquarters for the World's Largest Men's Underwear Company" (Bowling Green: Union Underwear, 1980); on earlier expansions, see *Park City Daily News*, 4 September 1959; and Bowling Green Chamber of Commerce Newsletter, May 1977, 4.

17. See, for example, Timothy Bartik, "Tennessee," in *The New Economic Role of American States: Strategies in a Competitive World Economy*, ed. R. Scott Fosler (New York: Oxford, 1991), 14–202.

18. *Park City Daily News* (Bowling Green, KY), 22 February 1980.

19. *Park City Daily News* (Bowling Green, KY), 1 April 1980.

20. *Park City Daily News* (Bowling Green, KY), 14 December 1986.

21. *Park City Daily News* (Bowling Green, KY), 13 September 1986.

22. *Park City Daily News* (Bowling Green, KY), 24 October 1986.

23. *Park City Daily News* (Bowling Green, KY), 4 December 1881, 5A.

24. *Park City Daily News* (Bowling Green, KY), 24 June 1993, 2A.

25. See *Moody's Industrial Manual, 1988*, 1257–1259.

26. Kenneth Labich, "Bill Farley Has a Dream Machine," *Fortune*, 27 May 1985, 115.

27. Bowling Green Chamber of Commerce, *Business News*, October 1987.

28. *Park City Daily News* (Bowling Green, KY), 25 September 1991, 2.

29. On Edwards, see Tyler Bridges, *Bad Bet on the Bayou: The Rise of Gambling in Louisiana and the Fall of Governor Edwin Edwards* (New York: Farrar, Straus and Giroux, 2001).

30. Nancy Haas, "Fruit's Salad Days," *Financial World*, 22 June 1993, 46.

31. See *Chicago Sun Times*, 21 July 1994, 56; and *Park City Daily News* (Bowling Green, KY), 3 March 1994.

32. *Park City Daily News* (Bowling Green, KY), 25 October 1995.

33. *Daily News Record*, 15 May 1996, 10.

34. For a good review of the benefits package, see *New York Times*, 19 March 2000, 3.1.

35. *New York Times*, 31 October 1995, D6.

36. *Park City Daily News* (Bowling Green, KY), 22 December 1995.

37. *Fruit of the Loom 1995 Annual Report*, 2 and 30.

38. *U.S. News and World Report*, 18 March 1996, 63.

39. *Park City Daily News* (Bowling Green, KY), 19 October 1995.

40. *Park City Daily News* (Bowling Green, KY), 21 January 1996.

41. *Louisville Courier-Journal*, 24 August 1997, 23.

42. *Louisville Courier-Journal*, 8 August 1997, 1.

43. *Atlanta Journal and Constitution*, 27 November 1997, 10G.

44. *(New Orleans) Times-Picayune*, 30 December 1997, A2.

45. *New York Times*, 12 November 1997, D4.

46. Ibid.

47. *Park City Daily News* (Bowling Green, KY), 9 January 1998.

48. *Park City Daily News* (Bowling Green, KY), 29 January 1998.

49. *Park City Daily News* (Bowling Green, KY), 31 March 1998.

50. *Park City Daily News* (Bowling Green, KY), 9 May 1998.

51. *Park City Daily News* (Bowling Green, KY), 16 April 1998.

52. Ibid.

53. Institute for Policy Studies, *Executive Excess '98: Fifth Annual Executive Compensation Survey*, 23 April 1998.

54. *Louisville Courier-Journal*, 22 August 1998; see also *Courier-Journal*, 1 August 1998; and *Park City Daily News*, 9 July 1998.

55. *Park City Daily News* (Bowling Green, KY), 4 August 1985.

56. *Park City Daily News* (Bowling Green, KY), 14 November 1985.

57. On the history of corporate public-relations efforts in this transition, see Stuart Ewen, *Captains of Consciousness: Advertising and the Social Roots of the Consumer Culture*, 25th anniversary ed. (New York: Basic Books, 2001).

58. On NAFTA generally, see William A. Orme Jr., *Understanding NAFTA: Mexico, Free Trade, and the New North America* (Austin: University of Texas Press,

1996), 290–297. For critiques of Clinton, see Steve Charnowitz, "Promoting Higher Labor Standards," *Washington Quarterly* 18 (Summer 1995): 171–173; and "International Trade and Social Welfare: The New Agenda—Transcript of January 7, 1994, Meeting of the Section on International Law of the American Association of Law Schools," *Comparative Labor Law Journal* 17 (Winter 1996): 352–353.

59. *New York Times*, 16 December 1993, D8.
60. *U.S. News and World Report*, 5 December 1994, 83.
61. *Washington Post*, 6 April 1995; on Clinton's policies, see Joseph Stiglitz, *The Roaring Nineties* (New York: Norton, 2003).
62. Deborah Lutterbeck, "Falling Wages: Why Salaries Keep Sinking When Corporate Profits Are Soaring," *Common Cause Magazine*, Fall 1995. On the concept of "Mugwumps," that is, upper-middle-class "good government" reformers, often indifferent to the working class, see David M. Tucker, *Mugwumps: Public Moralists of the Gilded Age* (Columbia: University of Missouri Press, 1998).
63. *Daily News Record*, 21 June 1995, 3.
64. *Daily News Record*, 25 July 1995, 4; see also continuing disputes, 22 December 1995, 12.
65. For more on Cooper and later work as CEO of Crowell and Moring, a legal consulting firm helping firms globalize, see: www.crowell.com/professionals/Doral.Cooper (accessed 21 August 2009).
66. *Philadelphia Inquirer*, 17 September 1996, A1.
67. *Boston Globe*, 25 May 1996, 1.
68. *Business Mexico*, 1 September 1997.
69. *Time*, 28 July 1997, 22.
70. *Louisville Courier-Journal*, 24 August 1997.
71. *Washington Post*, 14 November 1997, A27; the *New York Times*, 16 November 1997, 26, also credited the announced shutdowns at Fruit of the Loom for being a symbol that Congress could not ignore.
72. *Chicago Sun-Times*, May 1998, 1.
73. *Washington Post*, 29 October 1999, A29; see also 19 October 1999, E4.
74. Campaign contribution information taken from Political Money Line website, http.tray.com/cgi-win/x+pacpg.exe?DoFn=C0030346196 (accessed 6 July 2003).
75. *Time*, 1 November 1999, 51; on final passage see *Journal of Commerce*, 9 March 2000; *New York Times*, 15 April 2000, A1; and *Washington Post*, 5 May 2000, E1.
76. As *Canadian Business*, 28 May 2001, 58, stated, "It took a long time for Fruit of the Loom to clue in and move its own manufacturing offshore. By then

much of the damage was done—high labor costs caused the company to close several factories, triggering a public relations nightmare."

77. "Fruit of the Loom Company Overview," at http://www.fruit.com/company/index.htm (accessed 18 August 1999).

78. U.S. Securities and Exchange Commission, *Fruit of the Loom 10-K Report, 1999.*

79. *Park City Daily News* (Bowling Green, KY), 31 December 2001.

80. All quotes from Jonathan Moore, *The Factory* (Lexington, KY: KET Production, 2002).

81. Ibid.

82. For example, Union Underwear long ago had received military contracts at Campbellsville; see *Park City Daily News* (Bowling Green, KY), 22 July 1983.

83. *(Lexington, KY) Herald-Leader*, 7 March 2004.

84. *Park City Daily News* (Bowling Green, KY), 11 May 2004.

85. Ibid.

86. Using U.S. Census Bureau median household income from 1990 and 2000, the 1989 income in 1999 dollars was $24,596.34; at the time of the 2000 census it stood at $22,922.

87. According to U.S. Census Bureau, American Fact Finder data on Taylor County, the median household income in 2004 dollars declined from $32,118 in 1989 to $31,599; data at http://factfinder.census.gov/servlet/QTTable (accessed 26 October 2008); 1989–1990 data converted to 2004 dollars using Bureau of Labor Statistics Consumer Price Index calculations.

88. *Park City Daily News* (Bowling Green, KY), 9 April 2004.

89. *Park City Daily News* (Bowling Green, KY), 3 October 2001.

90. Ibid.

91. Ibid.

92. Sergio Zendejas and Pieter de Vries, *Rural Transformations Seen from Below: Regional and Local Perspectives from Western Mexico* (La Jolla, CA: Center for U.S.-Mexican Studies, University of California San Diego, 1995); and Kim Moody, "NAFTA and the Corporate Redesign of North America," *Latin American Perspectives* 22 (1995): 95–116.

93. *Park City Daily News* (Bowling Green, KY), 7 October 2001.

94. Francis Fukuyama, "The End of History," *The National Interest*, Summer 1989, 7.

95. For a general analysis of NAFTA and the neoliberal process described here, see, on Mexico, Kim Moody, "NAFTA and the Corporate Redesign of North America," *Latin American Perspectives* 22 (Winter 1995): 95–116. Joanna B. Swanger, "Review: Labor in the Americas: Surviving in a World of Shifting

Boundaries," *Latin American Research Review* 38 (2003): 147–166, has a discussion of a number of recent studies.

96. *New York Times*, 21 July 1995, A25.

97. *Washington Post*, 24 July 1995.

98. *Ottawa Citizen*, 30 April 1996, A5.

99. On the role of the Clinton administration and Latin American "free trade," see Greg Grandin, *Empire's Workshop: Latin America, the United States, and the Rise of the New Imperialism* (New York: Henry Holt, 2007), 193–195; Grandin takes the view of this book that Clinton implemented neoliberal trade policy.

100. *The (Kitchener-Waterloo, ON) Record*, 2 September 1995, A13.

101. See U.S. Department of Labor, Bureau of International Labor Affairs, *The Apparel Industry and Codes of Conduct: A Solution to the International Child Labor Problem?* (Washington, DC: U.S. Department of Labor, 1996).

102. Loyola Marymount University, Center for Ethical Concerns, *The Consumer and Sweatshops*, available at http://www.marymount.edu/news/garment-study/overview.html (accessed 15 July 2004).

103. *Journal of Commerce*, 16 October 1997, 6A.

104. Ibid.

105. *Toronto Star*, 26 February 1999.

106. *Boston Globe*, 26 November 2000, 1.

107. Maria Gillen, "The Apparel Industry Partnership's Free Labor Association: A Solution to the Overseas Sweatshop Problem or the Emperor's New Clothes?" *Journal of International Law and Politics* 32 (Summer 2000): 1096.

108. On the code, see: http://www.fairlabor.org/fla/go.asp?u=/pub/mp&Page=FLA Charter (accessed 6 June 2011).

109. *Toronto Star*, 26 February 1999.

110. *New York Times*, 18 November 2009, B1 and B4.

111. Nancy Birdsall, "Life Is Unfair: Inequality in the World," *Foreign Policy* (Summer 1998): 91, has a summary of these arguments.

112. Kurt A. Ver Beek, "Maquiladoras: Exploitation or Emancipation? An Overview of the Situation of Maquiladora Workers in Honduras," *World Development* 29 (September 2001): 1553–1567.

113. Gavin Wright, *Old South, New South: Revolutions in the Southern Economy since the Civil War* (New York: Basic Books, 1986), 147 and 270.

114. Carol Pier, "Deliberate Indifference: El Salvador's Failure to Protect Worker's Rights," *Human Rights Watch* 15 (December 2003): 2.

115. J. F. Hornbeck, *The U.S–Central American Free Trade Agreement (cafta): Challenges for Sub-Regional Integration*, U.S. Congress, Congressional Research Service Report, 28 October 2003 (RL31870), 19.

116. James A. Hodges, *New Deal Labor Policy and the Southern Cotton Textile Industry, 1933–1941* (Knoxville: University of Tennessee Press, 1986).

117. On the challenge of organizing, see William M. Adler, *Mollie's Job: A Story of Life and Work on the Global Assembly Line* (New York: Scribner, 2000), 173–203; and Wright, *Old South, New South*, 254.

118. For a description of Orion and San Pedro Sula, see Kitty Krupat, "From War Zone to Free Trade Zone," in *No Sweat: Fashion, Free Trade, and the Rights of Garment Workers*, ed. Andrew Ross (New York: Verso, 1997), 51–53.

119. Robert B. Reich, *The Work of Nations: Preparing Ourselves for 21st Century Capitalism* (New York: Alfred Knopf, 1991), 3.

120. On the labor problems implicit in the shift of production, see Henry J. Frundt, "Central American Unions in the Era of Globalization," *Latin American Research Review* 37 (2002): 14–29.

121. *The (Baton Rouge, MS) Advocate*, 3 September 1996, 1C.

122. Birdsall, "Life Is Unfair," 89.

123. Ver Beek, "Maquiladoras," 1561 and 1566.

124. Michael Mortimore, "When Does Apparel Become a Peril? On the Nature of Industrialization in the Caribbean," in *Free Trade and Uneven Development*, ed. Gary Gereffi, David Spener, and Jennifer Bair (Philadelphia: Temple University Press, 2002), 302.

125. Joel Mokyr, "Dear Labor, Cheap Labor, and the Industrial Revolution," in *Favorites of Fortune: Technology, Growth, and Economic Development since the Industrial Revolution*, ed. Patrice Higonnet, David S. Landes, and Henry Rosovsky (Cambridge, MA: Harvard University Press, 1991), 187–188.

126. Burton Bledstein, *The Culture of Professionalism: The Middle Class and the Development of Higher Education* (New York: Norton, 1976), 88–89.

127. For a review of the financial crisis in early 1998, see *Financial Times*, 13 February 1998, 28.

128. *Irish Times*, 4 September 1998, 54.

129. De'Ann Weimer, "A Killing in the Caymans?" *Business Week*, 11 May 1998, 50.

130. *Park City Daily News* (Bowling Green, KY), 10 May 1998.

131. This response was typical of the long-term southern adoption of faith in laissez-faire approaches to economic growth; see, for example, the long-ago study by Marian Irish, "Political Thought and Political Behavior in the South," *Western Political Quarterly* 13 (June 1960): 407–409.

132. Weimer, "A Killing in the Caymans?" 50.

133. This was said by Ian Cummings and Joseph Steinberg in a letter to shareholders of Lancadia National Corp., quoted in a story that also focused on Fruit of the Loom; Tim Reason, "Love It or Leave It? Will the Outcry over Inversions

Change the Way Overseas Income Is Taxed?" *CFO: The Magazine for Senior Financial Executives* (July 2002): 39–42.

134. *New England Health Care Employees Pension Fund v. Fruit of the Loom*, U.S. District Court for the Western District of Kentucky, Bowling Green Division, Civil Action No. 1:98-CV-99-M, 16 August 1999; see also *Louisville Courier-Journal*, 16 July 1998, 1B.

135. *Chicago Sun-Times*, 24 August 1998, 48.

136. *Daily News Record*, 19 May 1999, C2.

137. *Chicago Sun-Times*, 1 September 1999, 67.

138. *Business Week*, 13 September 1999, 50.

139. On the board's incompetence, see *New York Times*, 19 March 2000, 16.

140. *Park City Daily News* (Bowling Green, KY), 16 October 1999.

141. *Louisville Courier-Journal*, 18 December 1999.

142. *Park City Daily News* (Bowling Green, KY), 20 December 1999.

143. Ibid.

144. Stephen Cummings, *The Dixification of America: The American Odyssey into the Conservative Economic Trap* (Westport, CT: Praeger, 1998), 175–177.

145. *Park City Daily News* (Bowling Green, KY), 30 December 1999.

146. *Memphis Commercial Appeal*, 1 January 2000, B1.

147. *Atlanta Journal and Constitution*, 30 December 1999, 1.

148. *Daily News Record*, 17 March 2000, 10.

149. Ibid.

150. *Park City Daily News* (Bowling Green, KY), 6 April 2000.

151. Letter from Murray Borrello, Chair Task Force Technical Committee to Donlin, Recano and Co., U.S. Bankruptcy Court agent, *In re: Fruit of the Loom, Inc. et al.*, no. 99–04497 (PJW), 9 August 2000.

152. *Park City Daily News* (Bowling Green, KY), 3 May 2002.

153. *(Fort Smith, AR) Times Record*, 13 July 2003; *Park City Daily News* (Bowling Green, KY), 16 March 2000; *Birmingham Business Journal*, 25 February 2002; *Grand Forks (ND) Herald*, 29 July 2003.

154. *New York Times*, 11 May 2001.

155. *Charlotte (NC) Observer*, 6 May 2004.

156. Howard L. Preston, *The Future South: A Historical Perspective for the Twenty-first Century* (Urbana: University of Illinois Press, 1991), 206.

157. Jeffrey Leiter, "Reactions to Subordination: Attitudes of Southern Textile Workers," *Social Forces* 64 (June 1986): 952–955.

CHAPTER 7. EXPERTS AND LOCAL KNOWLEDGE

1. For a good summary of the literature from a community or local perspective, see Frank Fischer, *Citizens, Experts, and the Environment: The Politics of Local Knowledge* (Durham, NC: Duke University Press, 2000), 29–67; see especially Steven Brint, *In an Age of Experts: The Changing Role of Professionals in Politics and Public Life* (Princeton, NJ: Princeton University Press, 1994); and for a global perspective, Ulrich Beck et al., *Reflective Modernization* (Newbury Park, CA: Sage, 1994).

2. James C. Scott, *Domination and the Arts of Resistance: Hidden Transcripts* (New Haven: Yale University Press, 1990); for a wider discussion of the concept, see essays in Richard A. Horsley, ed., *Hidden Transcripts and the Arts of Resistance: Applying the Work of James C. Scott to Jesus and Paul* (Leiden: Brill, 2004).

3. Mary Beth Rogers, *Cold Anger: A Story of Faith and Power Politics* (Denton: University of North Texas Press, 1990).

4. A phrase associated with the French microbiologist and environmentalist René Dubos; see René Dubos and Barbara Ward, *Only One Earth: The Care and Maintenance of a Small Planet* (New York: Norton, 1972). Of course Dubos would not have seen the phrase as limiting a community such as St. Louis; see Carol L. Moberg, René Dubos, *Friend of the Good Earth* (Washington, DC: ASM Press, 2005).

5. *Chemical Week*, 4 July 1990, 9; 1 October 1997, 5; 18 February 1998, 28; *Adhesives Age*, November 1997; *Chemical Marketing Reporter*, 9 February 1998, 5.

6. William M. Adler, *Mollie's Job: A Story of Life and Work on the Global Assembly Line* (New York: Scribner, 2000), 242.

7. *Chattanooga Times*, 31 May 1996, A1.

8. *Chemical Week*, 3 July 1996, 49; he chaired the CMA's panel on "Responsible Care."

9. *Chemical Week*, 3 December 1997, 31. *Chattanooga Free Press*, 10 June 1998, describes how the Chemical Education Foundation gave Velsicol its Product Stewardship Award for its community advisory panels.

10. *New York Times*, 27 January 1987, A1.

11. *Chicago Tribune*, 28 February 1987, 3.

12. *Los Angeles Times*, 11 August 1987, 1; 2 October 1987, A8; *Washington Post*, 12 August 1987, A1; 2 October 1987; *San Diego Union-Tribune*, 12 August 1987, A3; *New York Times*, 12 August 1987, A15; *Crain's Chicago Business*, 19 October 1987, 28.

13. *Washington Post*, 9 June 1988, A17.
14. P. F. Infante, S. S. Epstein, W. A. Newton, "Blood Dyscrasias and Childhood Tumors and Exposure to Chlordane and Heptachlor," *Scandinavian Journal of Work Environmental Health* 4 (1978): 137–150; and S. S. Epstein, and D. Ozonoff, "Leukemias and Blood Dyscrasias Following Exposure to Chlordane and Heptachlor," *Teratogenic Carcinogenic Mutagen* 7 (1987): 527–540.
15. *Orange County Register*, 30 August 1989, A9.
16. From http://iscentre.org/ (accessed 9 November 2010); the isc's offices were in a private home near the Memphis plant of Velsicol.
17. See "List of Participants," Stockholm Convention on Persistent Organic Pollutants Review Committee, Fifth Meeting, 12–16 October 2009, Geneva, www.unep/pops/poprc.5/inf/24 (accessed 8 November 2010).
18. Ida Damgaard, Niels Skakkebaek, Jorma Toppari et al., "Persistent Pesticides in Human Breast Milk and Cryptorchidism," *Environmental Health Perspectives* 114 (2006): 1133–1138; Saiyad Habibullah, Aruna Dewan, Vijay Bhatnagar et al., "Effect of Endosulfan on Male Reproductive Development," *Environmental Health Perspectives* 111 (2003): 1958–1962.
19. *Memphis Commercial Appeal*, 12 April 1993, A1. Other stories reported on their acute toxicity and the difficulty of detecting them if used on fruits or vegetables imported into the country; see for example *New York Times*, 7 June 1990, A16; and *St. Petersburg Times*, 3 September 1989, 2D. For later studies, see K. H. Kilburn and J. C. Thornton, "Protracted Neurotoxicity from Chlordane Sprayed to Kill Termites," *Environmental Health Perspectives* 103 (1995): 691–694.
20. *Memphis Commercial Appeal*, 23 May 2006; see us epa, *Velsicol Chemical Corporation Hardeman County Landfill NPL Site Second Five Year Review Report* (Atlanta: us epa Region 4, 2006), 6–9, 36.
21. *Isabel v. Velsicol*, U.S. District Court for the Western District of Tennessee, Western Division, Case No. 04-2997-DV.
22. Having been delisted on 8 September 1983, it was relisted on 3 March 2010 following a 1 July 2009 "Governor's Concurrence" (from Michigan governor Jennifer Granholm) formally requesting in epa terminology that the site be put on the National Priorities List; see epa press release No. 10-OPA021, 5 March 2010, "EPA Adds St. Louis, Mich., Site to Superfund National Priorities List."
23. us epa, Environmental News Release No. 10-OPA052, 18 May 2010; the most comprehensive review by the state and federal governments of the extent of contaminants at the sites probably was the eight-binder Michigan Department of Environmental Quality, "Remedial Investigation," completed in November 2006.

24. Weston Solutions, "Draft Remedial Investigation for Operable Unit One (OU-I) Velsicol Superfund Site Velsicol Chemical Corporation Site St. Louis, MI," prepared for Michigan Department of Environmental Quality, Remediation, and Redevelopment Division, October 2003, Work Order No. 20083.500.001, 1–5.

25. Memphis Environmental Center, "Containment System Assessment (SA) Report, Former Michigan Chemical Plant Site, St. Louis, MI," prepared for Velsicol Chemical Corporation, 1 October 1997.

26. U.S. Public Health Service, Agency for Toxic Substances and Disease Registry, "Preliminary Health Assessment: Velsicol Chemical Corporation (St. Louis Plant Site) CERCLIS No. MID000722439 Gratiot County, Michigan," Office of Health Assessment, ATSDR, 9 November 1988, 6.

27. Ibid.

28. U.S. Army, Corps of Engineers, Waterways Experiment Station, "Information Summary, Area of Concern: Saginaw River and Saginaw Bay Miscellaneous Paper EL 91-7, March 1991, prepared for US EPA, Great Lakes National Program Office.

29. U.S. Department of Health and Human Services, Public Health Service, Agency for Toxic Substances and Disease Registry, Division of Health Assessment and Consultation, "Site Review and Update: Velsicol Chemical Mich. St. Louis, Gratiot County, Michigan CERCLIS No. MID000722439," 30 September 1993, 6.

30. For a discussion of site history, see US EPA, *Record of Decision for Operable Unit 2—Pine River; Velsicol Chemical Superfund Site, St. Louis, Michigan*, February 1999, 7–10.

31. Michigan Department of Environmental Quality, Surface Water Quality Division, "Staff: Status of the St. Louis Impoundment in the Vicinity of the Former Velsicol Chemical Company, Gratiot County, Michigan," MI/DEQ/ SWQ—96/092, October 1996, 2–3.

32. Edward C. Lorenz, "Containing the Michigan PBB Crisis, 1973–1992: Testing the Environmental Policy Process," *Environmental History Review* 17 (Summer 1993), 61.

33. On the value of local knowledge in policymaking, see essays in such works as Alan Bicker, Paul Sillitoe, and Johan Pottier, eds., *Investigating Local Knowledge: New Directions, New Approaches* (Burlington, VT: Ashgate, 2004); Stephen B. Brush and Doreen Stabinsky, *Valuing Local Knowledge: Indigenous People and Intellectual Rights* (Washington, DC: Island Press, 1996); and Frank Fisher, *Citizens, Experts, and the Environment: The Politics of Local Knowledge* (Durham, NC: Duke University Press, 2001).

34. On early TAGs, see Richard L. Hembra, "EPA's Superfund TAG Program: Grants Benefit Citizens but Administrative Barriers Remain," in U.S. Congress, House of Representatives, Committee on Public Works and Transportation, *Hearing before the Subcommittee on Investigations and Oversight*, 102nd Cong., 2nd sess., 1992.

35. The environmental-justice movement originated in the United Church of Christ's Commission for Racial Justice, *Toxic Wastes and Race in the United States* (New York: United Church of Christ, 1987), followed by a US EPA advisory commission, NEJAC, and the issuance in 1995 by US EPA, Office of Solid Waste and Emergency Response Directive 9230 0-28, *Guidance for Community Advisory Groups at Superfund Sites* in December 1995.

36. See *(Alma, MI) Morning Sun*, 16 September 1997, 1.

37. *(Alma, MI) Morning Sun*, 30 October 1997, 1; on prelude to meeting, see 27 October 1997, 1.

38. Letter from T. Stuart Hill, US EPA community involvement coordinator, to each volunteer, 24 November 1997; meeting held 17 December 1997.

39. *(Alma, MI) Morning Sun*, 6 November 1997, 1–2.

40. See Minutes, Pine River Superfund Citizen Task Force Meeting, 21 January 1998, in Pine River Files, Alma College Archives; the US EPA *Draft Community Advisory Group Tool Kit*, November 1997, and US EPA *Superfund Today*, May 1996, both supplied to groups to facilitate formation of CAGs. That the tool kit was a draft is a result of the Pine River group being one of the first CAGs; the CAG was incorporated on 18 February 1998.

41. See letter from Barb Shrum and Jim Hall, Co-Chairs, Pine River Superfund Task Force, Pine River Committee, to Beth Reiner, US EPA Project Manager, 29 July 1998, in Pine River Files, Alma College Archives.

42. Letter from Charles Hanson to Elizabeth Reiner and Gaylene Vasaturo, Associate Regional Counsel, 17 April 1998, in Pine River Files, Alma College Archives.

43. For description of community response, see letter from Elizabeth Reiner to Charles Hanson, 27 April 1998, in Pine River Files, Alma College Archives.

44. Letters from Edward Lorenz, CAG Chair, to Janet Reno, U.S. Attorney General; Carol Browner, US EPA Administrator; David Ulrich, US EPA Region 5 Administrator; Gail Ginsburg, US EPA Region 5 Counsel; Jennifer Granholm, Michigan Attorney General; Russell Harding, Director, Michigan Department of Environmental Quality, all 16 February 1999, in Pine River Files, Alma College Archives.

45. See minutes of Pine River Superfund Citizen Task Force meeting, 22 July 1998, in Pine River Files, Alma College Archives.

46. Treaty of Saginaw, 7 Stat. 203 (1819).

47. Letter from Kevin Chamberlain, Tribal Chief, to Charles Hanson, Velsicol Vice President, 15 July 1998.
48. See fax from US EPA Region 5, 7 August 1998, "You Are Invited to Attend a Native American Pipe Ceremony for the Pine River Sediment Removal Project," in Pine River Files, Alma College Archives.
49. The Saginaw Chippewa became involved in the machinations of Jack Abramoff; generally on the challenges of gaming and tribal politics, see Steven Light and Kathryn Rand, *Indian Gaming and Tribal Sovereignty: The Casino Compromise* (Lawrence: University Press of Kansas, 2005).
50. On tribe's efforts to expel Bill Snowden, see *Detroit News*, 16 August 2001.
51. Conflict resolution was encouraged by the passage of the Administrative Dispute Resolution Act in 1996, Public Law 104-320; however, many communities saw dispute resolution as a process that favored leaving pollution in place, since inherently it meant compromise between cleanup and pollution.
52. The US EPA retained Donald Nelson from Washington State University to facilitate Building Community through Consensus meetings; these were held not in St. Louis but in Alma, Michigan, and included primarily public officials and no CAG officers; see Memorandum, Cooperative Extension Washington State University, 15 December 1998, to Abdel Abdalla et al. for membership list; included in fourteen attendees: three EPA staff, two from state health department, one from state environmental agency, three city officials, one county official, and three citizens not in elected office; *(Alma, MI) Morning Sun*, 8 April 1999.
53. Dave Dempsey, *Ruin and Recovery* (Ann Arbor: University of Michigan Press, 2001), 115–117.
54. See, for example, the publication from EPA: *Guidance for Community Advisory Groups at Superfund Sites*, OSWER Directive 9230.0-28, EPA 540-K-96-001, December 1995, and US EPA, *About the Community Advisory Group Toolkit*, Solid Waste and Emergency Response (5204G), November 1997, 9–12, discussed incorporation. Because the Pine River group began at the start of the CAG process, the CAG Toolkit of November 1997 was listed as a "draft."
55. The EPA's *Guidance for Community Advisory Groups*, 21, said to name the group for the "name and location of site."
56. *(Alma, MI) Morning Sun*, 26–27 February 1999, 1–2.
57. Ibid.
58. See Minutes, Pine River Superfund Citizen Task Force Meeting, 16 February 2000, in Pine River Files, Alma College Archives.
59. Letter from Murray Borrello, CAG Technical Chair, to Donlin, Recano and Co. [agents U.S. Bankruptcy Court], *In re:* Fruit of the Loom et al., No. 99-04497 (PJW), 9 August 2000; follow-up claim filed Edward Lorenz, CAG

Chair, to Donlan Recano, 3 May 2001, in Pine River Files, Alma College Archives.

60. Letter from Edward Lorenz, CAG Chair, to Lois Gartner, US EPA, 20 August 2001; response, Gartner to Lorenz, 29 August 2001, in Pine River Files, Alma College Archives.

61. Letter from Edward Lorenz, CAG Legal Chair, to Bill Brighton, U.S. Department of Justice; Alan Tenenbaum, U.S. Department of Justice; William Hartwig, U.S. Fish and Wildlife Service; Steven Williams, U.S. Fish and Wildlife Service; Rob Rocker, National Oceanic and Atmospheric Administration; and Marguerita Matera, National Oceanic and Atmospheric Administration, 8 February 2002; sent immediately after CAG bankruptcy claim rejected by Fruit of the Loom, in Pine River Files, Alma College Archives.

62. U.S. Department of Justice, "Notice of Filing of Environmental Settlement In re Fruit of the Loom Inc.," *Federal Register* 67 (2 May 2002): 22108; see letter from Gwendolyn Wilkie, NOAA Attorney to Jane Keon, CAG Chair, 17 May 2002, who said, "First, I would like to thank you and the other members of the Pine River Superfund Citizen Task Force for contacting the National Oceanic and Atmospheric Administration (NOAA) . . .", in Pine River Files, Alma College Archives.

63. Letter from Jane Keon, CAG Chair, and Edward Lorenz, CAG Legal Chair, to U.S. Department of Justice, Assistant Attorney General of the Environmental and Natural Resource Division, 22 May 2002, requested a public hearing held in St. Louis, Michigan, in June 2002, in Pine River Files, Alma College Archives.

64. E-mail from Rita Harris to Jane Keon and Ed Lorenz, 9 July 2002, in Pine River Files, Alma College Archives.

65. Letter from Edward Lorenz, CAG Legal Chair, to Thomas Sansonetti, U.S. Department of Justice, Environment and Natural Resource Division, 26 June 2002, in Pine River Files, Alma College Archives, requested that the final draft recognize the CAG role, and that if there were excess moneys under the settlement, that they be specifically reserved to pay for natural-resource damage restoration; those changes were made in the U.S. Motion and Memorandum attached to the settlement text; see United States' Motion and Memorandum of Law in support of Motion to Approve Proposed Settlement Agreement under Environmental Laws, *In re: Fruit of the Loom, Inc. et al . . . Debtors*, No. 99-4497 (PWJ), U.S. Bankruptcy Court for the District of Delaware, 9 August 2002, 26, 28.

66. *Portland (ME) Press Herald*, 13 May 2004; this matter was being resolved through foreclosure on Farley's estates on the Maine coast.

67. Letter from Richard E. Benzie, District Engineer, Division of Water Supply, Michigan Department of Public Health to Jim Pavlik, Water Superintendent,

City of St. Louis, 4 August 1981; this was reported in the regional press as a great reassurance; see "St. Louis Water Reported Safe," *Gratiot County Herald*, 20 August 1981, and "Tests Indicate St. Louis City Water Safe," *(Alma, MI) Morning Sun*, 18 August 1981.

68. For a news account of the problem with the city aquifer, see for example *Detroit News*, 8 November 2005; for an account from the government perspective, see *The Remediator: The Newsletter of the Remediation and Redevelopment Division* [of the Michigan Department of Environmental Quality] (Spring 2006): 5.

69. *City of St. Louis, Michigan, vs. Velsicol et al.*, State of Michigan, 29th Judicial Circuit.

70. Settlement Agreement *City of St. Louis, MI v. Velsicol Chemical Corp. et al.*, U.S. District Court, Eastern District of Michigan, Northern Division, Case No. 1:07-cv-13683 (TLL) (CEB), 18 March 2011; for U.S. Intervention on side of Velsicol, see *United States Motion to Intervene in City of St. Louis v. Velsicol Chemical Corp. et. al.*, U.S. District Court, Eastern District of Michigan, Northern Division, Case No. 1:07-cv-13683, 14 December 2007; McConkie quoted in *(Alma, MI) Morning Sun* 31 March 2011, 1. .

71. For a discussion of practices such as Fruit, Velsicol and AIG followed related to the insurance, see Charles Murray, *Revolt in the Boardrooms: The New Rules of Corporate Power in America* (New York: HarperCollins, 2007), 12.

72. AIG became so upset at the trustees' requests, it sued; *American International Specialty Lines Insurance Co. v. NWI-I, Inc, Lepetomane II, Inc.*, U.S. District Court, Northern District of Illinois, Eastern Div. 05-C 6386, 8 November 2005; the trust countersued on 6 January 2006.

73. Velsicol was sold to Arsenal Capital Partners of New York on 11 October 2005; Arsenal, according to its website, http.arsenalcapital.com, was "a private equity firm that invests in lower middle market, niche manufacturing and service businesses. In statements that sounded like something from Farley's public-relations process, Arsenal said, "We look to partner with strong management teams to improve their competitiveness and accelerate growth."

74. In 2008, Arsenal split Velsicol, transferring several plants to Genovique Specialty Holdings.

75. *In the Matter of Velsicol Chemical Corp.*, US EPA Regions 4, 5, 2, Consent Agreement Region 4, Docket No. CERCLA 04-2005-3770; Region 5, Docket No. V-W-05-C-814; Region 2, Docket No. CERCLA 02-2005-2016.

76. E-mail from Allen Dearry, NIEHS to Ed Lorenz, 29 August 1999; American Chemical Society meeting held 22–26 August 1999, in Pine River Files, Alma College Archives.

77. Summary Statement from Frederick Tyson, NIEHS Special Emphasis Panel, to Wilfried Karmaus, Application No. 1-R01 ES10680-01, 25 April 2000, in Health Grants 2000, in Pine River Archives, Alma College.

78. Among the experts referenced in the proposal were Paul C. Stearn and Harvey V. Fineberg, eds., *Understanding Risk: Informing Decisions in a Democratic Society* (Washington, DC: National Academy Press, 1996); and Orywin Renn, Thomas Webler, and Peter Wiedemann, eds., *Fairness and Competence in Citizen Participation: Evaluating Models of Environmental Discourse* (Boston: Kluwer Academic, 1995); and various essays in Carl G. Herndl and Stuart C. Brown, eds., *Green Culture: Environmental Rhetoric in Contemporary America* (Madison: University of Wisconsin Press, 1996).

79. Frederick L. Tyson, NIEHS Special Emphasis Panel, 1 R01 ES1-680-01A1, to Wilfried Karmaus, 10–12 July 2001, 2, in Health Studies 2001 Grant Proposal, in Pine River Archives, Alma College.

80. Daniel S. Greenberg, "Creaky NIH Is Past Its Prime," *Baltimore Sun*, 12 July 1998, 3C; for more on Greenberg's views, see his *Science, Money, and Politics* (Chicago: University of Chicago Press, 2001), cited in CAG correspondence to NIEHS.

81. Letter from Edward Lorenz to Anne P. Sassaman, Director, Division of Extramural Research and Training, NIH, 4 September 2001, in NIEHS Gripe folder, in Pine River Archives, Alma College.

82. Letter from Senator Cark Levin to Ruth Kirschstein, Acting Director, National Institutes of Health, 17 September 2001, in NIEHS Gripes folder, Pine River Archives, Alma College.

83. On problems at NIEHS, see, for example, Dan Ferber, "NIEHS Toxicologist Receives a 'Gag Order,'" *Science*, 9 August 2002, 915–916.

84. See Virlyn W. Burse et al., "Brief Communication: Preliminary Investigation of the Use of Dried-Blood Spots for the Assessment of *in Utero* Exposure to Environmental Pollutants," *Biochemical and Molecular Medicine* 61 (1997): 236–239.

85. Letter from Edward Lorenz to James Haveman, Director, Michigan Department of Community Health, 20 January 2000, in Health–Infant Blood Spots folder, in Pine River Files, Alma College Archives.

86. Letter from Edward Lorenz to Assistant Administrator, ATSDR, 30 April 2004, in ATSDR Petition 2004–2005 file, in Pine River Files, Alma College Archives.

87. See for example "Physician's Fact Sheet, Velsicol Corporation Workers' Chemical Exposures and Health Issues," developed by the ATSDR office at the Michigan Department of Community Health, 17 October 2006, and presented to Gratiot County physicians at the grand rounds at the regional hospital, in Environmental Health Forum file, in Pine River Files, Alma College Archives.

88. Letter from Dana B. Barr, Chief, Pesticide Laboratory, Public Health Service, to Edward Lorenz, Pine River Group, 24 January 2005, in ATSDR Health Work 2005 File, in Pine River Files, Alma College Archives.

89. *Wall Street Journal*, 8 November 2005, A16.

90. On Koenig and the Annapolis Center, see http://www.martin-blanck.com/ harold_ koenig.html (accessed 27 May 2010); and http://www.Annapolis center.org (accessed 6 December 2005).

91. http://www.annapoliscenter.org (accessed 6 December 2005).

92. Ibid.

93. See for example Richard Tren and Hugh High, "Smoked Out: Anti-Tobacco Activism at the World Bank," available at http://www.forcesitaly.org/italy/ rubr3/smokeout.htm. On Tren's work on tobacco, see Hadii Mamudu, Ross Hammond, and Stanton Glantz, "Tobacco Industry Attempts to Counter the World Bank Report Curbing the Epidemic and Obstruct the WHO Framework Convention on Tobacco Control," *Social Science and Medicine* 67 (December 2008): 1690–1699.

94. *Capitalism Magazine*, 30 June 2002.

95. Letter to the editor from Richard Tren, *Daily Telegraph* (London), 2 May, 2003, 29.

96. Quoted in the Thai periodical *The Nation*, 11 May 2010, LexisNexis (accessed 29 May 2010).

97. Tren, *Excellent Powder*, 332.

98. Ibid., 224; note the source cited by Tren in footnote 443 had errors, making checking the source impossible.

99. See Paul Starobin, "Who Turned Out the Enlightenment?" *National Journal*, 29 July 2006, 20–26.

100. *Wall Street Journal*, 19 November 2005, ellipses in original.

101. Tren, *Excellent Powder*, 106.

102. On the general practice of lobbyists casting doubt on DDT regulations and other efforts to regulate grounded in scientific studies of harm, see Naomi Oreskes and Erik Conway, *Merchants of Doubt: How a Handful of Scientists Obscured the Truth on Issues from Tobacco Smoke to Global Warming* (New York: Bloomsbury Press, 2010); on anti–Rachel Carson lobbying, see especially 216–239.

103. Starobin, "Who Turned Out the Enlightenment?" 24.

104. Henry N. Pollack, *Uncertain Science . . . Uncertain World* (New York: Cambridge University Press, 2003), 23–42.

105. Thomas R. Dunlap, *DDT: Scientists, Citizens, and Public Policy* (Princeton, NJ: Princeton University Press, 1981), 235, said this in 1981; it was quoted approvingly in Tren, *Excellent Powder*, 333.

106. Joanne Silberner, "WHO Backs Use of DDT against Malaria," National Public Radio, *All Things Considered*, September 15, 2006.

107. Sheila Kaplan, "Great Lakes Danger Zones," Center for Public Integrity, at http://www.publicintegrity.org/GreatLakes/index.htm (accessed 21 February 2008).

108. Letter from Stupak and Dingell to Julie Gerberding, Director, CDC, 28 February 2008, in DDT Presenter–De Rosa, Chris File, in Pine River Files, Alma College Archives.

109. E-mail from Doris Cellarius to Ed Lorenz, 5 March 2008, in DDT Presenter–DeRosa, Chris File, in Pine River Files, Alma College Archives.

110. Recording of Eugene Kenaga International DDT Conference, held at Alma College, Pine River Archives, Alma College.

111. Brenda Eskenazi et al., "The Pine River Statement: Human Health Consequences of DDT Use," *Environmental Health Perspectives* 117 (September 2009): 1369.

112. "Scientific American News," 4 May 2009, at http://www.scientificamerican .com/article.cfm?id=ddt-use-to-combat-malaria&print=true (accessed 4 May 2009).

113. United Nations Environmental Program, Stakeholder Meeting to Review the Draft Business Plan to Promote a Global Partnership for Developing Alternatives to DDT, UNEP/POPS/DTBP.1/11, Geneva, 3–5 November 2008.

114. Paulo Coelho, *The Alchemist* (New York: Harper Collins, 1993), 121.

CHAPTER 8. CONSEQUENCES AND CONTROLS

1. United Press story, 4 March 1984, www.LexisNexis.com (accessed 17 November 1998), told of Greenpeace activists plugging the pipes at Marshall, Illinois.

2. On rise of neoliberalism, see David Harvey, *A Brief History of Neoliberalism* (New York: Oxford University Press, 2005), 39–63; William Hudson, *The Libertarian Illusion: Ideology, Public Policy, and the Assault on the Common Good* (Washington, DC: CQ Press, 2008); also Xavier Bonal, "The Neoliberal Educational Agenda and the Legitimation Crisis: Old and New State Strategies," *British Journal of Sociology Education* 24 (April 2003): 159–175.

3. On the mixed system that dominated U.S. political-economic policy in the middle of the twentieth century, see Edward D. Berkowitz and Kim McQuaid, *Creating the Welfare State: The Political Economy of 20th-Century Reform*, rev. ed. (Lawrence: University Press of Kansas, 1992).

4. David B. Robertson and Dennis R. Judd, *The Development of American Public Policy: The Structure of Policy Restraint* (Glenview, IL: Scott Foresman, 1989), 321–349, make the point that the only policy area where the United States led the developed world and did not practice its usual policy restraint was environmental policy.

5. Herbert Simon, *Administrative Behavior*, 3rd ed. (New York: The Free Press, 1976), 38–41, 74–77, 80–81, 240–44, discussed bounded rationality and the complicated relationship of values and facts; for a general review of incrementalism and reasons for its validity, see Jonathan Bendor, "A Model of Muddling Through," *American Political Science Review* 89 (December 1995): 819–840.

6. Harvey, *A Brief History of Neoliberalism*, 66–73. For a critical attack on WTO policy, see Naomi Klein, *The Shock Doctrine: The Rise of Disaster Capitalism* (New York: Picador, 2007), especially 352–352.

7. For example, see Donald B. Kraybill, ed., *The Amish and the State*, 2nd ed. (Baltimore: Johns Hopkins University Press, 2003), especially 87–108.

8. Marilee S. Grindle, "Reforming Land Tenure in Mexico: Peasants, Markets, and the State," in *The Challenge of Institutional Reform in Mexico*, ed. Riorden Roett (Boulder, CO: Lynne Rienner, 1995), 39–56.

9. On issues such as the juvenile death penalty, see Amnesty International, *United States of America: The Death Penalty and Juvenile Offenders* (New York: Amnesty International Publications, 1991); in March 2005 the U.S. Supreme Court finally overturned the juvenile death penalty in *Roper v. Simmons*, 543 U.S. 551 (2005).

10. Lance Compa, *Unfair Advantage: Workers' Freedom of Association in the United States under International Human Rights Standards* (Ithaca, NY: ILR Press, Cornell University, 2004), xix–xxii, discusses the post-2001 Supreme Court assault on workers' rights.

11. *New England Health Care Employees Pension Fund et al. v. Fruit of the Loom et al.*, 98cv00099, U.S. District Court for the Western District of Kentucky; see *Louisville Courier-Journal*, 22 March 2006, for settlement of the case and a companion one, *Fidel v. Farley*, 00cv48 in the same court.

12. Lawrence E. Mitchell, *Corporate Irresponsibility: America's Newest Export* (New Haven: Yale University Press, 2001), 166. It is important to clarify that not everything that goes by the name "investor activism" is the same; some contemporary "hedge funds" are very active, but not necessarily calling for the good of communities; see *Business Week*, 18 May 2007.

13. Thomas Berry, *The Great Work: Our Way into the Future* (New York: Bell Tower, 1999), 119.

14. As late as 1997, TIAA-CREF had 858,000 Class A shares of Fruit; College Retirement Equities Fund, *(CREF) Semi-Annual Report: Stock Account Financial Statement (Unaudited), Including Statements of Investments, June 30, 1997* (New York: CREF, 1997), 44.

15. Berry, *The Great Work*, 171.

16. Robert A. G. Monks and Nell Minow, *Power and Accountability* (New York: Harper Business, 1991), 184.

17. See Lily Qiu, "Which Institutional Investors Monitor? Evidence from Acquisition Activity," *Brown Economics Working Paper Series No. 2004–21* (June 2006): 1–4, 25–27.

18. John Perkins, *The Secret History of the American Empire* (New York: Penguin Group, 2007), 304–305.

19. International Financial Services, Hedge Funds April 2007, at http://www.ifsl. org.uk/pdf_handler.cfm?file=CBS_hedge_funds_2007&CFID=828181&C FToken+77793579 (accessed 11 July 2007).

20. Ibid., 2–3, shows that approximately 80 percent of hedge funds were controlled from the United States and Britain in 2006. On the contrasting Anglo-American model of treatment of stakeholders, see Will Hutton and Anthony Giddens, ed., *On the Edge: Living with Global Capitalism* (London: Jonathan Cape, 2000), 12–13.

21. See Charles Derber, *Corporation Nation: How Corporations Are Taking Over Our Lives and What We Can Do about It* (New York: St. Martin's Press, 1998), 250.

22. *New York Times*, 17 July 2005, BU-1, reported on the criticism of the Deutsche Bank analyst Bill Dreher, who criticized Costco for paying employees too well, thus depriving stockholders of their rightful return.

23. Mitchell, *Corporate Irresponsibility*, 208.

24. For a discussion of wealth creation, see Margaret M. Blair, *Ownership and Control: Rethinking Corporate Governance for the Twenty-First Century* (Washington, DC: Brookings Institution, 1995), 275–322.

25. Marina V.N. Whitman, *New World, New Rules: The Changing Role of the American Corporation* (Boston: Harvard Business School Press, 1999), 73–80, believes we have moved "from managerial to investor capitalism."

26. See, for example, Henry Dethloff and Keith Bryant, "Entrepreneurship," in *American Business History: Case Studies*, ed. Henry Dethloff and C. Joseph Pusateri (Arlington Heights, IL: Harlan Davidson, 1987), 4–21.

27. One term for this behavior might be the "long con"; see Matt Taibbi, *Graftopia, Bubble Machines, Vampire Squids, and the Long Con That Is Breaking America* (New York: Spiegel and Grau, 2010).

28. See, for example, "No Wonder C.E.O.'s Love Those Mergers," *New York Times*, 18 July 2004, 3.

29. With the U.S. economic collapse in 2008, the subject of executive bonuses became regular news, for example at AIG, the insurer of Fruit of the Loom and Velsicol; see Ralph Vartedian and James Oliphant, "The AIG Uproar," *Los Angeles Times*, 19 March 2009, A18. For an analysis of the larger trends in income, see Larry M. Bartels, *Unequal Democracy: The Political Economy of the New Gilded Age* (New York: Russell Sage Foundation, 2008).

30. John Kenneth Galbraith, *The Economics of Innocent Fraud: Truth for Our Time* (Boston: Houghton Mifflin, 2004), 27.

31. Robert H. Frank and Philip J. Cook, *The Winner-Take-All Society* (New York: The Free Press, 1995), 71.

32. For example, in a study in Mexico, Cedric Durand, "Externalities from Foreign Direct Investment in the Mexican Retailing Sector," *Cambridge Journal of Economics* 31 (2007): 395, found productivity declines and limitations on the adoption of new ideas within firms following globalization of a "developing world" industry.

33. Geoffrey R. D. Underhill, *Industrial Crisis and the Open Economy: Politics, Global Trade, and the Textile Industry in the Advanced Economies* (New York: St. Martin's Press, 1998), 16.

34. Cummings, 163, for example, discusses the auto industry's varied responses to import competition. On workers and quality, see for example the classic study of textile production, Tamara Hareven, *Family Time and Industrial Time: The Relationship between the Family and Work in a New England Industrial Community* (New York: Cambridge University Press, 1982), 302, which found that one third of worker grievances at the huge Amoskeag mills in the 1920s related to product quality.

35. In addition to Underhill, *Industrial Crisis and the Open Economy*, see for example the analysis of Robert E. Scott and Thea M. Lee, "The Cost of Trade Protection Reconsidered: U.S. Steel, Textiles, and Apparel," in *U.S. Trade Policy and Global Growth: New Directions in the International Economy*, ed. Robert A. Blecker (Armonk, NY: M. E. Sharpe, 1996), especially 126.

36. Underhill, *Industrial Crisis and the Open Economy*, 94.

37. See Denis O'Hearn, "Economic Globalization," in *From the Local to the Global: Key Issues in Development Studies*, ed. Gerard McCann and Stephen McCloskey (London: Pluto Press, 2003); and Aldrie Henry-Lee, "Convergence? The Lewis Model and the Rights-Based Approach to Development," *Social and Economic Studies* 54 (2005): 91–121—two examples of studies that critique simple claims that movement of production to the "developing world" generally improves prospects for those countries.

38. John Madeley, *Big Business, Poor Peoples: The Impact of Transnational Corporations on the World's Poor* (New York: Zed Books, 1999), 109. In Mexican maquiladoras there has grown up the domestic Mexican shelter corporation that at least retains some of the excess profits within the country.

39. Michael Mortimore, "When Does Apparel Become a Peril? On the Nature of Industrialization in the Caribbean." In *Free Trade and Uneven Development*, ed. Gary Gereffi, David Spener, and Jennifer Bair (Philadelphia: Temple University Press, 2002), 305.

40. *New York Times*, 12 November 1997, A28, is typical.

41. On the Centre, see http://iscentre.org/ (accessed 9 November 2010), with Velsicol's vice president Chuck Hanson serving as executive director and

attending various international pesticide-related meetings, such as those of the UN Environmental Program.

42. For a good general review of the issue of endocrine disruption, see Sheldon Krimsky, *Hormonal Chaos: The Scientific and Social Origins of the Environmental Endocrine Hypothesis* (Baltimore: Johns Hopkins University Press, 2002).

43. These behaviors contradict those advocated by the proponents of environmental accounting, not simply popular advocates such as Paul Hawken, *Natural Capitalism: Creating the Next Industrial Revolution* (Boston: Little, Brown and Co., 1999). Even specialized accounting works advocated the value of behaviors contrary to Velsicol's; see for example Grant Ledgerwood, Elizabeth Street, and Riki Therivel, *Implementing an Environmental Audit: How to Gain a Competitive Advantage Using Quality and Environmental Responsibility* (New York: Irwin Professional Publishing, 1994), 27–42.

44. Richard Tardanico and Mark B. Rosenberg, "Two Souths in the New Global Order," in *Poverty of Development: Global Restructuring and Regional Transformations in the U.S. South and Mexican South*, ed. Tardanico and Rosenberg (New York: Routledge, 2000), 5–6, and 19.

45. Agreeing with Cummings, *The Dixification of America*, on the need of technology in the American South, see David L. Carlton and Peter A. Coclanis, *The South, the Nation, and the World: Perspectives on Southern Economic Development* (Charlottesville: University of Virginia Press, 2003), 11.

46. Cummings, *The Dixification of America*, 154–155.

47. Marsha Witten, *All Is Forgiven: The Secular Message of American Protestantism* (Princeton, NJ: Princeton University Press, 1993), and later Alan Wolfe, *The Transformation of American Religion* (New York: Free Press, 2003) explain how U.S. churches are often more secular than sacred.

48. The Clayton Act of 1914, Title 15 USC, chapter 1, and the International Labor Organization's "Declaration of Philadelphia" maintain that labor is not a commodity.

49. John A. Ryan, *A Living Wage: Its Ethical and Economic Aspects*, introduction by Richard Ely (New York: Macmillan, 1906).

50. See Bruce Pietrykowski, "Fordism at Ford: Spatial Decentralization and Labor Segmentation at the Ford Motor Company, 1920–1950," *Economic Geography* 71 (October 1995): 389, summarizes the contradictions in this policy.

51. Nick Heffernan, *Capital, Class and Technology in Contemporary American Culture* (London: Pluto Press, 2000), 24–28, reviews the Marxist origins of the concepts of Fordism and post-Fordism.

52. Cathie Jo Martin, *Stuck in Neutral: Business and the Politics of Human Capital Investment Policy* (Princeton, NJ: Princeton University Press, 2000), 8–9.

53. Neil H. Jacoby, *Corporate Power and Social Responsibility: A Blueprint for the Future* (New York: Macmillan, 1973), 7–11.

54. The U.S. Department of Labor praised the new law; see "Fact Sheet: The Pension Protection Act of 2006: Ensuring Greater Retirement Security for American Workers," at http://www.whitehouse.gov/news/releases/2006/08/print/20060817.html (accessed 18 August 2006). Labor organizations, such as the National Education Association, http://www.nea.org/lac/pension/index.html?mode=print (accessed 18 August 2006), criticized the new law, saying, for example, "NEA opposes the bill because of deep concerns about financial and administrative burdens that could discourage private sector employers from offering or retaining defined benefit pension plans. . . . it would not be long before the states might seek to rid themselves of defined benefit plans as well."

55. James C. Cobb, *The Selling of the South: The Southern Crusade for Industrial Development, 1936–1990*, 2nd ed. (Urbana: University of Illinois Press, 1993), 58.

56. Michael Piore and Charles F. Sabel, *The Second Industrial Divide* (New York: Basic Books, 1984), 240.

57. Mitchell, *Corporate Irresponsibility*, 73.

58. Michael Novak, *Business as a Calling: Work and the Examined Life* (New York: The Free Press, 1996), 146–152, lists seven business responsibilities, which begin with "respect for the dignity of persons."

59. Joseph E. Stiglitz, *Globalization and Its Discontents* (New York: W.W. Norton, 2002), 247–250.

60. On greed, see Phyllis Tickle, *Greed: The Seven Deadly Sins* (New York: Oxford University Press, 2004); for example, all the readings in the common lectionary of the Christian churches for Year C, 13th Proper, Ecclesiastes 1:2, 12–14; 2:18–23; Psalm 49:1–12; Luke 12:13–21; and especially Colossians 3:1–11, refer to greed as the worst sin.

61. Galbraith, *The Economics of Innocent Fraud*, 51.

62. See for example "Hourly Pay in U.S. Not Keeping Pace with Price Rises," *New York Times*, 18 July 2004, 1; also see longer-term data in Lester Thurow, "Almost Everywhere: Surging Inequality and Falling Real Wages," in *The American Corporation Today*, ed. Carl Kaysen (New York: Oxford University Press, 1996), 383–412.

63. Mitchell, *Corporate Irresponsibility*, 8–10.

64. Robert Wuthnow, *God and Mammon in America* (New York: The Free Press, 1994), 126.

65. John A. Hall and Charles Lindholm, *Is America Breaking Apart?* (Princeton, NJ: Princeton University Press, 1999), 126.

66. Deborah Dougherty and Edward Bowman, "The Effects of Organizational Downsizing on Product Innovation," *California Management Review* 37 (Summer 1995): 28–44.

67. Hedrick Smith, *Rethinking America* (New York: Random House, 1995).

68. Mitchell, *Corporate Irresponsibility*, 278.

69. For a good review of its widespread acceptance, see Delia B. Comti, *Reconciling Free Trade, Fair Trade, and Interdependence: The Rhetoric of Presidential Economic Leadership* (Westport, CT: Praeger, 1998).

70. On "non-trade barriers" in the global textile industry, see Amy Glasmeier, Jeffery W. Thompson, and Amy J. Kays, "The Geography of Trade Policy: Trade Regimes and Location Decisions in the Textile and Apparel Complex," *Transactions of the Institute of British Geographers*, new series, 18 (1993); they also focus on U.S. dominance of trade barrier construction.

71. Thomas Donaldson and Thomas Dunfee, "Toward a Unified Conception of Business Ethics: Imaginative Social Contracts Theory," *Academy of Management Review* 19 (1994): 257.

72. Galbraith, *The Economics of Innocent Fraud*, 15.

73. Thomas L. Friedman, *The World Is Flat: A Brief History of the Twenty-First Century* (New York: Farrar, Straus and Giroux, 2005), 371–372.

74. Gavin Wright, *Old South, New South: Revolutions in the Southern Economy since the Civil War* (New York: Basic Books, 1986), 274.

75. Mitchell, *Corporate Irresponsibility*, 188.

76. Tardanico and Rosenberg, "Two Souths in the New Global Order," 37.

77. Fruit of the Loom came in for criticism following a coup in Honduras on 28 June 2009, when it failed to join with Nike and other apparel firms in calling for restoration of democracy; letter from Nike to Secretary of State Hillary Clinton, 27 July 2009, http//www.nikebiz.com/responsibility/2009Secretary ClintonHondurasLetter.html (accessed 21 August 2009).

78. A critical attack on university integrity is found in Chris Hedges, *Empire of Illusion: The End of Literacy and the Triumph of Spectacle* (New York: Nation Books, 2009), 89–114.

79. Thomas Berry, *The Great Work*, 73, has written, "As now functioning, the university prepares students for their role in extending human domination over the natural world. . . . Use of this power in a deleterious manner has devastated the planet."

80. In contrast to the U.S. free-market approach, the chemical-industry trade association in Britain has identified the positive impacts of regulation, which they see as stimulating innovation through challenges. For example, Chemical Industries Association, *Policy Priorities for the UK Chemical Industry* (London: Chemical Industries Association, 2007), 4, says, "The chemical industry operates in a strict regulatory environment. We support the need for this to

give confidence to the public and set the right safety, environmental, health and employment challenges."

81. Richard Goldthwaite, *The Economy of Renaissance Florence* (Baltimore: Johns Hopkins University Press, 2009); for a focus on the importance of "civic community" in modern northern Italy, see Robert Putnam, *Making Democracy Work: Civic Tradition in Modern Italy* (Princeton, NJ: Princeton University Press, 1994), 83–116.

82. Mario B. Mignone, *Italy Today: At the Crossroads of the New Millennium*, rev. ed. (New York: Peter Lang, 1998), 168–170, and for consumption data by region, table 11, 444.

83. Marvin B. Becker, *The Decline of the Commune*, vol. 1 of *Florence in Transition* (Baltimore: Johns Hopkins University Press, 1967), 281. For a later study of similar trends in Early Modern England, see Phil Withington, "Public Discourse, Corporate Citizenship, and State Formation in Early Modern England," *American Historical Review* 112 (October 2007): 1038, who is not as certain these types of developments provide a model for contemporary debates about civic life.

84. Becker, *The Decline of the Commune*, 232.

85. Ibid., 231–232.

86. Ibid., 231.

87. Mary L. Wingerd, *Claiming the City: Politics, Faith, and the Power of Place in St. Paul* (Ithaca, NY: Cornell University Press, 2001), 91.

88. "Revolt in the Northwest," *Fortune*, April 1936, 118–119.

89. For other case studies, see Xavier de Souza Briggs, *Democracy as Problem Solving: Civic Capacity in Communities across the Globe* (Cambridge, MA: MIT Press, 2008).

90. Wingerd, *Claiming the City*, 273.

91. Peter Spiro, *Beyond Citizenship: American Identity after Globalization* (New York: Oxford University Press, 2008), finds that the nation as a locale of citizenship has been undermined by globalization; it is "overinclusive." However, he misses the value of local citizenship linked emotionally to the nation.

92. Jackson Lears, "Looking Backward: In Defense of Nostalgia," *Lingua Franca* 7 (December-January 1998): 61.

93. Wingerd, *Claiming the City*, 273.

94. Charles Lindblom, "The Science of Muddling Through," *Public Administration Review* 19 (1959): 79–88, modified in his "Still Muddling, Not Yet Through," *Public Administration Review* 39 (1979): 517–526.

95. Becker, *The Decline of the Commune*, 231–232.

96. Charles Lindblom, *The Market System: What It Is, How It Works, and What to Make of It* (New Haven: Yale University Press, 2001), 236–250, refers to this response as the ways the market obstructs democracy.

97. See Robert Kuttner, *Everything for Sale: The Virtues and Limits of Markets* (New York: Alfred Knopf, 1997), 39–67, which describes how markets undermine civility, personal security, and liberty.

98. *(Alma, MI) Morning Sun*, 24 July 2005, 5A.

99. Much of the environmental literature misses the similarity between environmental conflict and the confrontations between workers, investors, and general communities with powerful interests. The environmental justice movement assumes there is uniqueness to the exploitation of the poor and minorities; this study sees similarities. On environmental justice, see Daniel Faber, *Capitalizing on Environmental Injustice: the Polluter-Industrial Complex in the Age of Globalization* (Lanham, MD: Rowman and Littlefield, 2008).

100. Becker, *The Decline of the Commune*, 232.

101. Albert O. Hirschman, *Exit, Voice, and Loyalty* (Cambridge, MA: Harvard University Press, 1970), 106–119, discusses this tendency.

Bibliography

MANUSCRIPT AND ARCHIVAL SOURCES

The research for this book has been facilitated by librarians and archivists at a number of institutions, especially the Bentley Historical Library at the University of Michigan, the Regenstein Library at the University of Chicago, Michigan State University Archives, the University of Notre Dame Library, Purdue University Library, Western Kentucky University, and especially Alma College, which maintains the Pine River Archives. There has been special help from the U.S. Archives, Chicago branch and staff of the Smithsonian Institution. Public libraries have been crucial to finding often obscure local records, ranging from the large libraries in Chicago, Indianapolis, Memphis, and Toledo to smaller community libraries in Bowling Green, Kentucky; Clarksville, Tennessee; Valley, Alabama; and St. Louis, Michigan. In addition, staff of Human Rights Watch were generous in giving time and information on employment issues in Central America.

INTERPRETATIVE WORKS

A number of sources have shaped the book's references to history, politics, the policy process, and especially community and individual interaction with government, social institutions, and experts. Included here are studies of institutions such as universities and churches, American exceptionalism, and the contrasting practices in other cultures, especially those works guiding the interpretation of the experience of Italy and other non-U.S. political cultures.

Beck, Ulrich, et al. *Reflective Modernization*. Newbury Park, CA: Sage, 1994.

Becker, Marvin B. *The Decline of the Commune*. Vol. 1 of *Florence in Transition*. Baltimore: Johns Hopkins University Press, 1967.

Bellah, Robert N., et al. *Habits of the Heart: Individualism and Commitment in American Life*. New York: Harper and Row, 1985.

Bendor, Jonathan. "A Model of Muddling Through." *American Political Science Review* 89 (December 1995): 819–840.

Berry, Thomas. *The Great Work: Our Way into the Future*. New York: Bell Tower, 1999.

Bicker, Alan, Paul Sillitoe, and Johan Pottier, eds. *Investigating Local Knowledge: New Directions, New Approaches*. Burlington, VT: Ashgate, 2004.

Bok, Derek. *Universities and the Marketplace: The Commercialization of Higher Education*. Princeton, NJ: Princeton University Press, 2003.

Bonal, Xavier. "The Neoliberal Educational Agenda and the Legitimation Crisis: Old and New State Strategies." *British Journal of Sociology Education* 24 (April 2003): 159–175.

Brint, Steven. *In an Age of Experts: The Changing Role of Professionals in Politics and Public Life*. Princeton, NJ: Princeton University Press, 1994.

Brush, Stephen B., and Doreen Stabinsky. *Valuing Local Knowledge: Indigenous People and Intellectual Rights*. Washington, DC: Island Press, 1996.

Cater, Douglass. *Power in Washington*. New York: Random House, 1964.

de Souza Briggs, Xavier. *Democracy as Problem Solving: Civic Capacity in Communities across the Globe*. Cambridge, MA: MIT Press, 2008.

Diggins, John P. *Ronald Reagan: Fate, Freedom, and the Making of History*. New York: Norton, 2008.

Ehrenhalt, Alan. *The Lost City: The Forgotten Virtues of Community in America*. New York: Basic Books, 1995.

Elshtain, Jean Bethke. *Democracy on Trial*. New York: Basic Books, 1995.

Etzioni, Amitai. *The Spirit of Community: Rights, Responsibilities, and the Communitarian Agenda* (New York: Crown Publishers, 1993).

Friedman, Thomas L. *The World Is Flat: A Brief History of the Twenty-First Century*. New York: Farrar, Straus and Giroux, 2005.

Gilder, George. *Wealth and Poverty*. New York: Basic Books, 1980.

Glendon, Mary Ann. *Rights Talk: The Impoverishment of Political Discourse*. New York: Free Press, 1991.

Goldthwaite, Richard. *The Economy of Renaissance Florence*. Baltimore: Johns Hopkins University Press, 2009.

Haber, Samuel. *Efficiency and Uplift: Scientific Management in the Progressive Era, 1890–1920*. Chicago: University of Chicago Press, 1973.

Hall, John A., and Charles Lindholm. *Is America Breaking Apart?* Princeton, NJ: Princeton University Press, 1999.

Hargrove, Erwin. *Jimmy Carter as President: Leadership and the Politics of the Public Good.* Baton Rouge: Louisiana State University Press, 1988.

Harvey, David. *A Brief History of Neoliberalism.* New York: Oxford University Press, 2005.

Hedges, Chris. *Empire of Illusion: The End of Literacy and the Triumph of Spectacle.* New York: Nation Books, 2009.

Hirschman, Albert O. *Exit, Voice, and Loyalty.* Cambridge, MA: Harvard University Press, 1970.

Hudson, William. *The Libertarian Illusion: Ideology, Public Policy, and the Assault on the Common Good.* Washington, DC: CQ Press, 2008.

Kingdon, John W. *America the Unusual.* New York: Worth Publishing, 1999.

Krimsky, Sheldon. *Science and the Private Interest: Has the Lure of Profits Corrupted Biomedical Research?* Lanham, MD: Rowman and Littlefield, 2003.

Kristeller, Paul Oskar. *Renaissance Thought: The Classic, Scholastic, and Humanist Strains.* New York: Harper and Row, 1961.

Kuttner, Robert. *The Squandering of America: How the Failure of Our Politics Undermines Our Prosperity.* New York: Knopf, 2007.

Lasch, Christopher. *The Culture of Narcissism: American Life in an Age of Diminishing Expectations.* New York: W.W. Norton, 1978.

Lindblom, Charles. "Still Muddling, Not Yet Through." *Public Administration Review* 39 (1979): 517–526.

Lindblom, Charles. "The Science of Muddling Through." *Public Administration Review* 19 (1959): 79–88.

Lipset, Seymour M. *American Exceptionalism: A Double Edged Sword.* New York: W.W. Norton, 1996.

Lowi, Theodore. *The End of Liberalism.* New York: W.W. Norton, 1969.

Lowi, Theodore. "Four Systems of Policy, Politics, and Choice." *Public Administration Review* 32 (July–August 1972): 298–310.

Maier, Charles. *Among Empires: American Ascendancy and Its Predecessors.* Cambridge, MA: Harvard University Press, 2006.

McConnell, Grant. *Private Power and American Democracy.* New York: Vintage Books, 1966.

Michels, Robert. *Political Parties: A Sociological Study of the Oligarchical Tendencies of Modern Democracy.* Translated by Eden and Cedar Paul. Glencoe, IL: The Free Press, 1962.

Mignone, Mario B. *Italy Today: At the Crossroads of the New Millennium.* Rev. ed. New York: Peter Lang, 1998.

Passmore, John. *Man's Responsibility for Nature.* London: Duckworth, 1974.

Patterson, Thomas E. *The Vanishing Voter: Public Involvement in an Age of Uncertainty.* New York: Knopf, 2002.

Perkins, John. *The Secret History of the American Empire.* New York: Penguin Group, 2007.

Phillips, Kevin. *The Politics of Rich and Poor: Wealth and the American Electorate in the Reagan Aftermath.* New York: Random House, 1992.

Putnam, Robert. *Making Democracy Work: Civic Tradition in Modern Italy.* Princeton, NJ: Princeton University Press, 1994.

Reeves, Thomas. *The Empty Church: Does Organized Religion Matter Anymore?* New York: Free Press, 1998.

Robertson, David B., and Dennis R. Judd. *The Development of American Public Policy: The Structure of Policy Restraint.* Glenview, IL: Scott Foresman, 1989.

Rogers, Mary Beth. *Cold Anger: A Story of Faith and Power Politics.* Denton: University of North Texas Press, 1990.

Rosenberg, Nathan, and Richard R. Nelson. "American Universities and Technical Advance in Industry." *Research Policy* 23 (1994): 323–348.

Rosenbloom, David H. "Public Administrative Theory and the Separation of Powers." *Public Administration Review* 43 (May-June 1983): 219–226.

Scott, James C. *Domination and the Arts of Resistance: Hidden Transcripts.* New Haven: Yale University Press, 1990.

Scott, James C. *Seeing Like a State: How Certain Schemes to Improve the Human Condition Have Failed.* New Haven: Yale University Press, 1998.

Shafer, Byron E., ed. *Is America Different? A New Look at American Exceptionalism.* New York: Oxford University Press, 1991.

Shepsie, Kenneth A. "Losers in Politics (and How They Sometimes Become Winners): William Riker's Heresthetic." *Perspectives on Politics* 1 (June 2003): 307–315.

Simon, Herbert. *Administrative Behavior.* 3rd ed. New York: The Free Press, 1976.

Slater, Phillip. *The Pursuit of Loneliness.* Boston: Beacon Press, 1970.

Smith, Hedrick. *Rethinking America.* New York: Random House, 1995.

Steigerwald, David. *The Sixties and the End of Modern America.* New York: St. Martin's Press, 1995.

Stone, Walter J. *Republic at Risk: Self-Interest in American Politics.* Pacific Grove, CA: Brooks Cole, 1990.

Tocqueville, Alexis de. *Democracy in America.* Edited by Phillips Bradley. New York: Alfred A. Knopf, 1956.

Tolchin, Susan J., and Martin Tolchin. *Dismantling America: The Rush to Deregulate.* New York: Oxford University Press, 1985.

Wald, Kenneth D. *Religion and Politics in the United States.* 4th ed. Lanham, MD: Rowman and Littlefield, 2003.

Walzer, Michael. *Spheres of Justice: A Defense of Pluralism and Equality.* New York: Basic Books, 1983.

West, Darrell. *The Rise and Fall of the Media Establishment.* New York: St. Martin's Press, 2001.

Withington, Phil. "Public Discourse, Corporate Citizenship, and State Formation in Early Modern England." *American Historical Review* 112 (October 2007): 1016–1038.

Witten, Marsha. *All Is Forgiven: The Secular Message of American Protestantism.* Princeton, NJ: Princeton University Press, 1993.

Wolfe, Alan. *The Transformation of American Religion.* New York: Free Press, 2003.

STATE AND REGIONAL POLITICS AND REGIONAL ECONOMIC AND TRADE POLICIES

In order to understand the historical, political, social, and economic contexts of the many communities in which the firms studied here had facilities, many specialized local and regional sources were helpful. Also listed here are some less geographically focused works on economic development and trade practices and their impacts. Since many subsidiaries of Fruit of the Loom located in the U.S. South and later in Central America, there are a number of studies of those regions that proved especially helpful. Because so much of this study focuses on St. Louis, Michigan, and the environmental problems in that community, there is a separate listing of studies related to that community in the last section of this bibliography.

Adler, William M. *Mollie's Job: A Story of Life and Work on the Global Assembly Line.* New York: Scribner, 2000.

Applebome, Peter. *Dixie Rising: How the South Is Shaping American Values, Politics, and Culture.* New York: Random House, 1996.

Bridges, Tyler. *Bad Bet on the Bayou: The Rise of Gambling in Louisiana and the Fall of Governor Edwin Edwards.* New York: Farrar, Straus and Giroux, 2001.

Burner, David. *The Politics of Provincialism: The Democratic Party in Transition, 1918–1932.* New York: W.W. Norton, 1967.

Carlton, David L., and Peter A. Coclanis. *The South, the Nation, and the World: Perspectives on Southern Economic Development.* Charlottesville: University of Virginia Press, 2003.

Charnowitz, Steve. "International Trade and Social Welfare: The New Agenda— Transcript of January 7, 1994, Meeting of the Section on International Law of the American Association of Law Schools." *Comparative Labor Law Journal* 17 (Winter 1996): 352–353.

Cobb, James C. *The Selling of the South: The Southern Crusade for Industrial Development, 1936–1990.* 2nd ed. Urbana: University of Illinois Press, 1993.

Comti, Delia B. *Reconciling Free Trade, Fair Trade, and Interdependence: The Rhetoric of Presidential Economic Leadership.* Westport, CT: Praeger, 1998.

Cummings, Stephen. *The Dixification of America: The American Odyssey into the Conservative Economic Trap.* Westport, CT: Praeger, 1998.

Daniel, Pete. *Lost Revolutions: The South in the 1950s.* Chapel Hill: University of North Carolina Press, 2000.

Daniel, Pete. *Standing at the Crossroads: Southern Life in the Twentieth Century.* New York: Hill and Wang, 1986.

Dempsey, Dave. *William G. Milliken: Michigan's Passionate Moderate.* Ann Arbor: University of Michigan Press, 2006.

Frundt, Henry J. "Central American Unions in the Era of Globalization." *Latin American Research Review* 37 (2002): 14–29.

Gillen, Maria. "The Apparel Industry Partnership's Free Labor Association: A Solution to the Overseas Sweatshop Problem or the Emperor's New Clothes?" *Journal of International Law and Politics* 32 (Summer 2000).

Glasmeier, Amy, Jeffery W. Thompson, and Amy J. Kays. "The Geography of Trade Policy: Trade Regimes and Location Decisions in the Textile and Apparel Complex." *Transactions of the Institute of British Geographers*, new series, 18 (1993).

Grandin, Greg. *Empire's Workshop: Latin America, the United States, and the Rise of the New Imperialism.* New York: Henry Holt, 2007.

Grindle, Marilee S. "Reforming Land Tenure in Mexico: Peasants, Markets, and the State." In *The Challenge of Institutional Reform in Mexico*, ed. Riorden Roett, 39–56. Boulder, CO: Lynn Rienner, 1995.

Henry-Lee, Aldrie. "Convergence? The Lewis Model and the Rights-Based Approach to Development." *Social and Economic Studies* 54 (2005): 91–121.

Heyrman, Christine L. *Southern Cross: The Beginnings of the Bible Belt.* Chapel Hill: University of North Carolina Press, 1998.

Hodges, James A. *New Deal Labor Policy and the Southern Cotton Textile Industry, 1933–1941.* Knoxville: University of Tennessee Press, 1986.

Hornbeck, J. F. *The U.S.–Central American Free Trade Agreement (cafta): Challenges for Sub-Regional Integration.* U.S. Congress, Congressional Research Service Report, 28 October 2003 (RL31870).

Hutton, Will, and Anthony Giddens, eds. *On the Edge: Living with Global Capitalism.* London: Jonathan Cape, 2000.

Irish, Marian. "Political Thought and Political Behavior in the South." *Western Political Quarterly* 13 (June 1960): 406–420.

Key, V. O., Jr. *Southern Politics in State and Nation.* New York: Alfred Knopf, 1950.

Kilian, Michael, Connie Fletcher, and F. Richard Ciccone. *Who Runs Chicago?* New York: St. Martin's Press, 1979.

Krupat, Kitty. "From War Zone to Free Trade Zone." In *No Sweat: Fashion, Free Trade, and the Rights of Garment Workers,* ed. Andrew Ross, 51–53. New York: Verso, 1997.

Lamphere, Louise. *From Working Daughters to Working Mothers: Immigrant Women in a New England Industrial Community.* Ithaca, NY: Cornell University Press, 1987.

Leifermann, Henry P. *Crystal Lee: A Woman of Inheritance.* New York: Macmillan, 1975.

Leiter, Jeffrey. "Reactions to Subordination: Attitudes of Southern Textile Workers." *Social Forces* 64 (June 1986): 952–955.

Marshall, F. Ray. *Labor in the South.* Cambridge, MA: Harvard University Press, 1967.

Martin, Andrew W. "Organizational Structure, Authority, and Protest: The Case of Union Organizing in the United States, 1990–2001." *Social Forces* 85 (March 2007): 1413–1436.

Mason, Lucy Randolph. *To Win These Rights: A Personal Story of the CIO in the South.* Westport, CT: Greenwood Press, 1970.

Minchin, Timothy J. *"Don't Sleep with Stevens!" The J.P. Stevens Campaign and the Struggle to Organize the South, 1963–1980.* Gainesville: University Press of Florida, 2005.

Mokyr, Joel. "Dear Labor, Cheap Labor, and the Industrial Revolution." In *Favorites of Fortune: Technology, Growth, and Economic Development since the Industrial Revolution,* ed. Patrice Higonnet, David S. Landes, and Henry Rosovsky, 177–200. Cambridge, MA: Harvard University Press, 1991.

Moody, Kim. "NAFTA and the Corporate Redesign of North America." *Latin American Perspectives* 22 (1995): 95–116.

Mortimore, Michael. "When Does Apparel Become a Peril? On the Nature of Industrialization in the Caribbean." In *Free Trade and Uneven Development,* ed. Gary Gereffi, David Spener, and Jennifer Bair, 287–306. Philadelphia: Temple University Press, 2002.

O'Hearn, Denis. "Economic Globalization." In *From the Local to the Global: Key Issues in Development Studies,* ed. Gerard McCann and Stephen McCloskey, 85–103. London: Pluto Press, 2003.

Orme, William A., Jr. *Understanding NAFTA: Mexico, Free Trade, and the New North America.* Austin: University of Texas Press, 1996.

Pier, Carol. "Deliberate Indifference: El Salvador's Failure to Protect Workers' Rights." *Human Rights Watch* 15 (December 2003).

Pohlmann, Marcus D., and Michael P. Kirby. *Racial Politics at the Crossroads*. Knoxville: University of Tennessee Press, 1996.

Pope, Liston. *Millhands and Preachers: A Study of Gastonia*. New Haven: Yale University Press, 1942.

Preston, Howard L. *The Future South: A Historical Perspective for the Twenty-first Century*. Urbana: University of Illinois Press, 1991.

Scott, Robert E., and Thea M. Lee. "The Cost of Trade Protection Reconsidered: U.S. Steel, Textiles, and Apparel." In *U.S. Trade Policy and Global Growth: New Directions in the International Economy*, ed. Robert A. Blecker, 108–135. Armonk, NY: M. E. Sharpe, 1996.

Spiro, Peter. *Beyond Citizenship: American Identity after Globalization*. New York: Oxford University Press, 2008.

Stiglitz, Joseph E. *Globalization and Its Discontents*. New York: W.W. Norton, 2002.

Swanger, Joanna B. "Review: Labor in the Americas: Surviving in a World of Shifting Boundaries." *Latin American Research Review* 38 (2003): 147–166.

Tardanico, Richard, and Mark B. Rosenberg. "Two Souths in the New Global Order." In *Poverty of Development: Global Restructuring and Regional Transformations in the U.S. South and Mexican South*, ed. Richard Tardanico and Mark B. Rosenberg, 3–18. New York: Routledge, 2000.

Tindall, George B. *The Emergence of the New South, 1913–1945*. Baton Rouge: Louisiana State University Press, 1967.

Tucker, David M. *Memphis since Crum: Bossism, Blacks, and Civic Reformers, 1948–1968*. Knoxville: University of Tennessee Press, 1980.

Tucker, David M. *Mugwumps: Public Moralists of the Gilded Age*. Columbia: University of Missouri Press, 1998.

Underhill, Geoffrey R. D. *Industrial Crisis and the Open Economy: Politics, Global Trade, and the Textile Industry in the Advanced Economies*. New York: St. Martin's Press, 1998.

Ver Beek, Kurt A. "Maquiladoras: Exploitation or Emancipation? An Overview of the Situation of Maquiladora Workers in Honduras." *World Development* 29 (September 2001): 1553–1567.

Wingerd, Mary L. *Claiming the City: Politics, Faith, and the Power of Place in St. Paul*. Ithaca, NY: Cornell University Press, 2001.

Wright, Gavin. *Old South, New South: Revolutions in the Southern Economy since the Civil War*. New York: Basic Books, 1986.

Zendejas, Sergio, and Pieter de Vries. *Rural Transformations Seen from Below: Regional and Local Perspectives from Western Mexico*. La Jolla, CA: Center for U.S.-Mexican Studies, University of California San Diego, 1995.

ENVIRONMENTAL POLICY

There are a number of studies of environmental issues and the development of the environmental policy process that helped shape this work. Especially helpful were some focused on the history of U.S. environmental policy and the interaction of economic and environmental policies and practices, and those that explain alternatives to the view that economic growth must include environmental exploitation. There are separate sections below on works related to science and risk assessment, ddt and other pesticides and Silent Spring, and the pbb mistake in St. Louis, Michigan.

Bliese, John R. E. *The Greening of Conservative America.* Boulder, CO: Westview Press, 2001.

Czarnezki, Jason J., and Adrianne K. Zahner. "The Utility of Non-Use Values in Natural Resource Damage Assessments." *Boston College Environmental Affairs Law Review* 32 (2005): 509–526.

Dempsey, Dave. *Ruin and Recovery.* Ann Arbor: University of Michigan Press, 2001.

Dubos, René, and Barbara Ward. *Only One Earth: The Care and Maintenance of a Small Planet.* New York: Norton, 1972.

Faber, Daniel. *Capitalizing on Environmental Injustice: The Polluter-Industrial Complex in the Age of Globalization.* Lanham, MD: Rowman and Littlefield, 2008.

Fischer, Frank. *Citizens, Experts, and the Environment: The Politics of Local Knowledge.* Durham, NC: Duke University Press, 2000.

Flippen, J. Brooks. *Nixon and the Environment.* Albuquerque: University of New Mexico Press, 2006.

Gray, Rob, and Jan Bebbington. *Accounting for the Environment.* 2nd ed. London: Sage Publications, 2001.

Hawken, Paul. *Natural Capitalism: Creating the Next Industrial Revolution.* Boston: Little, Brown and Co., 1999.

Hawken, Paul. *The Ecology of Commerce.* New York: HarperCollins, 1993.

Hays, Samuel P. *Beauty, Health, and Permanence: Environmental Politics in the United States, 1955–1985.* New York: Cambridge University Press, 1987.

Hays, Samuel P. "From Conservation to Environment: Environmental Politics in the United States since World War II." *Environmental Review* 6 (Fall 1982): 14–29.

Herndl, Carl G., and Stuart C. Brown, eds. *Green Culture: Environmental Rhetoric in Contemporary America.* Madison: University of Wisconsin Press, 1996.

Hunt, Stephen. "Winning Ways: Globalization and the Impact of the Health and Wealth Gospel." *Journal of Contemporary Religion* 15 (October 2000): 331–347.

Ingram, Helen M., and Dean E. Mann. "Environmental Policy from Innovation to Implementation." In *Nationalizing Government: Public Policies in America*, ed. Theodore Lowi and Alan Stone, 131–162. Washington, DC: Congressional Quarterly, 1984.

Kline, Benjamin. *First along the River: A Brief History of the U.S. Environmental Movement.* 2nd ed. San Francisco: Acada Books, 2000.

Krimsky, Sheldon. *Hormonal Chaos: The Scientific and Social Origins of the Environmental Endocrine Hypothesis.* Baltimore: Johns Hopkins University Press, 2002.

Layzer, Judith A. *The Environmental Case: Translating Values into Policy.* Washington, DC: CQ Press, 2002.

Ledgerwood, Grant, Elizabeth Street, and Riki Therivel. *Implementing an Environmental Audit: How to Gain a Competitive Advantage Using Quality and Environmental Responsibility.* New York: Irwin Professional Publishing, 1994.

Lewis, Jack. "Looking Backward: A Historical Perspective on Environmental Regulations." *EPA Journal* 14 (March 1988): 42–46.

Lovins, Amory, and L. Hunter Lovins. *Natural Capitalism: Creating the Next Industrial Revolution.* Boston: Little, Brown and Co., 1999.

Mazmanian, Daniel, and David Morell. *Beyond Superfailure: America's Toxics Policy for the 1990s.* Boulder, CO: Westview Press, 1992.

Moberg, Carol L., and René Dubos. *Friend of the Good Earth.* Washington, DC: ASM Press, 2005.

Ophuls, William. *Ecology and the Politics of Scarcity.* San Francisco: W. H. Freeman, 1977.

Portney, Paul R., ed. *Natural Resources and the Environment: The Reagan Approach.* Washington, DC: Urban Institute Press, 1984.

Probst, Katherine N., and David M. Konisky. *Superfund's Future: What Will It Cost?* Washington, DC: Resources for the Future, 2001.

Quarles, John. *Cleaning Up America: An Insider's View of the EPA.* Boston: Houghton Mifflin, 1976.

Renn, Orywin, Thomas Webler, and Peter Wiedemann, eds. *Fairness and Competence in Citizen Participation: Evaluating Models of Environmental Discourse.* Boston: Kluwer Academic, 1995.

Rosenbaum, Walter. *Environmental Politics and Policy.* 2nd ed. Washington, DC: Congressional Quarterly, 1991.

Rosenbaum, Walter. *The Politics of Environmental Concern.* New York: Praeger, 1973.

Shabecoff, Philip. *Earth Rising: American Environmentalism in the 21st Century.* Washington, DC: Island Press, 2000.

Shutkin, William A. *The Land That Could Be.* Cambridge, MA: MIT Press, 2001.

Vig, Norman J., and Michael E. Kraft, eds. *Environmental Policy in the 1980s: Reagan's New Agenda.* Washington, DC: Congressional Quarterly, 1984.

Weber, Edward P. *Pluralism by the Rules: Conflict and Cooperation in Environmental Regulation.* Washington, DC: Georgetown University Press, 1998.

ECONOMIC AND BUSINESS HISTORY

Many works on economic policy, corporate finance, and management have helped shape this study. Included here are those sources as well as some focused on social policy, labor standards, and globalization. Sources that specifically address the intersection of economic and environmental policy are grouped under the "Environmental Policy" heading. Others, with a focus on trade and development in the U.S. South and Central America, are listed under the "State and Regional Policy" heading.

Anders, George. *The Merchants of Debt.* New York: Basic Books, 1992.

Arnold, Kathleen R. *America's New Working Class: Race, Gender, and Ethnicity in a Biopolitical Age.* University Park: Pennsylvania State University Press, 2008.

Baker, George P., and George David Smith. *The New Financial Capitalists: Kohlberg Kravis Roberts and the Creation of Corporate Value.* New York: Cambridge University Press, 1998.

Barber, Benjamin. *Consumed: How Markets Corrupt Children, Infantilize Adults, and Swallow Citizens Whole.* New York: W.W. Norton, 2007.

Bartels, Larry M. *Unequal Democracy: The Political Economy of the New Gilded Age.* New York: Russell Sage Foundation, 2008.

Bartlett, Sarah. *The Money Machine.* New York: Warner Books, 1992.

Berkowitz, Edward D., and Kim McQuaid. *Creating the Welfare State: The Political Economy of 20th-Century Reform.* Rev. ed. Lawrence: University Press of Kansas, 1992.

Bernays, Edward L. *The Engineering of Consent.* Norman: University of Oklahoma Press, 1955.

Blackford, Mansel G., and K. Austin Kerr. *Business Enterprise in American History.* 2nd ed. Boston: Houghton Mifflin, 1990.

Blair, Margaret M. *Ownership and Control: Rethinking Corporate Governance for the Twenty-First Century.* Washington, DC: Brookings Institution, 1995.

Bledstein, Burton. *The Culture of Professionalism: The Middle Class and the Development of Higher Education.* New York: Norton, 1976.

Bobo, Kim. *Wage Theft in America.* New York: New Press, 2009.

Bruck, Connie. *The Predators' Ball.* New York: Penguin Books, 1988.

Burrough, Brian, and John Helyer. *Barbarians at the Gate.* New York: Harper and Row, 1990.

Carey, Alex. *Taking the Risk out of Democracy: Corporate Propaganda versus Freedom and Liberty.* Urbana: University of Illinois Press, 1997.

Charnowitz, Steve. "Promoting Higher Labor Standards." *Washington Quarterly* 18 (Summer 1995): 171–173.

Clark, Gordon L. "Limits of Statutory Responses to Corporate Restructuring Illustrated with References to Plant Closing Legislation." *Economic Geography* 67 (January 1991): 22–41.

Cohen, Daniel. *Globalization and Its Enemies.* Cambridge, MA: MIT Press, 2006.

Compa, Lance. *Unfair Advantage: Workers' Freedom of Association in the United States under International Human Rights Standards.* Ithaca, NY: ILR Press, Cornell University, 2004.

deLeon, Linda. "Ethics and Entrepreneurship." *Policy Studies Journal* 24 (Autumn 1996): 495–511.

Derber, Charles. *Corporation Nation: How Corporations Are Taking Over Our Lives and What We Can Do about It.* New York: St. Martin's Press, 1998.

Dethloff, Henry, and Keith Bryant. "Entrepreneurship." In *American Business History: Case Studies*, ed. Henry Dethloff and C. Joseph Pusateri, 4–21. Arlington Heights, IL: Harlan Davidson, 1987.

Donaldson, Thomas, and Thomas Dunfee. "Toward a Unified Conception of Business Ethics: Imaginative Social Contracts Theory." *Academy of Management Review* 19 (1994): 252–285.

Dougherty, Deborah, and Edward Bowman. "The Effects of Organizational Downsizing on Product Innovation." *California Management Review* 37 (Summer 1995): 28–44.

Drucker, Peter F. *Innovation and Entrepreneurship: Practice and Principles.* New York: Harper and Row, 1985.

Elkington, John. *Cannibals with Forks: The Triple Bottom Line of the Twenty-First Century Business.* Oxford: Capstone, 1997.

Ewen, Stuart. *Captains of Consciousness: Advertising and the Social Roots of the Consumer Culture.* 25th anniversary ed. New York: Basic Books, 2001.

Farrell, Larry C. *The Entrepreneurial Age: Awakening the Spirit of Enterprise in People, Companies, and Countries.* New York: Allworth Press, 2001.

Feldstein, Martin. *American Economic Policy in the 1980s.* Chicago: University of Chicago Press, 1994.

Frank, Robert H., and Philip J. Cook. *The Winner-Take-All Society.* New York: The Free Press, 1995.

Galambos, Louis. *Competition and Cooperation.* Baltimore: Johns Hopkins University Press, 1966.

Galbraith, John Kenneth. *The Economics of Innocent Fraud: Truth for Our Time.* Boston: Houghton Mifflin, 2004.

Galbraith, John Kenneth. *The New Industrial State.* Boston: Houghton Mifflin, 1967.

Hacker, Jacob. *The Great Risk Shift: The Assault on American Jobs, Families, Health Care and Retirement.* New York: Oxford University Press, 2006.

Heffernan, Nick. *Capital, Class and Technology in Contemporary American Culture.* London: Pluto Press, 2000.

Ikerd, John E. *Sustainable Capitalism: A Matter of Common Sense.* Bloomfield, CT: Kumarian Press, 2005.

Jacoby, Neil H. *Corporate Power and Social Responsibility: A Blueprint for the Future.* New York: Macmillan, 1973.

Kane, John, and Haig Patapan. "In Search of Prudence: The Hidden Problem of Managerial Reform." *Public Administration Review* 66 (September-October 2006): 711–725.

Kapp, K. William. *The Social Costs of Private Enterprise.* Cambridge, MA: Harvard University Press, 1950.

Klein, Naomi. *The Shock Doctrine: The Rise of Disaster Capitalism.* New York: Picador, 2007.

Kuttner, Robert. *Everything for Sale: The Virtues and Limits of Markets.* New York: Alfred Knopf, 1997.

Lewis, Michael. *Liar's Poker.* New York: Norton, 1989.

Lindblom, Charles. *The Market System: What It Is, How It Works, and What to Make of It.* New Haven: Yale University Press, 2001.

MacAvoy, Paul W., and Ira M. Millstein. *The Recurrent Crisis in Corporate Governance.* New York: Palgrave Macmillan, 2003.

Madeley, John. *Big Business, Poor Peoples: The Impact of Transnational Corporations on the World's Poor.* New York: Zed Books, 1999.

Martin, Cathie Jo. *Stuck in Neutral: Business and the Politics of Human Capital Investment Policy.* Princeton, NJ: Princeton University Press, 2000.

Mitchell, Lawrence E. *Corporate Irresponsibility: America's Newest Export.* New Haven: Yale University Press, 2001.

Moloney, Kevin. *Rethinking Public Relations: PR Propaganda and Democracy.* 2nd ed. New York: Routledge, 2000.

Monks, Robert A. G., and Nell Minow. *Power and Accountability.* New York: Harper Business, 1991.

Novak, Michael. *Business as a Calling: Work and the Examined Life.* New York: The Free Press, 1996.

Oreskes, Naomi, and Erik Conway. *Merchants of Doubt: How a Handful of Scientists Obscured the Truth on Issues from Tobacco Smoke to Global Warming.* New York: Bloomsbury Press, 2010.

Pietrykowski, Bruce. "Fordism at Ford: Spatial Decentralization and Labor

Segmentation at the Ford Motor Company, 1920–1950." *Economic Geography* 71 (October 1995): 383–401.

Piore, Michael, and Charles F. Sabel. *The Second Industrial Divide.* New York: Basic Books, 1984.

Qiu, Lily. "Which Institutional Investors Monitor? Evidence from Acquisition Activity." *Brown Economics Working Paper Series No. 2004–21* (June 2006): 1–27.

Reich, Robert B. *The Work of Nations: Preparing Ourselves for 21st Century Capitalism.* New York: Alfred Knopf, 1991.

Riker, William. *The Art of Political Manipulation.* New Haven: Yale University Press, 1986.

Roe, Mark. *Weak Owners, Strong Managers.* Princeton, NJ: Princeton University Press, 1994.

Ryan, John A. *A Living Wage: Its Ethical and Economic Aspects.* New York: Macmillan, 1906.

Schumpeter, Joseph. *Capitalism, Socialism, and Democracy.* 3rd ed. New York: Harper and Row, 1950.

Stewart, James B. *Den of Thieves.* New York: Simon and Schuster, 1991.

Stiglitz, Joseph. *The Roaring Nineties.* New York: Norton, 2003.

Taibbi, Matt. *Griftopia, Bubble Machines, Vampire Squids, and the Long Con That Is Breaking America.* New York: Spiegel and Grau, 2010.

Tickle, Phyllis. *Greed: The Seven Deadly Sins.* New York: Oxford University Press, 2004.

Vogel, David. *Fluctuating Fortunes: The Political Power of Business in America.* New York: Basic Books, 1989.

Weiss, Richard. *The American Myth of Success: From Horatio Alger to Norman Vincent Peale.* New York: Basic Books, 1969.

Whitman, Marina v.N. *New World, New Rules: The Changing Role of the American Corporation.* Boston: Harvard Business School Press, 1999.

Wuthnow, Robert. *God and Mammon in America.* New York: The Free Press, 1994.

SCIENTIFIC RISK

A number of works on risk to human health and the environment are helpful in this book. In addition, some that study more general science policy and politics and the role of risk in the United States are grouped here. Works focused on more general environmental policy are listed earlier. Those focused on ddt, other pesticides, and pbb are identified in the two sections that follow.

Ashford, Nicholas A. *Crisis in the Workplace: Occupational Disease and Injury.* Cambridge, MA: MIT Press, 1976.

Beck, Ulrich. *Risk Society: Towards a New Modernity.* Newbury Park, CA: Sage Publications, 1992.

Bosso, Christopher J. *Pesticides and Politics: The Life Cycle of a Public Issue.* Pittsburgh, PA: University of Pittsburgh Press, 1987.

Corbett, Thomas H. *Cancer and Chemicals.* Chicago: Nelson-Hall, 1977.

Dworsky, Leonard. *Pollution.* New York: Chelsea House Publishers, 1971.

Fischhoff, Baruch, Paul Slovic, and Sarah Uchtenstein. "How Safe Is Safe Enough? A Psychometric Study of Attitudes towards Technological Risks and Benefits." *Policy Sciences* 9 (1978): 127–151.

Goklany, Indur M. *The Precautionary Principle: A Critical Appraisal of Environmental Risk Assessment.* Washington, DC: Cato Institute, 2001.

Kates, Robert W. *Risk Assessment of Environmental Hazards.* Chichester, UK: John Wiley and Sons, 1978.

Lester, James P., and Ann Bowman, eds. *The Politics of Hazardous Waste Management.* Durham, NC: Duke University Press, 1983.

Lowrance, William W. *Of Acceptable Risk.* Los Altos, CA: William Kaufmann, Inc., 1976.

Markowitz, Gerald, and David Rosner. *Deceit and Denial: The Deadly Politics of Industrial Pollution.* Berkeley: University of California Press, 2002.

Omang, Joanne. "Perception of Risk: A Journalist's Perspective." In *The Analysis of Actual versus Perceived Risks,* ed. Vincent T. Covello et al., 267–271. New York: Plenum Press, 1983.

Perrow, Charles. *Normal Accidents: Living with High-Risk Technologies.* New York: Basic Books, Inc., 1984.

Pollack, Henry N. *Uncertain Science . . . Uncertain World.* New York: Cambridge University Press, 2003.

Reich, Michael R. "Environmental Politics and Science: The Case of PBB Contamination in Michigan." *American Journal of Public Health* 73 (March 1983): 307–311.

Stearn, Paul C., and Harvey V. Fineberg, eds. *Understanding Risk: Informing Decisions in a Democratic Society.* Washington, DC: National Academy Press, 1996.

Sunstein, Cass R. "Beyond the Precautionary Principle." *University of Pennsylvania Law Review* 151 (January 2003): 1003–1010.

Sunstein, Cass R. *Risk and Reason: Safety, Law, and the Environment.* New York: Cambridge University Press, 2002.

Viscusi, W. Kip. *Risk by Choice.* Cambridge, MA: Harvard University Press, 1983.

Wiener, Jonathan B. "Precaution in a Multi-Risk World." In *Human and Ecological*

Risk Assessment: Theory and Practice, ed. Dennis D. Paustenbach, 1509–1532. New York: J. Wiley, 2002.

Wildavsky, Aaron. *But Is It True? A Citizen's Guide to Environmental Health and Safety.* Cambridge, MA: Harvard University Press, 1995.

DDT AND RACHEL CARSON

So much of this book focuses on ddt and related persistent organic pollutants, and the campaign to identify their risks and the policies to address those risks, that this section includes the most helpful studies. Earlier sections of the bibliography contain general works on environmental and science policy. The last section of this bibliography identifies those sources that focus on pbb and the contamination in St. Louis, Michigan.

Acquavella, John F., Belinda Ireland, and Jonathan Ramlow. "Organochlorines and Breast Cancer." *Journal of the National Cancer Institute* 85 (November 1993): 1872–1875.

Alvarez, Walter C., and Samuel Hyman. "Absence of Toxic Manifestations in Workers Exposed to Chlordane." *AMA Archives of Industrial Hygiene and Occupational Medicine* 8 (November 1953): 480–483.

Baldwin, I. L. "Chemicals and Pests." *Science* 137 (28 September 1962): 1042–1043.

Brooks, Paul. *The House of Life: Rachel Carson at Work.* Boston: Houghton Mifflin, 1972.

Burse, Virlyn W., et al. "Brief Communication: Preliminary Investigation of the Use of Dried-Blood Spots for the Assessment of *in Utero* Exposure to Environmental Pollutants." *Biochemical and Molecular Medicine* 61 (1997): 236–239.

Carmen, G. C., et al. "Absorption of DDT and Parathion by Fruits." *Abstracts, 115th Meeting American Chemical Society* (1949): 30A

Carson, Rachel. *Silent Spring.* Boston: Houghton Mifflin, 1962.

Damgaard, Ida, Niels Skakkebaek, Jorma Toppari et al. "Persistent Pesticides in Human Breast Milk and Cryptorchidism." *Environmental Health Perspectives* 114 (2006): 1133–1138.

Daniel, Pete. *Toxic Drift: Pesticides and Health in the Post–World War II South.* Baton Rouge: Louisiana State University Press, 2005.

Davidow, B., and J. L. Radomski. "Isolation of an Epoxide Metabolite from Fat Tissues of Dogs Fed Heptachlor." *Journal of Pharmacology and Experimental Therapeutics* 107 (March 1953): 259–265.

DeWitt, James B. "Chronic Toxicity to Quail and Pheasants of Some Chlorinated Insecticides." *Journal of Agricultural and Food Chemistry* 4 (1956): 863–866.

Drinker, Cecil K., et al. "The Problem of Possible Systemic Effects from Certain Chlorinated Hydrocarbons." *Journal of Industrial Hygiene and Toxicology* 19 (September 1937): 283.

Dunlap, Thomas R. *DDT: Scientists, Citizens, and Public Policy.* Princeton, NJ: Princeton University Press, 1981.

Ekborn, Anders, Anders Wicklund-Glynn, Hans-Olov Adami. "DDT and Testicular Cancer." *The Lancet* 347 (24 February 1996): 553–554.

Epstein, S. S., and D. Ozonoff. "Leukemias and Blood Dyscrasias following Exposure to Chlordane and Heptachlor." *Teratogenic Carcinogenic Mutagen* 7 (1987): 527–540.

Eskenazi, Brenda, et al. "The Pine River Statement: Human Health Consequences of DDT Use." *Environmental Health Perspectives* 117 (September 2009): 1359–1367.

Fitzhugh, O. Garth, and Arthur A. Nelson. "The Chronic Oral Toxicity of DDT (2,2-bis p-chlorophenyl-1,1,1-tri-chloroethane)." *Journal of Pharmacology and Experimental Therapeutics* 89 (January 1947): 18–30.

Gannon, Norman, and J. H. Bigger. "The Conversion of Aldrin and Heptachlor to Their Epoxides in Soil." *Journal of Economic Entomology* 51 (February 1958): 1–2.

Garabrant, David H., et al. "DDT and Related Compounds and Risk of Pancreatic Cancer." *Journal of the National Cancer Institute* 84 (1992): 764–771.

Graham, Frank, Jr. *Since Silent Spring.* Greenwich, CT: Fawcett, 1970.

Habibullah, Saiyad, Aruna Dewan, Vijay Bhatnagar, et al. "Effect of Endosulfan on Male Reproductive Development." *Environmental Health Perspectives* 111 (2003): 1958–1962.

Infante, P. F., S. S. Epstein, and W. A. Newton. "Blood Dyscrasias and Childhood Tumors and Exposure to Chlordane and Heptachlor." *Scandinavian Journal of Work Environment and Health* 4 (1978): 137–150.

Jacobziner, Harold, and H. W. Raybin. "Poisoning by Insecticide (Endrin)." *New York State Journal of Medicine* 59 (15 May 1959): 2017–2022.

Jorgenson, J. Lisa. "Aldrin and Dieldrin: A Review of Research on Their Production, Environmental Deposition and Fate, Bioaccumulation, Toxicology, and Epidemiology in the United States." *Environmental Health Perspectives* 109 (March 2001): 114.

Kilburn, K. H., and J. C. Thornton. "Protracted Neurotoxicity from Chlordane Sprayed to Kill Termites." *Environmental Health Perspectives* 103 (1995): 691–694.

Lear, Linda. *Rachel Carson: Witness for Nature*. New York: Henry Holt, 1997.

Longnecker, Matthew P., et al. "Association between Maternal Serum Concentration of the DDT Metabolite DDE and Preterm and Small-for-Gestational Age Babies at Birth." *The Lancet* 358 (14 July 2001): 110–114.

Lorenz, Edward C. "Containing the Michigan PBB Crisis, 1973–1992: Testing the Environmental Policy Process." *Environmental History Review* 17 (Summer 1993): 49–68.

Malats, Nuria, Francisco X. Real, and Miquel Porta. "DDT and Pancreatic Cancer." *Journal of the National Cancer Institute* 85 (February 1993): 329.

McCay, Mary A. *Rachel Carson*. New York: Twayne Publishers, 1993.

Myers, C. S. "Endrin and Related Pesticides: A Review." *Pennsylvania Department of Health Research Report No. 45* (1958).

Nelson, Arthur A., and Geoffrey Woodard. "Severe Adrenal Cortical Atrophy (Cytotoxic) and Hepatic Damage Produced in Dogs by Feeding DDD or TDE." *A.M.A. Archives of Pathology* 48 (October 1949): 392.

Princi, Frank, and George H. Spurbeck. "A Study of Workers Exposed to the Insecticides Chlordan [*sic*], Aldrin, and Dieldrin." *AMA Archives of Industrial Hygiene and Occupational Medicine* 3 (January 1951): 64–72.

Riker, William. *The Theory of Political Coalitions*. New Haven: Yale University Press, 1962.

Roberts, Donald, and Richard Tren. *The Excellent Powder: DDT's Political and Scientific History*. Indianapolis: Dog Ear Publishing, 2010.

Rogan, Walter J., et al. "Polychlorinated Biphenyls (PCBs) and Dichlorodiphenyl Dichloroethene (DDE) in Human Milk: Effects on Growth, Morbidity, and Duration of Lactation." *American Journal of Public Health* 77 (October 1987): 1294–1297.

Schapsmeier, Edward L., and Frederick H. Schapsmeier. *Ezra Taft Benson and the Politics of Agriculture*. Danville, IL: Interstate Printers, 1975.

Shames, Lisa, Mark Munekata, and Malcolm Pike. "Re: Blood Levels of Organochlorine Residues and Risk of Breast Cancer." *Journal of the National Cancer Institute* 86 (November 1994): 1642–1643.

Smith, Ray F., et al. "Secretion of DDT in Milk of Dairy Cows Fed Low Residue Alfalfa." *Journal of Economic Entomology* 41 (1948): 759–763.

Southgate, M. Therese. "Review of Silent Spring." *Journal of the American Medical Association* 182 (10 November 1962): 704.

Von Oettingen, Wolfgang F. *The Halogenated Aliphatic, Olefinic, Cyclic, Aromatic, and Aliphatic-Aromatic Hydrocarbons: Including the Halogenated Insecticides, Their Toxicity and Potential Dangers*. Washington, DC: U.S. Dept. of Health, Education, and Welfare, Public Health Service Publication No. 414, 1955.

Whitten, Jamie L. *That We May Live*. Princeton, NJ: D. Van Nostrand, 1966.

Wolff, Mary S., et al. "Blood Levels of Organochlorine Residues and Risk of Breast Cancer." *Journal of the National Cancer Institute* 85 (April 1993): 648–652.

ST. LOUIS AND THE PBB CRISIS

This final section of the bibliography contains works focused on the St. Louis operations of Michigan Chemical and Velsicol and the PBB crisis of the 1970s.

Anderson, Henry A., et al. "Investigation of the Health Status of Michigan Chemical Corporation Employees." *Environmental Health Perspectives* 23 (April 1978): 187–191.

Blanck, Heidi M., et al. "Age at Menarche and Tanner Stage in Girls Exposed *in utero* and Postnatally to Polybrominated Biphenyls." *Epidemiology* 11 (2000): 641–647.

Cary, Mike, and Jon Thompson. "Your Money and Your Life." *Seven Days*, 8 December 1978, 21–24.

Chen, Edwin. *PBB: An American Tragedy*. Englewood Cliffs, NJ: Prentice Hall, Inc., 1979.

Chumlea, W. C., et al. "Age at Menarche and Racial Comparisons in U.S. Girls." *Pediatrics* 111 (2003): 110–113.

Coyer, Bryan, and Don Schwerin. "Bureaucratic Regulation and Farmer Protest in the Michigan PBB Contamination Case." *Rural Sociology* 46 (Winter 1981): 718–721.

Cronon, William. *Nature's Metropolis*. New York: W.W. Norton, 1991.

Ebert, Daniel W., Janet Elashoff, and Kenneth Wilcox. "Possible Effect of Neonatal Polybrominated Biphenyl Exposure on the Developmental Abilities of Children." *American Journal of Public Health* 73 (March 1983): 286–288.

Egginton, Joyce. *The Poisoning of Michigan* (East Lansing: Michigan State University Press, 2009).

Henderson, Alden K., et al. "Breast Cancer among Women Exposed to Polybrominated Biphenyls." *Epidemiology* 6 (September 1965): 544–546.

Hesse, John L., and Richard A. Powers. "Polybrominated Biphenyl (PBB) Contamination of the Pine River, Gratiot and Midland Counties, Michigan." *Environmental Health Perspectives* 23 (April 1978): 24–25.

Hoque, Ashraful, et al. "Cancer among a Michigan Cohort Exposed to Polybrominated Biphenyls in 1973." *Epidemiology* 9 (July 1998): 373–377.

MacAvoy, Paul W., and Ira M. Millstein. *The Recurrent Crisis in Corporate Governance.* New York: Palgrave Macmillan, 2003.

Moskow, Michael. "Disruptions in Global Financial Markets: The Role of Public Policy." *Economic Perspectives (Third Quarter 2000).* Chicago: Federal Reserve Bank of Chicago, 2000.

Plouffe, J. F., et al. "Abnormal Lymphocyte Responses in Residents of a Town with a Cluster of Hodgkin's Disease." *Clinical Experimental Immunology* 35 (1979): 163–170.

Schwartz, Edward M., and William Rae. "Effect of Polybrominated Bipenyls (PBB) on Developmental Abilities of Children." *American Journal of Public Health* 73 (March 1983): 277–280.

Schwartz, Richard S., Jeffrey P. Callen, and Joseph Silva. "A Cluster of Hodgkin's Disease in a Small Community: Evidence for Environmental Factors." *American Journal of Epidemiology* 108 (1978): 19–26.

Seagull, Elizabeth. "Developmental Abilities of Children Exposed to Polybrominated Biphenyls (PBB)." *American Journal of Public Health* 73 (March 1983): 281–285.

Touzeau, Lois. *Our Side of the Story.* New York: Vantage Press, 1985.

Weimer, Arthur M. "An Economic History of Alma, Michigan." Ph.D. dissertation, University of Chicago, 1934.

Xintaras, Charles et al. *NIOSH Health Survey of Velsicol Pesticide Workers: Occupational Exposure to Leptophos and Other Chemicals.* Cincinnati: Division of Biomedical and Behavioral Science, NIOSH, U.S. Dept. of Health, Education, and Welfare, 1978.

Index